AF371259

Hotspots of Subterranean Biodiversity

Hotspots of Subterranean Biodiversity

Editors

Tanja Pipan
David C. Culver
Louis Deharveng

MDPI • Basel • Beijing • Wuhan • Barcelona • Belgrade • Manchester • Tokyo • Cluj • Tianjin

Editors

Tanja Pipan
ZRC SAZU Karst Research
Institute
Slovenia

David C. Culver
American University
USA

Louis Deharveng
Sorbonne Université
France

Editorial Office
MDPI
St. Alban-Anlage 66
4052 Basel, Switzerland

This is a reprint of articles from the Special Issue published online in the open access journal *Diversity* (ISSN 1424-2818) (available at: https://www.mdpi.com/journal/diversity/special_issues/HotspotsSubterranean_Biodiversity).

For citation purposes, cite each article independently as indicated on the article page online and as indicated below:

LastName, A.A.; LastName, B.B.; LastName, C.C. Article Title. *Journal Name* **Year**, *Volume Number*, Page Range.

ISBN 978-3-0365-2359-0 (Hbk)
ISBN 978-3-0365-2360-6 (PDF)

Cover image courtesy of Didier Rigal
The cover image shows a crab *Cancrocaeca xenomorpha* from a cave of Sulawesi, Indonesia.

Contents

About the Editors

Tanja Pipan received her Ph.D. in Biology from the University of Ljubljana in 2003. She is currently Research Advisor at ZRC SAZU Karst Research Institute and Associate Professor of Biology at University of Nova Gorica. She has research interests in the biology of shallow subterranean habitats, ecology of the epikarst copepod fauna, patterns of subterranean biodiversity, and karst ecosystem function. She is largely responsible for the recognition of epikarst as a significant subterranean habitat. She has published over 50 papers and has authored three books. Pipan is country coordinator for the International and European Long Term Ecological Research Program. In 2016, she received the ZRC SAZU Gold Award: a prize for outstanding scientific research.

Louis Deharveng received his Ph.D. in Biology from the University of Toulouse (France) in 1982. He is currently Research Director Emeritus of the CNRS based at Museum National d'Histoire Naturelle in Paris. His research interest is in the taxonomy and evolution of springtails and the biodiversity of the hypogean environment. He has published more than 200 papers and led two European projects on these research topics. He has organized or co-organized several international projects and symposia on biodiversity between 2001 and 2019, especially in France, Indonesia, China, Vietnam and South Africa, and set up several biospeological expeditions in Southeast Asia and the Pacific. He is a member of the International Society for Subterranean Biology and the chair of the Cave Invertebrate Specialist Group at IUCN.

David C. Culver received his Ph.D. in Biology from Yale University in 1970. He is currently Professor of Environmental Science Emeritus at American University, Washington, DC, with broad research interests in subterranean biology, especially biodiversity, biogeography, and shallow subterranean habitats. He has published well over 100 research papers on subterranean biology, and is the author or editor of eight books on speleobiology. He has organized or co-organized international symposia on conservation and protection of karst (1997), mapping subterranean biodiversity (2001), epikarst (2003), karst ecosystems (2007), and carbon in karst (2013), among others. He is an Honorary Life Member of both the National Speleological Society and the International Society for Subterranean Biology, and a member of the Explorers Club.

Preface to "Hotspots of Subterranean Biodiversity"

Caves have long intrigued biologists because of the unique life forms they harbor—species without eyes or pigment, often with elongated, spindly appendages and other non-optic sensory hypertrophies. Caves and their inhabitants were once thought to be rare, but current research shows that neither is truly rare. Worldwide, there are well in excess of 100,000 caves and 5000 species specialized for caves and other aphotic, subterranean habitats. These species hold a special fascination for evolutionary biologists because of their bizarre morphology, and they present challenges to students of biodiversity because sampling cave fauna represents a considerable logistical challenge and because species ranges are typically very small due to the constraints of subterranean dispersal. More than 20 years ago, David Culver, together with the Slovenian speleobiologist, Boris Sket, assembled the first list of individual caves and karst wells (20 in total) with more than 20 specialized species. Since that time, this list of hotspot sites has grown to 22 caves and karst wells, each with 25 or more aquatic or terrestrial species (and in a few cases both) specialized for caves and wells.

In this Special Issue, detailed accounts are given of 14 of the 22 hotspot sites in Australia, Bermuda, Mexico, France, Romania, Slovenia, Spain, Sulawesi, and the United States. For each of these sites, a detailed taxonomic consideration is presented, as well as detailed description of the site, usually with a cave map. These descriptions should be of interest, not only to speleobiologists, but to anyone interested in biodiversity in general. Although the pattern is still incompletely known, the terrestrial pattern is a global one, with hotspots throughout the world. Aquatic hotspots are more constrained, somewhat concentrated in central Europe (six out of ten cases).

Our thanks go the publisher, especially Emma Li, the Managing Editor for *Diversity*, and to all the authors who contributed to this Special Issue—Hotspots of Subterranean Biodiversity—printed here.

Tanja Pipan, David C. Culver, Louis Deharveng
Editors

Editorial

Hotspots of Subterranean Biodiversity

Tanja Pipan [1,*], Louis Deharveng [2,*] and David C. Culver [3,*]

[1]   Research Centre of the Slovenian Academy of Sciences and Arts, Karst Research Institute, Novi trg 2, SI-1000 Ljubljana, Slovenia

[2]   Institut de Systématique, Évolution, Biodiversité (ISYEB), UMR7205, CNRS, Muséum national d'Histoire naturelle, Sorbonne Université, EPHE, 45 rue Buffon, 75005 Paris, France

[3]   Department of Environmental Science, American University, 4400 Massachusetts Ave. NW, Washington, DC 20016, USA

*   Correspondence: tanja.pipan@zrc-sazu.si (T.P.); louis.deharveng@mnhn.fr (L.D.); dculver@american.edu (D.C.C.)

Received: 18 May 2020; Accepted: 21 May 2020; Published: 25 May 2020

Abstract: Worldwide, caves and groundwater habitats harbor thousands of species modified and limited to subterranean habitats in karst. Data are concentrated in Europe and USA, where a number of detailed analyses have been performed. Much less is known with respect to global patterns due to a lack of data. This special issue will focus on and discuss the global patterns of individual hotspot caves and groundwater habitats.

Keywords: α-diversity; biogeography; biospeleology; cave biology; caves; hotspots; invertebrates; subterranean biodiversity

---

1. Introduction

The denizens of caves and other habitats without light hold a special interest for students of biodiversity [1]. The fauna have bizarre and very distinctive morphology, including the lack of or reduction in eyes and pigment, elongated appendages, and often a ghost-like appearance [1–4]. How these species from disparate groups, including both vertebrates and invertebrates, evolved this convergent morphology remains a subject of intense research, both in the laboratory [5–7] and in the field [8,9]. Whatever the details, cave fauna are a magnificent example of convergent evolution and adaptation [10,11].

What is less well known is that there are thousands of species specialized for life in darkness, especially in caves. While speciation can occur as a result of isolation in caves, followed by subterranean dispersal and more isolation, opportunities for subterranean dispersal are restricted [4,12,13], and the specialized subterranean fauna is the result of many hundreds if not thousands of separate colonizations from the surface to the subsurface. These numbers are sufficient to expect predictable geographic patterns to emerge from their analysis [14].

While there are now several datasets on the georeferenced occurrence of specialized subterranean species from both Europe and the United States with over 2000 records [15–18], these records were hard-won and without parallel in other continents and regions. Brazil is the best-known region outside of Europe and the United States and the number of records is much smaller, and mostly for as yet undescribed species [19,20]. Exploration and access to caves is often formidable, and the number of caves is very large. For example, 45,000 caves were known from the USA in 1999 [21], and that number has grown considerably since that time. Of course, not all caves need to be sampled to obtain a good estimate of species richness for a region, but a rather large number do. This is because β-diversity dominates α-diversity [22,23]. One study of the European terrestrial fauna found that approximately 100 caves were needed to have accumulation curves of species numbers approach an asymptote [24].

The availability of large numbers of records for non-cave subterranean habitats is almost non-existent, with two exceptions. One is the analysis of epikarst communities in Slovenia [25], a habitat that is indirectly sampled through the continuous sampling of water dripping from cave ceilings [26]. The other is the larger scale sampling of porous aquifers in Europe as part of the PASCALIS project (Protocol for the ASsessment and Conservation of Aquatic Life in the Subsurface) [27]. There are also several studies that combine cave and non-cave subterranean aquatic diversity [28,29]. This sampling deficiency in subterranean habitats not directly accessible to humans has been named the "Racovitzan impediment" [30].

In spite of these impediments, we know a great deal about cave biodiversity patterns in Europe and, to a lesser extent, the United States. In Europe, there is a ridge of high species richness for both aquatic and terrestrial species at a latitude of approximately 45° N along the spine of the Pyrenees and through the Dinaric karst of Slovenia, Croatia, Serbia, and Montenegro [16,24]. Species range sizes follow Rapoport's Rule and increase with increasing latitude [16], and species turnover is high over short distances [16,23]. Explanations of patterns are highly scale-dependent. For example, the ridge of high species richness is largely concordant with areas of high productivity [24], but productivity is unimportant in determining hotspots within hotspots [15]. The explanations for regional differences have proven to be complex, often involving a combination of energy availability, spatial heterogeneity, and history, and the explanations vary from region to region, being spatially non-stationary [31].

Determining the global patterns of subterranean species richness has remained elusive. At one level, the possibilities of determining the patterns seem difficult at best and a long way in the future. If thousands of records are needed to describe and understand the European pattern, global studies with tens of thousands of records would be needed. However, there may be a shortcut at hand that at least holds promise of a global picture, albeit incomplete. It starts with the observation by Gibert and Deharveng [22] in their classic paper on subterranean biodiversity that regional diversity is a good predictor of local diversity and vice versa. This was buttressed by later findings that species accumulation curves rarely crossed, and thus the regional qualitative pattern could be captured by a relatively small number of samples [24,32]. Culver and Sket [33] took this idea to its logical extreme, and considered only caves and karst wells with the highest species richness, originally finding 20 sites with 20 or more species specialized for subterranean life. While the coverage of large numbers of caves in a relatively small area was (and is) limited to Europe and the United States, they reasoned that at least a few outstanding caves, extensively sampled, were known from most large karst areas. Since the publication of the first hotspot list in 2000, knowledge of the global cave fauna has grown exponentially. Species lists are available for several tropical countries [20,34,35], and a number of caves throughout the tropics and sub-tropics are now well sampled [36]. In a later update [37], the bar was raised to either 25 terrestrial or 25 aquatic species. In 2019, there were 24 examples of hotspot caves known [38]—16 from the temperate zone, 5 from the subtemperate zone, and 3 from the subtropics. The tropics have at least five caves with 20 or more species specialized for subterranean life, the original hotspot criterion [37]. The demonstration of hotspot caves in the tropics also raises some other issues. One is how to treat undescribed species, which is the case for the majority of tropical species [39]. If they are ignored, then tropical and subtropical caves will appear to be artificially depauperate. If they are fully counted, the likelihood that some of these species are not valid [40] is ignored. Of course, described species may also be wrongly thought to be limited to subterranean habitats. In addition, the tropical cave fauna often has a component specialized on guano but never found outside of caves. These guanobionts typically show less eye and pigment loss and less appendage elongation [38,39]. Should they be discounted as cave specialists? If so, then once again tropical and subtropical caves may appear to be artificially depauperate.

This special issue of *Diversity* aims to bring together information on hotspot caves and karst ground-watered habitats for in-depth analyses and comparisons. First and foremost, there will be a species list for each of the hotspot caves, information that is strangely unpublished for many of the hotspot caves. This is especially important given the controversy around the ecological status of cave species [41]. Deharveng and Bedos [41] pointed out that considerable confusion exists in the literature about the terms troglobiont—which should be used only for species not found outside of caves, irrespective of their morphology—and troglomorph [42,43], species with reduced eyes and pigment and elongated appendages. The two are not identical, a problem that arises not only with guanobionts but also with all species without troglomorphic features occurring in caves [44]. The special issue will also provide a physical setting for the caves and groundwater habitats, including their hydrogeological and environmental context, their use by humans, the nature of the karst in which they are situated, and the knowledge on nearby cave biodiversity. An interesting side note is that a number of hotspot caves are or have been commercial caves, including Mammoth Cave in Kentucky, USA; the Postojna-Planina Cave System in Slovenia; and Vjetrenica in Bosnia and Hercegovina. Meanwhile, some groundwater hotspots are industrially exploited for water consumption. Finally, there will also be a summary of what we know, the patterns of distribution, and future research directions. For more information, contact any of the editors.

Conflicts of Interest: The authors declare no conflict of interest.

References

1. Fenolio, D. *Life in the Dark: Illustrating Biodiversity in the Shadowy Haunts of Planet Earth*; Johns Hopkins University Press: Baltimore, MD, USA, 2016.
2. Juberthie, C. *Encyclopaedia Biospeologica*; Decu, V., Ed.; Societé Internationale de Biospéologie: Moulis, France, 1994; Volume 3.
3. Botosaneanu, L. *Stygofauna Mundi*; E.J. Brill: Leiden, The Netherlands, 1986.
4. Culver, D.C.; Pipan, T. *The Biology of Caves and Other Subterranean Habitats*, 2nd ed.; Oxford University Press: Oxford, UK, 2019.
5. Keene, A.C.; Yoshizawa, M.; McGaugh, S.E. *Biology and Evolution of the Mexican Cavefish*; Academic Press: Waltham, MA, USA, 2016.
6. Wilkens, H.; Strecker, U. *Evolution in the Dark. Darwin's Loss without Selection*; Springer Nature: Berlin, Germany, 2017.
7. Rohner, N.; Jarosz, D.F.; Kowalko, J.E.; Yoshizawa, M.; Jeffery, W.R.; Borowsky, R.L.; Lindquist, S.; Tabin, C.J. Cryptic variation in morphological evolution: HSP90 as a capacitor for loss of eyes in cavefish. *Science* **2013**, *342*, 1372–1375. [CrossRef] [PubMed]
8. Poulson, T.L. Cave adaptation in amblyopsid fishes. *Am. Midl. Nat.* **1963**, *70*, 257–290. [CrossRef]
9. Culver, D.C.; Kane, T.C.; Fong, D.W. *Adaptation and Natural Selection in Caves. Gammarus minus as a Case Study*; Harvard University Press: Cambridge, MA, USA, 1995.
10. Pipan, T.; Culver, D.C. Convergence and divergence in the subterranean realm: A reassessment. *Biol. J. Linn. Soc.* **2012**, *107*, 1–14. [CrossRef]
11. Culver, D.C.; Pipan, T. Shifting paradigms of the evolution of cave life. *Acta Carsologica* **2015**, *44*, 415–425. [CrossRef]
12. Trontelj, P. Vicariance and dispersal in caves. In *Encyclopedia of Caves*, 3rd ed.; White, W.B., Culver, D.C., Pipan, T., Eds.; Academic Press: Waltham, MA, USA, 2019; pp. 1103–1109.
13. Eme, D.; Malard, F.; Konecny-Dupré, L.; Lefébure, T.; Douady, C.J. Bayesian phylogenetic inferences reveal contrasted colonization dynamics among European groundwater isopods. *Mol. Ecol.* **2013**, *22*, 5865–5899. [CrossRef]
14. Zagmajster, M.; Christman, M.C. Mapping subterranean biodiversity. In *Encyclopedia of Caves*, 3rd ed.; White, W.B., Culver, D.C., Pipan, T., Eds.; Academic Press: Waltham, MA, USA, 2019; pp. 678–685.
15. Bregović, P.; Zagmajster, M. Understanding hotspots within a global hotspot—Identifying the drivers of regional species richness patterns in terrestrial subterranean habitats. *Insect Conserv. Biodivers.* **2016**, *9*, 268–281. [CrossRef]
16. Zagmajster, M.; Eme, D.; Fišer, C.; Galassi, D.; Marmonier, P.; Stoch, F.; Cornu, J.; Malard, F. Geographic variation in range size and beta diversity of groundwater crustaceans: Insights from habitats with low thermal seasonality. *Glob. Ecol. Biogeogr.* **2014**, *23*, 1135–1145. [CrossRef]

17. Christman, M.C.; Doctor, D.H.; Niemiller, M.L.; Weary, D.J.; Young, J.A.; Zigler, K.S.; Culver, D.C. Predicting the occurrence of cave-inhabiting fauna based on features of the Earth surface environment. *PLoS ONE* **2016**, *11*, e0160408. [CrossRef]

18. Deharveng, L.; Stoch, F.; Gibert, J.; Bedos, A.; Galassi, D.; Zagmajster, M.; Brancelj, A.; Camacho, A.; Fiers, F.; Martin, P.; et al. Groundwater biodiversity in Europe. *Freshw. Biol.* **2009**, *54*, 709–726. [CrossRef]

19. Trajano, E.; Bichuette, M.E. Diversity of Brazilian subterranean invertebrates, with a list of troglomorphic data. *Subterr. Biol.* **2010**, *7*, 1–16.

20. Souza-Silva, M.; Ferreira, R.L. The first two hotspots of subterranean biodiversity in South America. *Subterr. Biol.* **2016**, *19*, 1–21. [CrossRef]

21. Culver, D.C.; Hobbs III, H.H.; Christman, M.C.; Master, L.L. Distribution map of caves and cave animals in the United States. *J. Cave Karst Stud.* **1999**, *61*, 139–140.

22. Gibert, J.; Deharveng, L. Subterranean ecosystems: A truncated functional diversity. *Bioscience* **2002**, *52*, 473–481. [CrossRef]

23. Malard, F.; Boutin, C.; Camacho, A.I.; Ferreira, D.; Michel, G.; Sket, B.; Stoch, F. Diversity patterns of stygobiotic crustaceans across multiple spatial scales in western Europe. *Freshw. Biol.* **2009**, *54*, 756–776. [CrossRef]

24. Culver, D.C.; Deharveng, L.; Bedos, A.; Lewis, J.J.; Madden, M.; Reddell, J.R.; Sket, B.; Trontelj, P.; White, D. The mid-latitude biodiversity ridge in terrestrial cave fauna. *Ecography* **2006**, *29*, 120–128. [CrossRef]

25. Pipan, T.; Culver, D.C.; Papi, F.; Kozel, P. Partitioning diversity in subterranean invertebrates: The epikarst fauna of Slovenia. *PLoS ONE* **2018**, *13*, e0185991. [CrossRef]

26. Pipan, T. *Epikarst—A Promising Habitat*; Založba ZRC: Ljubljana, Slovenia, 2005.

27. Gibert, J.; Culver, D.C. Assessing and conserving groundwater biodiversity: An introduction. *Freshw. Biol.* **2009**, *54*, 639–648. [CrossRef]

28. Hof, C.; Brändle, M.; Brandl, R. Latitudinal variation of diversity in European freshwater animals is not concordant across habitat types. *Glob. Ecol. Biogeogr.* **2008**, *17*, 539–546. [CrossRef]

29. Stoch, F.; Galassi, D.M.P. Stygobiotic crustacean species richness: A question of numbers, a matter of scale. *Hydrobiologia* **2010**, *653*, 217–234. [CrossRef]

30. Ficetola, C.F.; Canadoli, C.; Stoch, F. The Racovitzan impediment and the hidden biodiversity of unexplored environments. *Conserv. Biol.* **2019**, *33*, 214–216. [CrossRef]

31. Zagmajster, M.; Malard, F.; Eme, D.; Culver, D.C. Subterranean biodiversity patterns from global to regional scales. In *Cave Ecology*; Moldovan, O.T., Kováč, L., Halse, S., Eds.; Springer: Cham, Switzerland, 2018; pp. 195–228.

32. Dole-Olivier, M.J.; Castellarini, F.; Coineau, N.; Galassi, D.M.P.; Martin, P.; Mori, N.; Valdecasas, A.; Gibert, J. Towards an optimal sampling strategy to assess groundwater biodiversity: Comparison across six regions of Europe. *Freshw. Biol.* **2009**, *54*, 777–796. [CrossRef]

33. Culver, D.C.; Sket, B. Hotspots of subterranean biodiversity in caves and wells. *J. Cave Karst Stud.* **2000**, *62*, 11–17.

34. Palacios-Vargas, J.G.; Reddell, J.R. Actualización del inventario cavernicola (estigobiontes, estigófilos y troglobios) de México. *Mundos Subterráneos* **2013**, *24*, 33–95.

35. Trajano, E.; Gallão, J.E.; Bichuette, M.E. Spots of high diversity of troglobites in Brazil: The challenge of measuring subterranean diversity. *Biodivers. Conserv.* **2016**, *25*, 1805–1828. [CrossRef]

36. Deharveng, L.; Bedos, A. Biodiversity in the tropics. In *Encyclopedia of Caves*, 3rd ed.; White, W.B., Culver, D.C., Pipan, T., Eds.; Academic Press: Waltham, MA, USA, 2019; pp. 146–162.

37. Culver, D.C.; Pipan, T. Subterranean ecosystems. In *Encyclopedia of Biodiversity*, 2nd ed.; Levin, S.A., Ed.; Elsevier: Amsterdam, The Netherlands, 2013; Volume 7, pp. 49–62.

38. Deharveng, L.; Bedos, A. The cave fauna of southeast Asia. Origin, evolution, and ecology. In *Subterranean Ecosystems*; Wilken, H., Culver, D.C., Humphreys, W.F., Eds.; Elsevier: Amsterdam, The Netherlands, 2000; pp. 603–632.

39. Deharveng, L.; Lips, J.; Rahmali, C. Focus on guano. In *The Natural History of Santo*; Bouchet, T., Le Guyader, H., Pascal, O., Eds.; Museum National d'Histoir Naturelle: Paris, France, 2011; pp. 300–306.

40. Culver, D.C.; Trontelj, P.; Zagmajster, M.; Pipan, T. Paving the way for standardized and comparable subterranean biodiversity studies. *Subterr. Biol.* **2012**, *10*, 43–50. [CrossRef]

41. Deharveng, L.; Bedos, A. Diversity of terrestrial invertebrates in subterranean habitats. In *Cave Ecology*; Moldovan, O.T., Kováč, L., Halse, S., Eds.; Springer: Cham, Switzerland, 2018; pp. 107–172.
42. Christiansen, K.A. Proposition pour la classification des animaux cavernicoles. *Spelunca Mem.* **1962**, *2*, 76–78.
43. Christiansen, K.A. Morphological adaptation. In *Encyclopedia of Caves*, 2nd ed.; White, W.B., Culver, D.C., Eds.; Academic Press: Waltham, MA, USA, 2012; pp. 517–528.
44. Culver, D.C.; Pipan, T. *Shallow Subterranean Habitats. Ecology, Evolution, and Conservation*; Oxford University Press: Oxford, UK, 2014.

Article

Ojo Guareña: A Hotspot of Subterranean Biodiversity in Spain

Ana Isabel Camacho [1,*] and Carlos Puch [2]

1   Dpto. Biodiversidad y Biología Evolutiva, Museo Nacional de Ciencias Naturales (CSIC), José Gutiérrez Abascal 2, 28006 Madrid, Spain
2   Grupo Espeleológico Edelweiss, Club Cántabro de Exploraciones Subterráneas, Víctor de la Serna 26, 28016 Madrid, Spain; carlospuch@gmail.com
*   Correspondence: mcnac22@mncn.csic.es

Abstract: Ojo Guareña Natural Monument in Burgos (Spain) is an important and large karstic system. It consists of more than 110 km of surveyed galleries, and it has rich sources of organic material from the surface and permanent water circulation. It is the fourth largest cave system in the Iberian Peninsula, and one of the 10 largest in Europe. Ojo Guareña also ranks 23rd among the world's largest caves. To date, only volcanic caves in the Canary Islands, in which between 28 and 38 subterranean species occur, are considered subterranean diversity hotspots in Spain. Here, we provide the first list of subterranean taxa present in Ojo Guareñ, which is comprised of 54 taxa that includes 46 stygobiotic and eight troglobiotic species (some still unidentified at the species level), revealing Ojo Guareña as the largest known subterranean biodiversity hotspot in Spain and Portugal. In addition, we provide a list of an additional 48 taxa, 34 stygophiles and 14 troglophiles, found in the system, whose ecological status could change with detailed biological studies, which may change the number of strictly subterranean species present in the system. Indeed, at present, these numbers are provisional as they correspond to a small part of this sizeable cave system. The biodiversity of large areas of the system remains unknown as these areas have yet to be explored from the biological point of view. In addition, a large number of samples of both terrestrial and aquatic fauna are still under study by specialists. Furthermore, evidence of cryptic species within Bathynellacea (Crustacea) indicates an underestimation of biodiversity in the karstic system. Despite these limitations, the data available reveal the typical uneven distribution of subterranean aquatic fauna, and suggest that the great heterogeneity of the microhabitats in this wide and highly connected karstic extension led to the great richness of aquatic subterranean species.

Keywords: hotspot caves; Ojo Guareña natural monument; stygobionts; troglobionts; subterranean biodiversity

Citation: Camacho, A.I.; Puch, C. Ojo Guareña: A Hotspot of Subterranean Biodiversity in Spain. *Diversity* **2021**, *13*, 199. https://doi.org/10.3390/d13050199

Academic Editors: Tanja Pipan, David C. Culver and Louis Deharveng

Received: 31 March 2021
Accepted: 6 May 2021
Published: 8 May 2021

1. Introduction

Iberian Peninsula, together with Balearic Islands, is one of the regions in Europe with the greatest development of karst areas. Numerous caves, many of them large (more than 3 km of development and/or more than 300 m in depth), have been surveyed in the country [1]. Despite this, very little data on the subterranean fauna within these caves are available. Potential biospeleologists are discouraged by the difficulties of access and progression in the caves (e.g., large vertical shafts, meanders, narrow passages, etc.), the inaccessibility to man of much of the underground environment (flooded galleries, mesovoid shallow substratum-MSS-, hyporheic sites), and the low density of specimens in accessible populations (epikarst, especially). Thus, most samplings of these ecosystems have been limited and sporadic, and mainly focused on the study of specific taxonomic groups using capture techniques suited for them. Due to the great sampling deficiencies and the scarce biospeleological tradition in our country, there are very few inventories of the region's underground fauna beyond some lists of a few terrestrial or aquatic taxa in the more well-known caves. The only caves considered in the literature as subterranean diversity

Diversity **2021**, *13*, 199. https://doi.org/10.3390/d13050199   https://www.mdpi.com/journal/diversity

hotspots in Spain are four volcanic ones located in the Canary Islands: in these, between 28 and 36 troglobiotic species and 38 stygobiotic and interstitial species (volcanic anchialine habitats of Lanzarote) have been recorded [2,3]. Culver and Pipan (2013) also mention possible diversity hotspots in the Cantabrian Mountains in northern Spain, which contain the largest number of caves in the country [4]; however, more studies of this region are needed to confirm this hypothesis.

The Ojo Guareña Natural Monument (OGNM) karstic system is situated on the southern rim of the Cantabrian Mountains. It lies within a hypothetical zone (between ca 42° and 46° N) in Europe characterized as having "high biodiversity for terrestrial cave fauna" [5]. According to Gulden [6], OGNM is the fourth largest cave system in the Iberian Peninsula, and one of the 10 largest in Europe; it also ranks 23rd on the list of the world's largest caves. With more than 110 km of surveyed galleries, it is a vast cave system that has rich sources of organic material from the surface and also permanent groundwater flow (phreatic water). It is one of the few places in Spain where the terrestrial and aquatic environments have been sampled repeatedly and extensively [7]. However, we are far from knowing its true diversity. All of the "shortfalls" impeding our knowledge of biodiversity that are commonly mentioned at the global level (e.g., Linnean, Wallacean, Darwinian, Prestonian) [8] also apply at local scales. A more recently described shortfall, the "Racovitzan impediment" [9], is of particular importance in the case of the subterranean biodiversity of the OGNM: what remains unexplored, from a biological point of view, cannot be known.

The most accessible (and thus the best sampled) area of the OGNM extends from the main entrance at Palomera Cave to Museo de Cera Hall (covering about 3 km of passages). Moreover, this area contains sumps from the Guareña River and water that filters in from the Trema River. However, many of the specimens collected in this area and throughout the monument remain unidentified at the species level (taxonomic shortfall). Cryptic species of crustaceans have also been discovered in the area [10], indicating that the data presented here represent an underestimation of the diversity of the subterranean fauna. In addition, the biological composition of many kilometers of underground passages has yet to be surveyed, and the accumulation curves of stygobitic species [7,11] of the best sampled areas do not reach saturation, demonstrating our incomplete knowledge of the true biodiversity of this cave system's fauna.

In addition to its biological assemblages, the OG endokarst has unique geological, geomorphological, hydrological, paleoclimatic, archaeological, and paleontological characteristics that make it worthy of protection. It has a continuous and non-fragmented surface that is subjected to natural evolution with little human intervention and that can maintain the physical and biological characteristics of the system, ensuring the continued functioning of natural processes. Moreover, the absence of resource extraction by agricultural, forestry, hydraulic or mining (natural or artificial) activities reduce the likelihood that the functioning and resources of the ecosystems within the system, and the aesthetics of the landscape, would be significantly altered.

2. A Bit of History

The main cave of the system (OG), through its main entrances (Palomera Cave and Dolencias Shaft), has been known since the 1940s [4]. Although the cave did not initially arouse great interest among the scientific community—beyond simple sporadic visits and specific samplings—, interest later increased as the importance of its biological species [12–15], archaeological sites [16], and hydrogeological and geomorphological peculiarities became more evident. Due to these features, various public administrations began to establish general protection regimes for the karst system; eventually, in 1996, it was incorporated as a natural monument in the Network of Protected Natural Spaces in Spain and, more recently, in the European Ecological Network Natura 2000 [7]. In addition, given the uniqueness of its geology, it was included in a publication on the "Points of Geological Interest of the Cantabrian Coast" (1983). However, despite these acknowledgments, very little scientific study has been conducted in the system. In fact, until just this century, faunal sampling of the system had been sparse, and the findings almost anecdotal.

Here, we provide some details of past explorations of this natural monument in order to put into context the knowledge that exists of this vast karst system. The presence of numerous archaeological remains from the Paleolithic and Neolithic, and from the Iron and Middle ages, testifies that OG has been known since ancient times. However, geologists did not become interested in it until 1933 [17]. In 1956, the Edelweiss Cave Club (GEE), from Burgos, began to explore the large galleries that converge at the main entrance (development = 8 km). In 1958, with the collaboration of various speleological groups (Spanish and foreigners), the "OG 58" campaign was held, resulting in 8900 m being surveyed and produced the first plan of the cave. A complete topographic survey of the system continues to this day, and in parallel, exploration inside the main cave and the rest of the caves intensifies. By 1986 [18], surveys of new sectors, only accessible when the water is low, together with the union of other caves, considerably increased the known development of the complex to 89,071 m. Currently, 110 km in an area encompassing 15 km^2 have been explored and surveyed.

Biological sampling of OG was occasional during the second half of the twentieth century, though it is worth highlighting the initiative of Professor Eugenio Ortiz, director of the Museo Nacional de Ciencias Naturales de Madrid (MNCN), who, from 1968 to 1975, tried to establish an underground laboratory at the system (that never materialized) and an associated biological station to be sponsored by the Consejo Superior de Investigaciones Científicas (CSIC). Among the aquatic faunal samplings carried out during those years, Ortiz observed the presence of numerous specimens of the isopod *Stenasellus virei buchneri* Stammer, 1936 (Figure 1) in a pond near the Palomera entrance. He also collected specimens of Bathynellacea and some other groups; however, they were never studied or, at least, the results were never published. Other samples collected by Ortiz, especially terrestrial ones, were later studied by specialists of different fauna groups (e.g., oligochaetes, opilions, mites, myriapods, and beetles). Later (1975), T. Antón, a member of the Edelweiss Cave Club, and X. Bellés collected and studied some other terrestrial fauna, which led to several publications [19–26]. In 1984, Notenboom [27,28] reported on some aquatic faunal samples from one of the resurgences of the system (La Torcona). Camacho and Puch also conducted sporadic samplings of aquatic fauna in the cave during the 1980s and 1990s. Thanks to these works, which were performed without institutional support, the cave's fauna began to be inventoried [29]. As of 1993, a total of 81 taxa were known in the cave, of which 71 were identified to the species level (63 terrestrial and 8 aquatic). Of these, only a dozen of the well-identified species could be considered as strictly subterranean fauna [29]. From 2002 to 2004, OGNM was one of four areas under study as part of a project on European stygofauna (PASCALIS) [11]. The project areas included various caves, springs, wells, unsaturated zones (hyporheic system of epigean rivers) and porous aquifers. Finally, the Junta de Castilla y León and the MNCN of the CSIC signed agreements to carry out faunistic studies of the cave (from 2002 to 2004 and 2006 to 2009) prior to its limited opening to tourism. Small parts of the terrestrial [30] and aquatic ecosystems [31] were studied prior to any potential effects that tourism could have on the faunal composition. The results of all of these investigations have contributed to the lists of species constituting the main objective of this publication.

In this study, we provide the first lists of stygobiotic and troglobiotic (and stygophilic and troglophilic) taxa present in OGNM, which were compiled on the basis of the results of the aforementioned investigations. Although only a small fraction of this vast karst system has been surveyed, and some gaps in knowledge remain, we show that it is a hotspot of subterranean biodiversity with great biological potential.

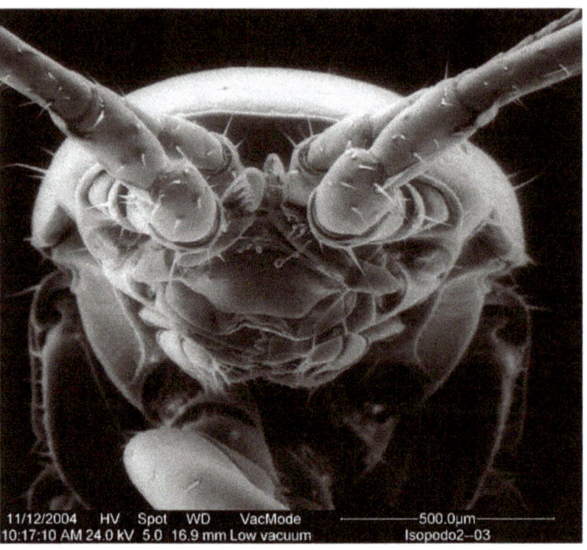

Figure 1. Isopod *Stenasellus virei buchneri*. Photo ESEM, MNCN, Madrid.

3. Materials and Methods

The study area (OGNM) is located in the Sotoscueva region (Burgos, Northern Spain), within the southernmost limit of the Cantabrian Mountains. OGNM is a paradigmatic example of a complete karst system, with a water absorption zone in the northwestern and central parts, an extensive network of galleries of more than 110 km in the main cave and evacuation points located at the southeastern limit of the system. It develops in limestones and dolomites of the Upper Coniacian (about 130 m of thickness), which lean on clay limestones and marls of the Middle Coniacian–Turonian (Upper Cretaceous) that act as an impermeable substrate (Figure 2). The surface landscape is arranged in the form of a geosyncline with a WNW–ESE alignment, giving rise to the characteristic slopes that stand out in the northern part of the area [4,18,32]. The rivers Guareña, to the north, and Trema, to the west, course underground when they encounter limestones and dolomites and resurface, after a long, partially known route, in La Torcona, a cave resurgence near the confluence of the Trema River with La Hoz Stream (Figure 3).

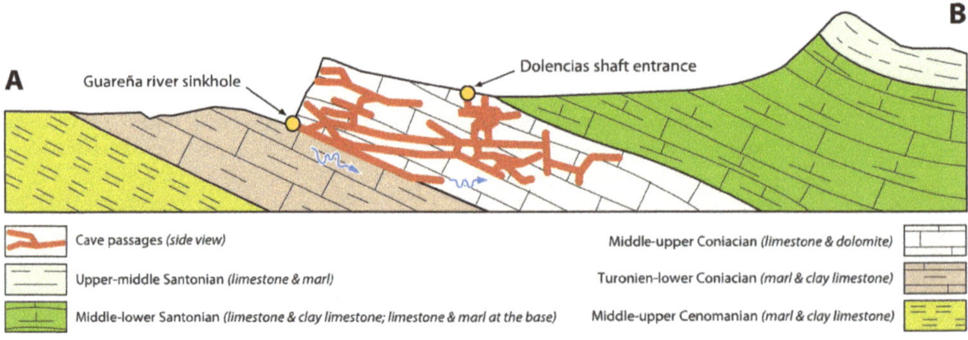

Figure 2. Geological section approximately North-South, corresponding to line A–B in Figure 3.

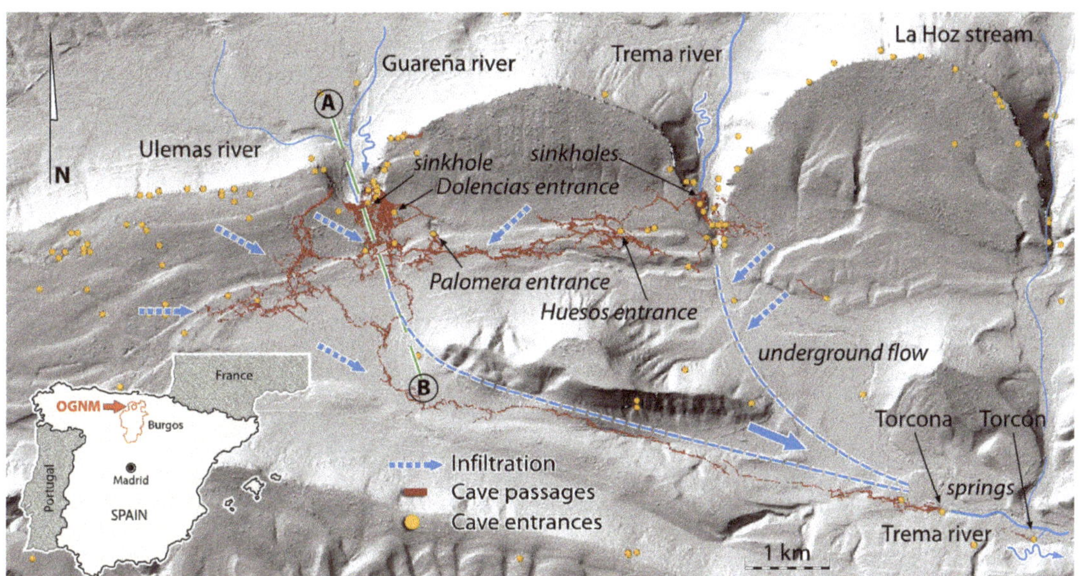

Figure 3. Location of caves, underground galleries, flows, and drainages on a digital elevation model of the OGNM terrain.

The main cave (OG) has 13 entrances: Palomera Cave, Dolencias Shaft, Huesos Shaft, Rizuelos Shaft, Cornejo Cave, Cuatro Pisos Cave, San Bernabé caves, Moro Cave, Trema river sinkholes, La Mina Cave, Guareña river sinkhole (the "Ojo" or Eye), Villallana Cave, and Torcona cave spring (main resurgence of the system) (Table S1) (Figure 3). The last two are connected to the main cave by long flooded passages. The general structure is comprised of a maze of galleries, mainly horizontal, that are arranged along the W–E and N–S axes. In the western part of the cave, the galleries are mainly oriented in a NE–SW direction. Other caves of the system that are not yet connected to the main cave include Kubía (surveyed length 550 m), Último Sumidero (350 m), Prado Vargas (130 m), Kaite (585 m), Covanería (320 m), and Jaime (650 m). A little further away from the main cave lies Las Yeguas Cave (1900 m). In 1981, the Trema river sinkholes (on the right bank of the river) and La Mina Cave (on the left bank) became connected through two conduits that traverse underneath the bottom of the Trema river canyon. The hypogeal course of this river is only known in the space between two sumps (300 m). In Cornejo Cave, the progression of the underground stream stops in a semi-flooded squeeze.

In short, the current ensemble of galleries developed in a 15 km^2 area (Figure 3). Of these, 30 km are concentrated in an area of just 1 km^2, underneath Alto de San Bernabé Hill, just above and to the south of the Guareña river sinkhole. The galleries are distributed in six largely overlapping levels, creating a three-dimensional maze known as "the West Daedalus". The upper levels are almost clogged inactive passages. The intermediate ones present a seasonal hydrological activity that can be very important, depending on the weather. The lower levels are the permanent beds of the Guareña and Trema rivers, to the north and west, respectively, and of Villamartín Stream, to the northwest, feeding a large underground aquifer installed around the axis of the syncline. After a long, partially known path, these water sources reappear in La Torcona, a cave near the eastern limit of the massif, and El Torcón, an inaccessible flooded spring. Hydrological recharge occurs by infiltration of rainwater and snowmelt through permeable formations, as well as by feeding from small hillside springs and by karst sinks that induce flow losses along epigeal channels. The most spectacular cases of recharging in the entire unit are found at the Guareña river sinkhole, at the base of the San Bernabé escarpment, through a penetrable sinkhole ("el Ojo"), and at the Trema river sinkhole in its own channel, shortly before the village of Cornejo. The entire endorheic basin represents an area of about 27 km^2. After a relatively rapid underground flow (1.5 km/day in low water conditions and 4 to 5 km/day in flooding), the flow drains towards the Trema River, on the southeastern limit of the karst. Although much is known about the behavior of the underground waters that come from the Guareña River, very little is known about those that come from the Trema River.

3.1. Sampling and Studies of the Fauna

As mentioned above, the OGNM was only occasionally sampled prior to the first decade of the twenty-first century, when the most intensive and systematic samplings, to date, occurred. The taxa lists compiled here are derived from previous studies of the material obtained from all known samplings, though with a bias towards the more recent ones. Below, we briefly mention some of the sampling techniques used, as well as the extent of sampling.

3.1.1. Samples of Aquatic Fauna

Although we do not know the number or specific locations of the samples collected or the specific sampling techniques used by Ortiz during his exploration of the cave, we do have detailed information on the systematic samplings carried out during this century. Between 2002 and 2009, 344 samples of aquatic fauna were collected at OGNM [31]. Sampling was carried out in the main cave (OG), plus 12 other caves and 12 springs, and in the hyporheic environments of the epigean rivers (Table S1):

(a) Main Cave (OG), about 3 km of main gallery (Palomera to Museo de Cera). A total of 59 sites were sampled (see Table S1 for site names): OG01 to OG16, periodically sampled (methods described below), and OG17 to OG59, sampled once or twice using hand nets (Figure 4). A total of 244 samples was studied.

Figure 4. Part of the survey of the main cave, indicating the sampling sites.

(b) Other caves in the system that were sampled on one or two occasions: Rizuelos, Kaite 2, Jaime, La Mina, Prado Vargas, García, San Bernabé, San Miguel river sinkhole, Las Llanas, Cornejo, Racino, and Redonda. A total of 31 samples were studied.

(c) Springs sampled only once: Torcona, Pozo del Infierno, Cornejo, La Mea, Villa, Salce, Cubío, La Calzada, Jordana, Avellanos, Mazo 1, and Mazo 2. A total of 24 samples were studied.

(d) Hyporheic and interstitial environments of the surrounding rivers and streams that are part of the OGNM aquifer, sampled on one or two occasions and at various points: Guareña, Ulemas, Trema, Trueba, Engaña, Nela, and La Hoz. A total of 43 samples were studied.

(e) Wells sampled only once: Torcón and Villabascones. Only two samples were studied.

Most samplings were restricted to the epikarstic zone of the caves [31]. However, sampling also occurred in some other areas including a couple of underground rivers in the main cave (Erizos and Guareña), the phreatic zone in Aburrimiento gallery (a long southernmost passage), and the resurgences of La Torcona and Villallana.

Eight sampling methods were used [33,34]: (1) traps with bait in free water (ponds and gours) and located within the substrate on riverbanks; (2) direct captures with a manual aspirator; (3) drip collection; (4) the Karaman-Chappuis method (on sandy shores of underground pools and epi- and hypogeal rivers); (5) removing the substrate and filtering the water with a hand net (ponds, gours, potholes) (Figure 5); (6) Bou-Rouch pumping (in La Torcona and in epigean rivers); (7) kicking (in the benthos of the Guareña and Erizos river streams); and (8) Cvetkov nets or other types of phreatobiological nets (in deep lakes and wells). All sampling nets (0.100 mm mesh size) were designed ad-hoc in order to adapt them for the different habitats and types of fauna studied. With the mesh size used, most adult forms of crustaceans and interstitial fauna can be collected.

Figure 5. Sampling in "Sala Edelweiss" OG09. Photo C. Puch.

The samples collected between 2002 and 2004 and between 2007 and 2009 were processed in the MNCN laboratories in Madrid under the supervision of A.I. Camacho [31]. As a great deal of the material collected (mainly ostracods, amphipods and copepods) are still being studied and identified by specialists, some of the entries on the stygobiotic species list have only been identified to the family or genus level. In most cases, samples were identified using classical (morphological) taxonomic methods. For samples of Bathynellacea, molecular analyses of nuclear (18S) and mitochondrial (COI) gene fragments were also carried out for taxonomic purposes. For this, we used DNA extraction and PCR amplification techniques previously developed for this type of fauna [10,35,36].

3.1.2. Samples of Terrestrial Fauna

The sampling techniques used in the earlier sporadic sampling visits to the cave are not known. We assume that most samplings were made de visu, a low yield technique, hence the scarcity of available material and data on the few troglobiotic species known at that time. From 2002 to 2004, samples of terrestrial fauna were collected using three techniques:

(1) Baited traps, after being set, were collected periodically (30, 60, 90, and 120 days after setting) in four selected spots in OG, along the main route from Palomera to Museo de Cera, over a distance of about 3 km. Samples collected outside the main cave (OG), from the access doline to the Palomera entrance, [29] were also included in the study.

(2) Sediment samples were collected at four locations in OG at six time points throughout the year, in an attempt to collect cave-dwelling fauna that are not attracted to the baited traps. Sediment samples of the access doline to Palomera, taken over four occasions, were also processed.

(3) Manual aspirators were used to capture samples de visu on every visit to OG.

The collected material was processed in the laboratories of the Department of Animal Biology I of the Faculty of Biology at the Complutense University of Madrid, under the supervision of Pérez-Zaballos [30]. Fauna were extracted from the sediment using the Berlese technique, and taxa were separated into individual groups. In a few cases, specimens could be identified only to the order or family level, and most specimens remain under study by taxonomists specializing in the different terrestrial groups. Some of the material has been deposited in the collections housed at the MNCN (Madrid, Spain).

4. Results

Overall, only partial results have been obtained from the samplings: most of the material collected, particularly the terrestrial specimens, are still under study, with their identification to a specific level still pending.

A total of 299 taxa has been identified in OGNM (Table 1), of which 112 are terrestrial and 187 are aquatic [30,31]. A total of 54 of these taxa constitutes subterranean fauna: 46 stygobiotics and 8 troglobiotics. However, another 48 taxa comprising 34 stygophilics and 14 troglophilics (Table S2) were also identified. In many cases, the ecological status of cave animals is difficult to determine [37] and therefore we feel their inclusion is warranted. In the future, the surface/subterranean status of some of these species could change as more information about their ecology and biology comes to light.

Table 1. Number of taxa known in the Ojo Guareña Natural Monument.

TAXA	Total Species	Stygobiotic	Troglobiotic	Stygophilic	Troglophilic
	Aquatic/Terrestrial	Species	Species	Species	Species
Rotifera	1/—	0	—	1	—
Cnidaria	1/—	0	—	0	—
Nematoda	1/?	—	0	—	0
Turbellaria	5/—	0	—	0	—
Hirudinea	3/—	0	—	1	—

Table 1. *Cont.*

TAXA	Total Species Aquatic/Terrestrial	Stygobiotic Species	Troglobiotic Species	Stygophilic Species	Troglophilic Species
Tardigrada	13/3	0	0	0	0
Oligochaeta	49/9	5	2	8	0
Mollusca					
Gastropoda	11/4	2	1	1	0
Bivalvia	5/0	0	—	0	—
Crustacea					
Cladocera	1/0	0	—	0	—
Ostracoda	22/0	8	—	9	—
Copepoda	37/0	9	—	12	—
Amphipoda	8/0	4	—	0	—
Isopoda	3/1	2	0	1	1
Bathynellacea	7/0	7	—	—	—
Limnohalacarida	1/—	1	—	0	—
Hydrachnidia	19/—	8	—	1	—
Oribatida	—/54	—	0	—	5
Araneae	—/1	—	0	—	1
Opiliones	—/2	—	0	—	1
Pseudoscorpionida	—/1	—	0	—	1
Myriapoda	—/3	—	1 Chilopoda	—	2 Diplopoda
Collembola	—/1	—	1	—	0
Diplura	—/2	—	1	—	0
Zygentoma	—/1	—	0	—	0
Coleoptera	—/13	—	2	—	3
Diptera	—/12 (Families)	—	0	—	0
Hemiptera	—/1 (Order level)	—	0	—	0
Lepidoptera	—/1 (Order level)	—	0	—	0
Orthoptera	—/1 (Order level)	—	0	—	0
Psocoptera	—/1 (Order level)	—	0	—	0
Psiphonaptera	—/1 (Order level)	—	0	—	0
TOTAL: 299	187/112	46	8	34	14

4.1. Aquatic Fauna

Table 1 shows the distribution of the 187 aquatic taxa in OGNM by faunal group. Crustacea is the largest group, with 78 species, 30 of which are stygobiotics (some without a specific name). Groups with the largest species numbers include Copepoda, Ostracoda and Bathynellacea. Only five of the 49 aquatic species of Oligochaeta are stygobionts. In the case of the Acari, nine of the 20 aquatic species are stygobiotics. The two remaining stygobiotics species are Mollusca (two gastropod species). Similar to the case of the stygobiotics, most of the 34 stygophilic species belong to the Copepoda (12 species), Ostracoda (9), and Oligochaeta (8) (Table 1).

Although this work focuses on the subterranean species present in OGNM, it is important to highlight the repeated presence of the up to 187 taxa in the samples, and their unequal distributions and abundances among the different subterranean aquatic habitats and sampling points. As an important energetic resource of the ecosystem, this aquatic fauna may be able to provide information on the diversity and disparity of populations in different localities [31].

Below, we provide a list of stygobiont taxa found in OGNM. All of the samples have been reviewed by expert taxonomists; thus, we consider the list as valid, without specifying whether the "sp." notation refers to a new species, an already known species, or, possibly, a cryptic species. The specific site(s) from which each taxa was collected in OGNM is also included.

List of stygobiont species found in the Ojo Guareña National Monument, and their specific sites (sampling sites in the main cave, through the Palomera entrance: OG01–OG59, Figure 4).

Oligochaeta

Gianius navarroi Rodriguez and Achurra, 2010. OG10 and OG14.
Delaya navarrensis (Delay, 1973). OG21.
Stylodrilus mariae Achurra, Rodriguez and Erséus, 2015. OG07 and OG08.
Trichodrilus tenuis Habre, 1960. OG14.
Trichodrilus sp. 1. OG09.

Mollusca

Palaospeum septentrionale (Rolan and Ramos, 1995). Sima Jaime Cave, De la Hoz Creek, and Torcona Spring.
Spiralix (Burgosia) burgensis Boeters, 2003. Sima Jaime Cave, Trema River, and Trueba River.

Hydrachnidia

Axonopsis (Paraxonopsis) vietsi Motas and Tanasachi, 1947. Trueba River.
Frontipodopsis reticulatifrons Szalay, 1945. OG08, OG10, and Engaña River.
Frontipodopsis sp. Lasía River.
Kongsbergia sp. Trueba and Lasía rivers.
Neoacarus hibernicus Halbert, 1944. Nela River.
Protzia sp. Trueba River.
Pseudotorrenticola sp. Nela River.
Sperchonopsis sp. Lasía River.

Limnohalacarida (Halacaridae)

Soldanellonyx chappuisi Walter, 1917. OG: 07–10, 12, 14, 15, 17, 19, 30, 31, 34, 36–38, and 50; caves: Rizuelos, García, San Miguel, Cornejo and Redonda; springs: Torcona and Calzada; rivers: Trema, Trueba, Engaña, Nela, and De la Hoz and Torcon Well.

Copepoda, Cyclopoida

Acanthocyclops cf. *venustus* (Norman and Scott, 1906). Engaña and Nela rivers.
Acanthocyclops n. sp. OG: 10, 31, 34, 37, and 38.
Graeteriella (G.) unisetigera (Graeter, 1910). OG02 and OG07.
Speocyclops infernus (Kiefer, 1930). OG: 01, 08, 09, 14, 16–19, 21, 31, 38, and Huesos 2 and 4.
Speocyclops sebastianus Kiefer, 1937. Trema River.
Speocyclops spelaeus Kiefer, 1937. Caves: García, San Bernabé, Las Llanas, and Racino.

Copepoda, Harpacticoida

Bryocamptus (Rheocamptus) pyrenaicus (Chappuis, 1923). Caves: Kaite 2, García, Cornejo, Racino, and Redonda; Salce Spring.
Paracamptus sp. SA. Nela River.
Parastenocaris sp. OG: 07–09, 12, 14, 16, 17, 20, 21, 31–38, 41–43, 45, 46, 51–54, and Huesos 2; Villa Spring and De la Hoz and De la Cueva creeks.

Ostracoda

> Candoninae cf. gen. sp. 1 "Rounded". OG: 07, 09, 12, 17, 19, 21, 39; spring: Torcona, Cornejo and Calzada; Nela River.
> Candoninae "Trapezoid" 1. OG: 09, 19, 21; caves: Jaime, La Mina, Las Llanas; Guareña River.
> Candoninae cf. gen. sp. 2 "Trapezoid" 2. OG09.
> Candoninae cf. gen. sp. 2 "Trapezoid" 3. OG: 1, 18, 37, 38.
> Candoninae "Triangular" 1. OG: 15, 30–32 and Trema River.
> Candoninae "Triangular" 2. OG16 and Sima Jaime Cave.
> Candoninae "Triangular" 3. Torcona Spring and Trema River.
> *Cypria* sp. OG: 7, 9, 10, 35, 40, 42, 43, 51, 55–59, Huesos 2 and 3; San Miguel Cave; springs: Cornejo, Villa, Salce, Cubio, and Jordana; rivers: Guareña and Lasía.

Amphipoda

> *Haploginglymus* sp. 1. OG38; Torcona Spring; river: Trema, Trueba, and Nela; Torcon well.
> *Niphargus* sp. OG37 and 38; Cubio Spring and Trueba River.
> *Pseudoniphargus burgensis* Notenboom, 1986. Torcona Spring.
> *Pseudoniphargus* n. sp. 1. Torcona Spring.

Isopoda

> *Cantabroniscus primitivus* Vandel, 1965. OG21.
> *Stenasellus virei buchneri* (Stammer, 1936). OG: 02, 07, 09, 10, 14, 15, 17, 21, 39, 40, 44, 53, 55, and 57; caves: García, San Bernabé, San Miguel, Las Llanas, and Redonda and Salce Spring.

Bathynellacea

> *Iberobathynella burgalensis* Camacho, 2005. OG53
> *Iberobathynella cornejoensis* Camacho, 2005. Redonda and Racino caves and Trema River.
> *Iberobathynella guarenensis* Camacho, 2003. OG57 ("Erizos River").
> *Vejdovskybathynella edelweiss* Camacho, 2007. OG01, 09, 16, 17, 38, 50, 57, and Huesos 3 and 4; Racino and Mina caves and Cubio Spring
> *Vejdovskybathynella* n. sp. 1 (cryptic species "*edelweiss*"). OG57 ("Erizos River").
> *Vejdovskybathynella* n. sp. 2 (cryptic species "*edelweiss*"). Redonda and Sima Jaime caves.
> Bathynellidae n. gen. n. sp. OG09, 14, Torcona Spring and Prado Vargas Cave.

In OG, 25 stygobiotic species have been found across the 59 sampling sites, with many of the species being found in OG09 (12 species), followed by OG07, OG08, and OG10 (11 species each) and OG14 (seven species) (Figure 4).

Of the five stygobiotic species of Oligochaeta found in OG, *Gianius navarroi* described from OG [38], has only been found in two small ponds, OG10 and OG14. The two species of Mollusca live in other caves, springs and rivers but not in OG. The eight stygobiotics of Hydrachnidia live in the rivers, though one, *Frontipodopsis reticulatifrons*, has also been found in OG8 and OG10, two small epikarstic ponds that are filled with water from diffuse drips. The stygobiotic species of Limnohalacaridae is widely distributed throughout the principal cave (in many of the sites), and in all of the other types of habitats (multiple springs and rivers, and one of the two wells).

Crustaceans, a major animal group found in groundwater worldwide [2], are very well represented in OGNM by 78 taxa, 30 of which are stygobionts (Table 1). Copepods are present in almost all ponds and pools in OG. Of the 37 copepod species found, nine are stygobiotics that live in several caves and rivers and in a couple of springs. Twenty-two identified species of ostracods are distributed around OGNM. Of these, eight are stygobiotics that inhabit all the different types of sites sampled. Species of three of the five genera of Amphipoda found at the system are stigobiotics. They live mainly in springs and rivers, and also in two sites of OG. Of the two stygobiotic isopod species found, *Cantabroniscus primitivus* has only been found in the "siete lagos" area (OG21). The other species, *Stenasellus virei buchneri* (Figure 1), has been found in many sites in OG and in some of the other caves and springs. In addition to these isopod species, we found *Proasellus* cf *ortizi*, (Table S2), which we conservatively consider a stygophilic, despite its cave-like morphology. This species has been consistently found in OG12, although at varying abundances, and in three springs (Avellanos, Mea, and Jordana).

Bathynellacea, a strictly stygobiotic group of crustaceans, is represented in OGNM by seven species belonging to two families: Parabathynellidae and Bathynellidae (Figures 6 and 7). Four of the species, three belonging to Parabathynellidae (*Iberobathynella* Schminke, 1973 genus) and one to Bathynellidae (*Vejdovskybathynella* Serban and Leclerc, 1984 genus), have already been formally described [12–14,39]. The second genus found belonging to Bathynelidae is pending description. On the basis of molecular data, we discovered two cryptic species of *Vejdovskybathynella* within the family Bathynellidae [40]. These species are currently under further study. To our knowledge, no other karst area in the world has as many known species of Bathynellacea living together. The specimens of Bathynellidae found in Erizos River and Redonda Cave were initially identified as *Vejdovskybathynella edelweiss* (Figure 6), which is known from seven sites in OG, and from other caves of the system (Huesos, Racino, La Mina, and Cubio Spring). It is the most widely distributed species in OGNM. However, the specimens from Erizos River show a genetic divergence in the COI gene of 14.8% with the nominal species, and 17% with the specimens from Redonda Cave, which, in turn, show a 15.2% divergence with the nominal species [40]. These results imply that the specimens from Erizos and from Redonda each constitute a distinct species, and that there are actually three morphologically similar species of *Vejdovskybathynella* in the sampled sites. We expect this case will not be an isolated event, suggesting the biodiversity of the system is far greater than has been estimated. The distribution of the bathynellaceans is patchy. Species of both families have been found together in Redonda Cave and Erizos River (the only site from which *I. guarenensis* has been described). *Vejdovskybathynella edelweiss* and the new genus coexist in OG09. *Iberobathynella burgalensis* (Figure 7) only lives in gours in OG53. *Iberobathynella cornejoensis*, which has not been found in the main cave, inhabits a couple of small caves of the complex including Redonda (near the Trema sinkholes, where it has also been found) and Racino (to the west).

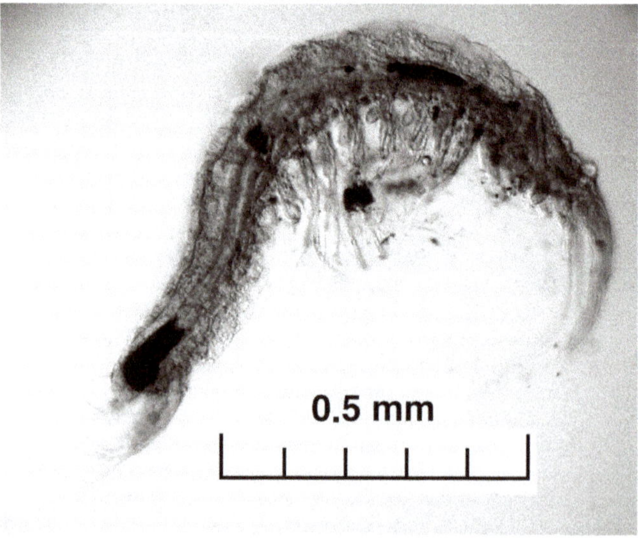

Figure 6. Habitus of *Vejdovskybathynella edelweiss* Camacho, 2007.

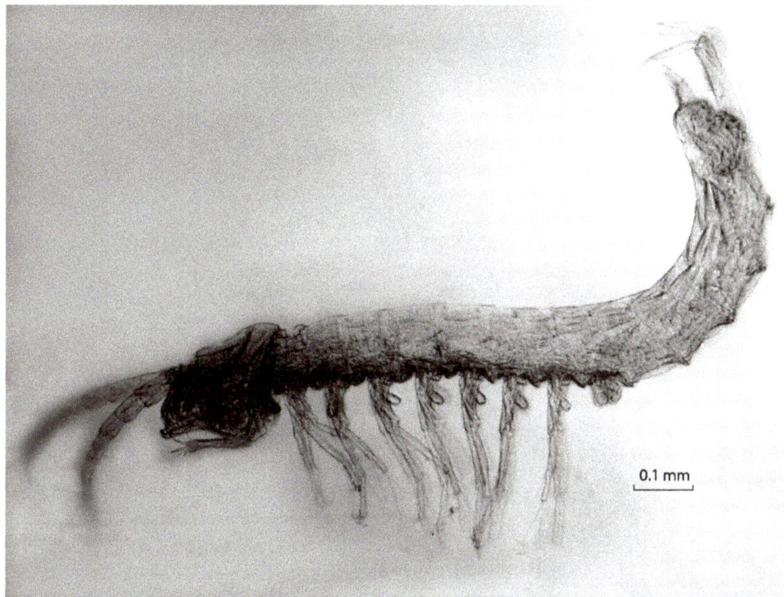

Figure 7. Habitus of *Iberobathynella burgalensis* Camacho, 2005.

4.2. Terrestrial Fauna

The terrestrial fauna in OGNM is poor compared with those in other systems: only 8 troglobiotic and 14 troglophilic species have been identified (Table S2). This is more likely due to a lack of sampling and study rather than to an absence of species. Given that we still have some samples waiting to be identified to the species level, we expect this number to increase.

Between 1968 and 1975, Ortiz sampled the terrestrial and aquatic fauna in Palomera. The material collected from two clayey areas of Palomera and from zones with organic residues throughout OG were subsequently studied by specialists [22–25] and gave rise to two monographs, one on terrestrial oligochaetes (eight species but only two troglobionts) [41] and another on oribatid mites (43 species but no troglobionts; three were new to Spain, and two were new to science) [42].

Other previous studies of terrestrial fauna in OG include those by Bellés and Antón [18]; Prieto and Gómez [43], who studied some terrestrial molluscs collected by the Niphargus Cave Club. Demange and Serra [44] described the only myriapod known from OG. Rambla cites the presence of a troglobiont opilion in OG [45] that Prieto and Zubiaga later studied [46]. Español [47] studied 10 coleopterans collected by Ortiz and found six species, one of them new to science (*Trechus ortizi*). The presence of some of these species in OG were confirmed through the samplings carried out between 2002 and 2003; however, most of the specimens from these later campaigns have yet to be identified by specialists.

The list of well-identified troglobiotic species is short, and specific information on the sites in which they have been found is lacking.

List of troglobiotic species found in the main cave of the Ojo Guareña National Monument:

Oligochaeta:

Aporrectodea rosea troglodyta (Álvarez, 1971).
Orodrilus paradoxoides (Álvarez, 1971).

Mollusca:

Zospeum schaufussi Frauenfeld, 1862.

Arachnida, Opiliones:

> *Litocampa zaldivarae* Sendra, Salgado and Monedero, 2003.

Myriapoda, Chilopoda:

> *Lithobius deroutae sexusbispiniger* Demange and Serra, 1978.

Collembola:

> *Verhoeffiella* cf *hispanicus* (Bonet, 1931).

Coleoptera:

> *Speocharis sharpi* Escalera, 1898.
> *Trechus ortizi* Español, 1970.

5. Discussion and Conclusions

Only a small fraction of the terrestrial ecosystem of OGNM has been studied: knowledge of the species that occur in the complex stems from sporadic samplings conducted during the second half of the twentieth century and between 2002 and 2003, with many of the specimens collected still unstudied. Therefore, a few data available about the troglobiotic fauna in the system makes any discussion irrelevant. By contrast, the abundance of information on the subterranean aquatic fauna of OGNM makes their discussion important, despite the fact that only a small part of the system's aquatic environment has been studied. Over a total of six years (2002–2004 and 2006–2009), a series of selected sites from the epikarst of the main gallery had been regularly sampled, with many other areas of the vast subterranean territory sporadically sampled.

The information available for the stygofauna goes beyond that provided by a still photo taken during sampling. Repeated samplings at different times of the annual cycle and over successive years (thus considering changes in environmental conditions or variability of drought-flood cycles) are necessary to find a stygofauna that is formed, in general, by "rare" species (i.e., those that are scarce or difficult to locate) that live in small populations in places with little to no access.

The obtained data reveal the typical patchy distribution of subterranean aquatic fauna [48]. The great heterogeneity of microhabitats, in a highly connected, vast continuous karstic extension, could explain the high richness of subterranean species observed in OGNM, as has been demonstrated in other cases [49,50]. Habitat heterogeneity and the spatiotemporal dynamics of this subterranean aquatic ecosystem are also evidenced by the great variation of species and communities observed, even over short periods of time and at a local scale of a few meters [31], a result common for subterranean environments [51–53].

Despite numerous samplings, covering the different microhabitats (gours, drips, ponds, lakes, and the interstitial medium of hypogean streams), large sections of OGNM remain unexplored, from both a topological (e.g., permanently flooded sections) and biological perspective. The inability to sample the entirety of the karst system implies an underestimation of its species richness.

The subterranean aquatic fauna in most karst areas in Europe has been fairly well studied, with around 1800 stygobiotic species known to date [54]. In Spain, slightly more than 200 are known from very few areas (representing 11% of all known species in Europe) [11]. In OGNM, of the 187 aquatic taxa found (see Table 1), 46 are stygobiotics, accounting for almost a quarter of all stygobiotic species known in our country. However, many of these taxa have been identified to only the genus level (e.g., for some taxa of Oligochaeta, Copepoda, Amphipoda, and Acari), or the subfamily level (e.g., Candoninae (Ostracoda)), and many specimens remain unidentified, suggesting the number of stygobiotic species will likely increase as more studies are published. We do not yet know the extent of endemism or the number of new species in OGNM, as many specimens from all of the major groups remain under study and have not yet been identified to a specific level.

Crustaceans constitute a major animal group in groundwater [2]: around 3400 species are known worldwide, 1200 of which are known to occur in Europe [54]. We identified 30 stygobiotic crustacean species in OGNM. Remarkably, we found seven species of

Bathynellacea in OGNM. This is a unique case as in no other karst system has this many species been found in coexistence. Moreover, the seven species are distributed among three genera: two known ones (*Iberobathynella* Schminke, 1973, and *Vejdovskybathynella* Serban and Leclerc, 1984, each with three species) and a third (Bathynellidae n. gen. n. sp.) that is new to science. Specimens of these species have been found, within the main cave, in five pools and the Erizos River, and in five of the other caves, two springs, and the Trema River. The two cryptic species of *Vejdovskybathynella*, *Vejdovskybathynella* sp. 1 and *Vejdovskybathynella* sp. 2, have also been found in two nearby caves outside OGNM, Río Chico, and Imunía [55]. The other five species of Bathynellacea are endemic to OGNM. A possible scenario is that the ancestral populations of these species, which were likely widely distributed in the area, became isolated due to droughts and vicariance. Although Bathynellacea lack active forms of dispersal (they do not have swimming larvae, like other crustaceans) and are generally not very mobile, a later (passive) dispersal event may have led to their genetic isolation and consequent speciation, resulting in the unusually high number of species coexisting, though in a patchy way, in a relatively small area. However, phylogeographic studies of these species are needed to confirm any of these hypotheses.

The largest number of stygobiotic species found in OGNM occur in the main cave (OG), with 25 species. A large number of these species is found at several sites: 12 species in OG09; 11 in OG07, OG08, and OG10; and seven in OG14. In addition, in general, the species were repeatedly found in all samplings, although in different proportions [31]. The 25 stygobiotics, along with the seven known troglobiotics, make OG a subterranean biodiversity hotspot within the wider OGNM hotspot (with its 54 subterranean species).

The number of stygobiont species in OGMN is striking. In no other area of Spain and Portugal has such a level of subterranean biodiversity been reported (see Table 1), even in the four hotspot caves in the Canary Islands (Felipe Reventón, Viento, Sobrado, and La Corona lava tube), which have between 28 and 36 troglobiotic species the first three and 38 stygobiotic species La Corona [2,3].

To improve our knowledge of the diversity of a region, it is essential to have large inventories of well-identified, georeferenced species in order to conduct standardized geospatial analyses that can be compared among regions and at different spatial scales [2,56]. Although the available data clearly demonstrate that the subterranean biodiversity in OGNM is relatively high, especially compared with that of other caves in the world [57,58]. The balance of sites and specific samples studied shows the relevance that the small known fraction can have compared with all that remains to be explored and studied from a biological perspective.

Among the common shortfalls of biodiversity data, the "Racovitzan impediment" is of great importance in the case of the OGNM. The sampling deficiency is large: there are numerous habitats not directly accessible to humans that, therefore, have not been sampled to date. Thus, we lack information about the species they may harbor. In addition, the lag between sampling efforts and the taxonomic identification of all specimens makes it difficult to obtain accurate data on the overall biodiversity of the system and its distribution. Although we are likely underestimating the real diversity of subterranean fauna in OGNM due to these issues, we confirm the prediction made by Culver and Pipan [2]: a subterranean hotspot is indeed present in northern Spain. Here, we provide the first list of stygobiotic and troglobiotic taxa present in OGNM, as well as demonstrate that, despite some knowledge gaps, this complex is a hotspot of subterranean biodiversity. Moreover, the diversity observed to date for OGNM implies its high rank among hotspot caves in not only Europe, but the world.

Future lines of research should focus on resolving the ecological status of the 48 taxa provisionally considered as stygophilics (34) or troglophilics (14) (see Table S2); discovering potential new cryptic species among the fauna through molecular analyses, such as those used to reveal cryptic species in the families Bathynellidae [55] and Parabathynellidae [59]; and identifying all remaining material to the lowest taxonomic level possible. These studies, along with a greater sampling effort, will all contribute to increase our knowledge of the subterranean biodiversity of this exceptionally conserved karst ecosystem.

Supplementary Materials: The following materials are available online at https://www.mdpi.com/article/10.3390/d13050199/s1: Figure S1, *Stenasellus virei buchneri* (Stammer, 1936) in OG14; Table S1, Georeferenced list of sampled sites on the Ojo Guareña Natural Monument; Table S2, List of stygophile and troglophile species and the sites where they have been found in Ojo Guareña Natural Monument.

Author Contributions: Both authors participated in all phases of the realization and writing of the paper. Both authors have read and agreed to the published version of the manuscript.

Funding: This work was supported by project PID2019-110243GB-100 (Ministerio de Ciencia e Innovación/FEDER).

Institutional Review Board Statement: Not applicable.

Informed Consent Statement: Not applicable.

Data Availability Statement: All data are available from the authors upon request.

Acknowledgments: We express our gratitude to J. Fernandez and B. Sanchez-Chillón for reviewing their bibliographic and specimen collections, respectively, for this study. We thank the Spanish PASCALIS team, as well as Fortunato Lázaro, Ana M. de Juan, and Jesús Robador for their participation in the samplings. Thanks to Pilar Rodríguez and Sanda Lepure for reviewing and correcting the list of oligochaetes and copepods, respectively. Thanks also to Consuelo Temiño and Beatriz Cabezas for their help from 2002–2010 under the agreements of the Junta de Castilla y León and the CSIC, and to the MNCN and the CSIC for their support in carrying out this work. And lastly, thanks to Miguel Rioseras for providing us with the surveys and cartographic data and Melinda Modrell, who helped us with the English translations.

Conflicts of Interest: The authors declare no conflict of interest.

References

1. Puch, C. La Montaña subterránea. *Pyrenaica* **2009**, *235*, 331–333.
2. Culver, D.; Pipan, T. Subterranean Ecosystems. In *Encyclopedia of Biodiversity*, 2nd ed.; Levin, S.A., Ed.; Academic Press: Waltham, MA, USA, 2013; Volume 7, pp. 49–62. [CrossRef]
3. Martínez, A.; González, B.C. Volcanic Anchialine Habitats of Lanzarote. In *Cave Ecology*; Ecological Studies 235; Moldovan, O.T., Kovác, L., Halse, S., Eds.; Springer: Cham, Switzerland, 2018; pp. 195–228.
4. Puch, C. *Grandes Cuevas y Simas de España*; Espelo Club de Grácia-Federación Española de Espeleología: Barcelona, Spain, 1998; p. 816.
5. Culver, D.C.; Deharveng, L.; Bedos, A.; Lewis, J.J.; Madden, M.; Reddell, J.R.; Sket, B.; Trontelj, P.; White, D. The mid-latitude biodiversity ridge in terrestrial cave fauna. *Ecography* **2006**, *29*, 120–128. [CrossRef]
6. Gulden, B. Official list of the National Speleological Society (USA). Available online: http://www.caverbob.com/wlong.htm (accessed on 2 February 2021).
7. Camacho, A.I.; Temiño, C.; Cabeza, B.; Puch, C. El Monumento Natural de Ojo Guareña (Burgos, España): Un "hotspot" de biodiversidad acuática subterránea. In *Cuevas: Patrimonio, Naturaleza, Cultura y Turismo*; Durán, J.J., Carrasco, F., Eds.; Asociación de Cuevas Turísticas Españolas: Madrid, Spain, 2010; pp. 621–636, ISBN 13:978-84-614-4630-8.
8. Wilson, E.O. Biodiversity research requires more boots on the ground. *Nat. Ecol. Evol.* **2017**, *1*, 1590–1591. [CrossRef]
9. Ficetola, G.F.; Canedoli, C.; Stoch, F. The Racovitzan impediment and the hidden biodiversity of unexplored environments. *Conser. Biol.* **2018**, *33*, 214–216. [CrossRef]
10. Camacho, A.I.; Dorda, B.A.; Rey, I. Undisclosed taxonomic diversity of Bathynellacea (Malacostraca: Syncarida) in the Iberian Peninsula revealed by molecular data. *J. Crust. Biol.* **2012**, *32*, 816–826. [CrossRef]
11. Deharveng, L.; Stoch, F.; Gibert, J.; Bedos, A.; Galassi, D.; Zagmajster, M.; Brancelj, A.; Camacho, A.; Fiers, F.; Martin, P.; et al. Groundwater biodiversity in Europe. *Freshw. Biol.* **2009**, *54*, 709–726. [CrossRef]
12. Camacho, A.I. Four new species of groundwater crustaceans (Syncarida, Bathynellacea, Parabathynellidae) endemic to the Iberian Peninsula. *J. Nat. Hist.* **2003**, *37*, 2885–2907. [CrossRef]
13. Camacho, A.I. One more piece in the genus puzzle: A new species of *Iberobathynella* Schminke, 1973 (Syncarida, Bathynellacea, Parabathynellidae) from the Iberian Peninsula. *Graellsia* **2005**, *61*, 123–133. [CrossRef]
14. Camacho, A.I. The first record of the genus *Vejdovskybathynella* Serban and Leclerc, 1984 (Syncarida, Bathynellacea, Bathynellidae) in the Iberian Peninsula: Three new species. *J. Nat. Hist.* **2007**, *41*, 2817–2841. [CrossRef]
15. Camacho, A.I.; Torres, T.; Puch, C.J.; Ortiz, J.E.; Valdecasas, A.G. Small-scale biogeographical pattern in groundwater Crustacea (Syncarida, Parabathynellidae). *Biodivers. Conserv.* **2006**, *15*, 3527–3541. [CrossRef]

16. Ortega, A.I.; Ruiz, F.; Martín, M.A.; Benito, A.; Vidal, M.; Bermejo, L.; Karampaglidis, T. Prehistoric human tracks in Ojo Guareña Cave System (Burgos, Spain): The Sala and Galerías de las Huellas. In *Reading Prehistoric Human Tracks*; Pastoor, A., Lenssen-Erz, T., Eds.; Springer: Berlin/Heidelberg, Germany, 2021; pp. 1–21. ISBN 10-3030604055; ISBN 13-978-3030604059.
17. Sáenz, C. Notas acerca de la estratigrafía del Supracretáceo y del Numulítico en la cabecera del Nela y zonas próximas. *Bol. R. Soc. Esp. Hist. Nat.* **1933**, *XXXIII*, 159–185.
18. GEE (Grupo Espeleológico Edelweiss). Monografía sobre Ojo Guareña. *Kaite* **1986**, *4–6*, 1–415.
19. Bellés, X. Ptinidos recogidos en cavidades subterráneas ibéricas (Col. Ptinidae). *Speleon* **1976**, *22*, 145–147.
20. Bellés, X. Notas sobre *Speocharis minus* Jeannel, 1909 y otros catópidos recogidos en cuevas de la provincia de Burgos. *Graellsia* **1977**, *31*, 115–124.
21. Bellés, X. Nuevos datos sobre la Fauna de Ojo Guareña. *Ixiltasun Izkutuak*, 1986; Unpublished.
22. Bellés, X. *Fauna Cavernícola i Intersticial de la Península Ibérica i les Illes Balears*; Monografies Cientifiques, 4; CSIC: Mallorca, Spain, 1987; p. 207.
23. Salgado, J.M. Origenes e distribuçao geográfica des Bathysciinae (Col. Catop.) cantábricos (Grupo *Speocharis*). *Ciénc. Biol.* **1976**, *1*, 105–130.
24. Salgado, J.M. Nuevos datos sobre Entomofauna Cavernícola de la zona de Carranza (Vizcaya). *Kobie* **1977**, *7*, 127–138.
25. Vives, E. Coleópteros cavernícolas nuevos o interesantes de la Península Ibérica y Baleares. *Speleon* **1976**, *22*, 159–169.
26. Vives, M. Noves localitats de Trichoniscidae cavernicolas de la fauna espanyola (Crustacís: Isopodes: Oniscoides). *Com. VI. Simp. Espeleol. Biospeleol.* **1977**, 97–101.
27. Notenboom, J.; Meijers, I. Investigaciones sobre la fauna de las aguas subterráneas de España: Lista de estaciones y primeros resultados. In *Verslagen en Technische Gegevens*; Instituut voor Taxonomische Zoölogie (Zoologische Museum), Universiteit von Amsterdam: Amsterdam, The Netherlands, 1985; Volume 42, pp. 1–43.
28. Notenboom, J. Introduction to the Iberian Groundwater amphipods. *Limnetica* **1990**, *6*, 165–176.
29. Camacho, A.I. *Informe Biológico: Ojo Guareña (Sotoscueva-Burgos)*, 1993; Unpublished. 29.
30. Camacho, A.I.; Valdecasas, A.G.; Rodríguez, J.; Puch, C. *Biodiversidad Faunística del Complejo Kárstico de Ojo Guareña: Evaluación de la Influencia de la Presión Humana en Algunas de Las Poblaciones de Invertebrados en el Monumento Natural de Ojo Guareña (Burgos)*, Informe Final, Marzo. 2005; Unpublished. 110
31. Camacho, A.I.; Puch, C. *Colonización, éxito Evolutivo y Biodiversidad Faunística del Complejo Kárstico de Ojo Guareña en el Monumento Natural de Ojo Guareña (Burgos)*, Informe Final. 2010; Unpublished. 240.
32. Olmo, P.d.; Ramírez del Pozo, J. *Mapa Geológico 1:50.000 n° 84 (Espinosa de los Monteros)*; IGME: Madrid, Spain, 1978.
33. Camacho, A.I. Sampling the subterranean biota. Cave (aquatic environment). In *The Natural History of Biospeleology*; Camacho, A.I., Ed.; Monografías del Museo Nacional de Ciencias Naturales, 7; CSIC: Madrid, Spain, 1992; pp. 135–168. ISBN 84-00-07280-4.
34. Malard, F.; Dole-Olivier, M.J.; Mathieu, J.; Stoch, F.; Brancelj, A.; Camacho, A.I.; Fiers, F.; Galassi, D.; Gibert, J.; Lefebure, T.; et al. *Sampling Manual for the Assessment of Regional Groundwater Biodiversity. PASCALIS Project (V Framework Programme. Key Action 2: Global Change, Climate and Biodiversity. 2.2.3 Assessing and Conserving Biodiversity)*; Malard, F., Ed.; Pascalis: Villeurbanne, France, Contract n°: N°EVK2-CT-2001-00121; 2002; p. 110. Available online: http//www.pascalis-project.org (accessed on 31 March 2021).
35. Camacho, A.I.; Dorda, B.A.; Rey, I. Identifying cryptic speciation across groundwater populations: First COI sequences of Bathynellidae (Crustacea, Syncarida). *Graellsia* **2011**, *67*, 7–12. [CrossRef]
36. Camacho, A.I.; Dorda, B.A.; Rey, I. Old and new taxonomic tools: Description of a new genus and two new species of Bathynellidae from Spain with morphological and molecular characters. *J. Nat. Hist.* **2013**, *47*, 1393–1420. [CrossRef]
37. Deharveng, L.; Bedos, A. Diversity of terrestrial invertebrates in subterranean habitats. In *Cave Ecology*; Moldovan, O.T., Kovác, L., Halse, S., Eds.; Springer: Cham, Switzerland, 2018; pp. 107–172.
38. Rodriguez, P.; Achurra, A. New species of aquatic oligochaetes (Annelida: Clitellata) from groundwaters in karstic areas of northern Spain, with taxonomic remarks on *Lophochaeta ignota* Stolc, 1886. *Zootaxa* **2010**, *2331*, 21–29. [CrossRef]
39. Camacho, A.I. Expanding the taxonomic conundrum: Three new species of groundwater crustaceans (Syncarida, Bathynellacea, Parabathynellidae) endemic to the Iberian Peninsula. *J. Nat. Hist.* **2005**, *39*, 1819–1838. [CrossRef]
40. Camacho, A.I.; Dorda, B.A.; Rey, I. Integrating DNA and morphological taxonomy to describe a new species of the family Bathynellidae (Crustacea, Syncarida) from Spain. *Graellsia* **2013**, *69*, 179–200. [CrossRef]
41. Álvarez, J. Biospeleología de la cueva Ojo Guareña. Oligoquetos terrícolas. *Bol. R. Soc. Esp. Hist. Nat. (Biol.)* **1971**, *69*, 11–18.
42. Pérez-Iñigo, C. Biospeleología de la cueva Ojo Guareña. Acaros oribátidos. *Bol. R. Soc. Esp. Hist. Nat. (Biol.)* **1969**, *67*, 143–160.
43. Prieto, C.E.; Gómez, B.J. Primeros datos de *Zospeum* (Mollusca, Gastropoda, Ellobiidae) para la provincia de Burgos. *Com. 2° Simp. Reg. Espeleol. Burgos* **1984**, 143–147.
44. Demange, J.M.; Serra, A. Etude des rapports de longueur des articles des P.15 de quelques *Lithobius* cavernicoles de l'Espagne et des Pyrénées françaises. Description d'une espèce et une sous-espèce nouvelles (Chilopoda, Lithobiomorpha). *Speleon* **1978**, *24*, 39–54.
45. Rambla, M. Contribución al estudio de los Opiliones de la Fauna Ibérica. Las especies del grupo *nemastoma bacilliferum*, Simon, 1879 en la Península Ibérica (Opiliones, Fam. Nemastomatidae). *Publ. Inst. Biol. Apllc.* **1968**, *45*, 33–56.
46. Prieto, C.E.; Zubiaga, A. El género *Ischyropsalis* C.L. Koch (Ischyropsalididae, Opiliones) en la provincia de Burgos. *Mem. 2° Simp. Reg. Espeleol. Burgos* **1984**, 15–19.

47. Español, F. Un nuevo *Trechus* cavernícola del norte de Burgos (Col. Trechidae). *Speleon* **1970**, *XVII*, 53–57.
48. Mösslacher, F. Evolution, Adaption und Verbreitung. In *Grundwasseroökologie*; Griebler, C., Mösslacher, F., Eds.; UTB-Facultas Verlag: Wien, Austria, 2003; pp. 253–310.
49. Culver, D.C.; Christman, M.C.; Sket, B.; Trontelj, P.P. Sampling adequacy in an extreme environment: Species richness patterns in Slovenian caves. *Biodivers. Conserv.* **2004**, *13*, 1209–1229. [CrossRef]
50. Bregovic, P.; Zagmajster, M. Understanding hotspots within a global hotspot—Identifying the drivers of regional species richness patterns in terrestrial subterranean habitats. *Insect Conserv. Biodivers.* **2016**, *9*, 268–281. [CrossRef]
51. Gibert, J.; Deharveng, L. Subterranean ecosystems: A truncated functional biodiversity. *Bioscience* **2002**, *52*, 473–481. [CrossRef]
52. Griebler, C.; Mösslacher, F. Grundwasser—eine ökosystemare Betrachtung. In *Grundwasserökologie*; Griebler, C., Mösslacher, F., Eds.; UTB-Facultas Verlag: Wien, Austria, 2003; pp. 253–310.
53. Hahn, H.J. The GW-Fauna-Index: A first approach to a quantitative ecological assessment of groundwater habitats. *Limnologica* **2006**, *36*, 119–137. [CrossRef]
54. Stoch, F.; Galassi, D.M.P. Stygobiotic crustacean species richness: A question of numbers, a matter of scale. *Hydrobiologia* **2010**, *653*, 217–234. [CrossRef]
55. Camacho, A.I.; Mas-Peinado, P.; Dorda, B.A.; Casado, A.; Brancelj, A.; Knight, L.R.F.D.; Hutchins, B.; Bou, C.; Perina, G.; Rey, I. Molecular tools unveil an underestimated diversity in a stygofauna family: A preliminary world phylogeny and updated morphology of Bathynellidae (Crustacea: Bathynellacea). *Zool. J. Linn. Soc.* **2018**, *183*, 70–96. [CrossRef]
56. Zagmajster, M.; Malard, F.; Eme, D.; Culver, D.C. Subterranean biodiversity patterns from global to regional scales. In *Cave Ecology, Ecological Studies 235*; Moldovan, O.T., Kovác, L., Halse, S., Eds.; Springer: Cham, Switzerland, 2018; pp. 195–228.
57. Deharveng, L.; Bedos, A. The cave fauna of southeast Asia. Origin, evolution, and ecology. In *Subterranean Ecosystems*; Wilken, H., Culver, D.C., Humphreys, W.F., Eds.; Elsevier: Amsterdam, The Netherlands, 2000; pp. 603–632.
58. Pipan, T.; Deharveng, L.; Culver, D.C. Hotspots of Subterranean Biodiversity. *Diversity* **2020**, *12*, 209. [CrossRef]
59. Camacho, A.I. Diversity, morphological homogeneity and genetic divergence in a taxonomically complex group of groundwater crustaceans: The little known Bathynellacea (Malacostraca). *Bull. Soc. d'Hist. Nat. Toulouse* **2019**, *154*, 105–160.

Article

The Chemoautotrophically Based Movile Cave Groundwater Ecosystem, a Hotspot of Subterranean Biodiversity

Traian Brad [1,2,†], **Sanda Iepure** [1,2,†] and **Serban M. Sarbu** [3,4,*,†]

1   "Emil Racoviță" Institute of Speleology, str. Clinicilor nr. 5-7, 400006 Cluj-Napoca, Romania; traian.brad@academia-cj.ro (T.B.); sanda.iepure@academia-cj.ro (S.I.)
2   Institutul Român de Știință și Tehnologie, str. Virgil Fulicea nr. 3, 400022 Cluj-Napoca, Romania
3   "Emil Racoviță" Institute of Speleology, str. Frumoasă nr. 31, 010986 București, Romania
4   Department of Biological Sciences, California State University, Chico, CA 95929, USA
*   Correspondence: serban.sarbu@yahoo.com
†   All the three authors have the same contribution.

Abstract: Movile Cave hosts one of the world's most diverse subsurface invertebrate communities. In the absence of matter and energy input from the surface, this ecosystem relies entirely on in situ primary productivity by chemoautotrophic microorganisms. The energy source for these microorganisms is the oxidation of hydrogen sulfide provided continuously from the deep thermomineral aquifer, alongside methane, and ammonium. The microbial biofilms that cover the water surface, the cave walls, and the sediments, along with the free-swimming microorganisms, represent the food that protists, rotifers, nematodes, gastropods, and crustacean rely on. Voracious water-scorpions, leeches, and planarians form the peak of the aquatic food web in Movile Cave. The terrestrial community is even more diverse. It is composed of various species of worms, isopods, pseudoscorpions, spiders, centipedes, millipedes, springtails, diplurans, and beetles. An updated list of invertebrate species thriving in Movile Cave is provided herein. With 52 invertebrate species (21 aquatic and 31 terrestrial), of which 37 are endemic for this unusual, but fascinating environment, Movile Cave is the first known chemosynthesis-based groundwater ecosystem. Therefore, Movile Cave deserves stringent attention and protection.

Keywords: Movile Cave; Romania; subterranean biodiversity; chemoautotrophically based; groundwater ecosystem

Citation: Brad, T.; Iepure, S.; Sarbu, S.M. The Chemoautotrophically Based Movile Cave Groundwater Ecosystem, a Hotspot of Subterranean Biodiversity. *Diversity* **2021**, *13*, 128. https://doi.org/10.3390/d13030128

Academic Editor: Michael Wink

Received: 24 February 2021
Accepted: 12 March 2021
Published: 17 March 2021

1. Introduction

Movile Cave is an underground void, which has no natural opening to the surface. It is located on the outskirts of the town of Mangalia (SE Romania), 2 km from the Black Sea shore [1] and it was intercepted 18 m below the surface by an artificial geological survey shaft dug in June 1986. The cave is developed in Sarmatian limestones (12.5 MY), which were covered by Quaternary deposits (clays and loess), approximately 2.5 million years ago [2], and it was formed by classical karst dissolution processes combined with sulfuric acid speleogenesis (SAS), a process mediated by sulfur-oxidizing bacteria [3]. Additional geographical description information and geological data is presented in Sarbu et al. 2019 [1]. The cave appears as a horizontal maze, with a total development of 240 m, and consists of two levels (Figure 1). The upper level is dry and it lacks speleothems. The atmosphere in the upper level is warm (21 °C) and it contains 19% dioxygen (O_2) and 1% carbon dioxide (CO_2). The lower level is 40 m long and partially flooded, with the Lake Room and three Air-Bells as the only aerated zones. The O_2 concentration decreases gradually in the Air-Bells to 7%, while the CO_2 concentration increases to 3.5% due to the activity of microorganisms, which form biofilms [4,5]. In the Air-Bells, the O_2 dissolved in water originates from the cave atmosphere and it is rapidly used for the oxidation of hydrogen sulfide (H_2S) and methane (CH_4). Below a depth of 1 mm, the water becomes completely anoxic [6]. The water is relatively stagnant at the surface in the Lake Room and in the nearby Air-Bells, while some flow (i.e., $5\,l\,s^{-1}$) was detected at depths over 1 m in the flooded cave passages [3].

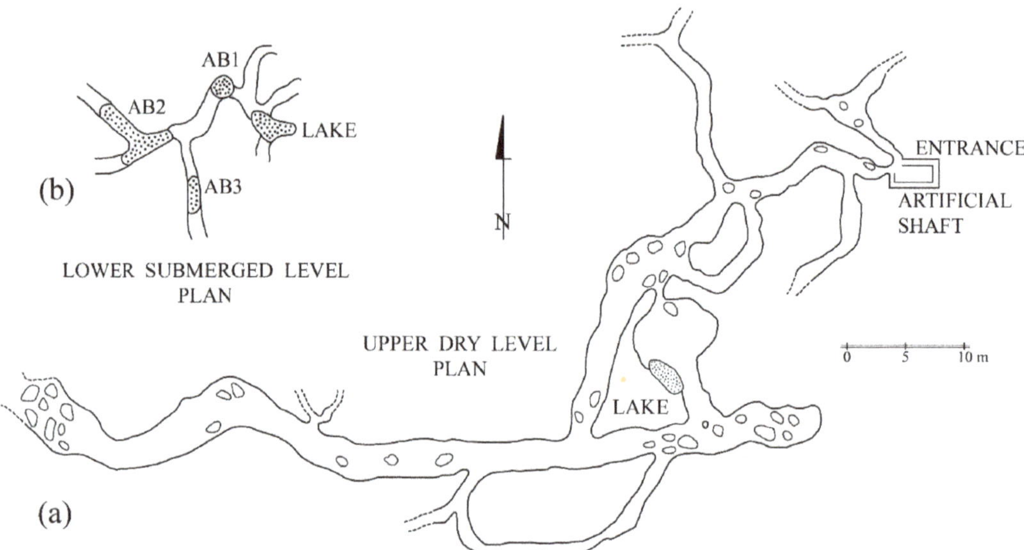

Figure 1. Plan of Movile Cave (**a**) with depiction of the water surfaces (dotted areas) in Lake Room and the three Air-Bells (AB) (**b**).

Movile Cave is the first subterranean groundwater ecosystem described to be solely based on chemosynthesis [7]. Reduced chemical compounds such as H_2S, CH_4 and ammonium (NH_4^+) are supplied continuously and in large amounts by the thermomineral water that ascends along geological faults from an artesian aquifer [5] located at a depth of 180 to 200 m in Mesozoic limestones [2]. Movile Cave is a window of access to a groundwater aquifer that occupies a surface area of approximately 50 to 100 km^2 and supports a particular hypogean ecosystem. Other windows of access to the aquifer are the old hand-dug wells in the town of Mangalia and in the surrounding villages, as well as several thermal sulfidic springs located along the sea-shore, at distances of 1 to 3 km from Movile Cave. Several of the endemic species that inhabit Movile Cave were also encountered in these springs and wells.

Along with CH_4 and NH_4, H_2S represents the energy source for a wide variety of microorganisms [8] that use O_2, nitrate, sulfate, and ferric iron (Fe^{3+}) as electron acceptors [9–11]. The carbon fixed in situ by microorganisms, either swimming freely or congregated in thick biofilms, represents the food base of the trophic webs in Movile Cave. Aquatic invertebrates that graze on microorganisms, roam at the water surface where O_2 is available in small concentrations. Terrestrial invertebrates are rarely found in the upper dry sections of the cave, but they are unusually abundant in the lower, partially flooded cave level (Figure 1), in the Lake Room and the Air-Bells, where isopods, pseudoscorpions, millipedes, and insects feed on nearly any organic debris or graze on soil microorganisms, while predatory chilopods and spiders chase their prey, consisting of isopods, collembola, beetles or other spiders.

Analogous chemosynthesis-based cave ecosystems, such as those in the Frasassi caves (Italy) [12–14], Melissotrypa Cave (Greece) [15], Ayyalon Cave (Israel) [16,17], and Tashan Cave (Iran) [18–20], display comparable diversities of cave dwelling invertebrates with large numbers of endemisms (Table 1). These ecosystems also depend primarily on in situ carbon fixation by chemoautotrophic microorganisms using the H_2S present in water. The natural resources in these ecosystems sustain, just like in Movile Cave, the growth of diverse and complex microbial communities, that form biofilms.

Table 1. The number of invertebrate species encountered in sulfidic cave ecosystems analogous to Movile Cave. All endemic species encountered in these sulfidic ecosystems are restricted to caves.

Cave	Species Present	Endemic Species
Movile Cave (Romania)	52	37
Frasassi caves (Italy)	56	16
Melissotrypa Cave (Greece)	30	8
Ayyalon Cave (Israel)	8	7
Tashan Cave (Iran)	3	3

These ecosystems can be appropriate examples for the ecological theory regarding the diversity-driven speciation [21] and convergent or parallel evolution. On the contrary, in caves with energy input from the surface, thus scarcer and less diverse food resources, the diversity of cave organisms is significantly lower.

The purpose of this study is to provide an up-to-date list of invertebrate cave-dwelling species living in the peculiar Movile Cave ecosystem and to draw attention to the scientific importance of these species and their fascinating habitat.

2. Movile Cave Fauna

2.1. Aquatic Fauna

In Movile Cave, the food base for aquatic invertebrates is produced autochthonously, and consists of microorganisms that thrive in the sulfide-rich water and sediments and use the energy resulting from the oxidation of the reduced chemical compounds from the thermo-mineral water [8].

The microorganisms swim freely in water as bacterioplankton, or they gather in thick biofilms floating at the water surface or they attach to rock surfaces (Figure 2). These represent a copious food source for the numerous consumers thriving in this groundwater ecosystem [7]. Various types of Archaea [10], sulfur-oxidizing bacteria [8], methanotrophs [9,10,22–24], or nitrifying and denitrifying bacteria [10], have been identified in Movile Cave. Representatives of a newly described strain of *Thiovulum* swim actively at the water surface and gather in loose veils [25]. Epi-symbiotic strains of *Thiothrix* sp. live on the bodies of aquatic amphipod crustaceans [11]. A diverse fungal community, associated with microbial mats and submerged sediments, is also present [26]. Underneath the floating mats formed by microorganisms, a diverse community of aquatic invertebrates feast on the abundant and assorted food provided [1,7].

Flagellate, ciliate, and amoebozoan protists and rotifers are the smallest grazers in this peculiar groundwater ecosystem [27]. They feed on bacterioplankton and microbial biofilms. Microorganisms, either pro- or eukaryotic, are consumed by the abundant meiofauna consisting of rotifers, nematodes, polychaetes, and copepod and ostracod (Figure 3C) crustaceans. Among the aquatic invertebrates, the nematodes (*Panagrolaimus* sp. and *Poikilolaimus* sp.) are important prey for cyclopoid copepods *Eucyclops greateri scythicus* [28]. Groups of hundreds of Moitessierid gastropods *Heleobia dobrogica* (Figure 3E) gather at the edge of the water, along the lake banks feeding on microbial biofilms [29]. Crustaceans of the Ostracoda, Copepoda, Amphipoda, and Isopoda (Figure 3B) swim at the water surface or slink on the submerged cave walls, and feed on smaller organisms they come across (Table 2). Eyeless and unpigmented water scorpions (*Nepa anophthalma*, Hemiptera), one of the top predators in this aquatic environment (Figure 3D), hide cautiously under the water surface, between the lake walls asperities, and wait for prey consisting of amphipods and isopods. Leeches (*Haemopis caeca*) are also considered top predators in this aquatic ecosystem (Figure 3A). They swim elegantly in the mass of water, diving deeper sometimes, and approach the lake banks where they predate on earthworms (*Helodrilus* sp.) that live in high numbers in the sediments. Flat worms (*Dendrocoelum obstinatum*) glide on sediments in shallow water near the lake banks, never deeper than 1 cm, where some dissolved O_2 is still present, or swim at the water surface [30]. They graze on microorganisms, or they can predate on worms and crustaceans [1].

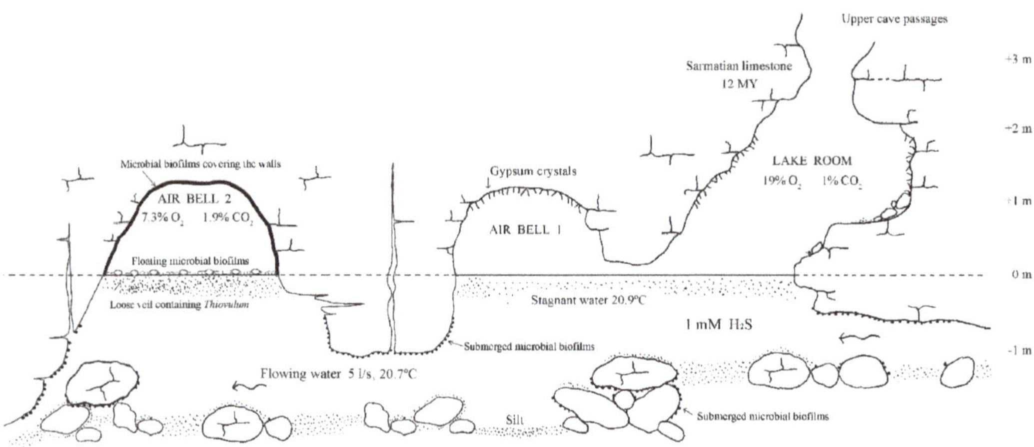

Figure 2. Longitudinal profile through the submerged cave gallery (modified after Sarbu and Popa, 1992). Some water flow is present at depths greater than 1 m, while at the surface, the water is rather stagnant and supports the growth of microorganisms in thick or loose biofilms. Anaerobic microbes attach to the cave walls, rocks, and sediments.

The trophic chains in Movile Cave are likely simple compared to the complex food webs in aboveground ecosystems with more complex interactions. The in-situ food production, the diversity of microorganisms and invertebrates in Movile Cave, are remarkably rich [1] for a cave environment. Conversely, the majority of caves resulting from epigenetic karstic dissolution processes, where the base of the food web is represented by input of allochthonous food of photoautotrophic origin, host lower numbers of cave-adapted species. Using stable isotope analysis, Sarbu et al. [5,7] provided a first diagram of the aquatic food web of Movile Cave. Its base is represented by the microbial biofilms formed mainly as a result of H_2S and CH_4 oxidation. The primary consumers are terrestrial grazers (*Archiboreoiulus serbansarbui*, *Trachelipus troglobius* and *Armadillidium tabacarui*), and aquatic grazers (*Helodrilus* sp., *Niphargus racovitzai*, *Niphargus dancaui*). These are predated upon by the secondary consumers and top predators (aquatic) *Nepa anophthalma* and *Haemopis caeca*; (terrestrial) *Medon dobrogicus*, *Agraecina cristiani* and *Cryptops speleorex*.

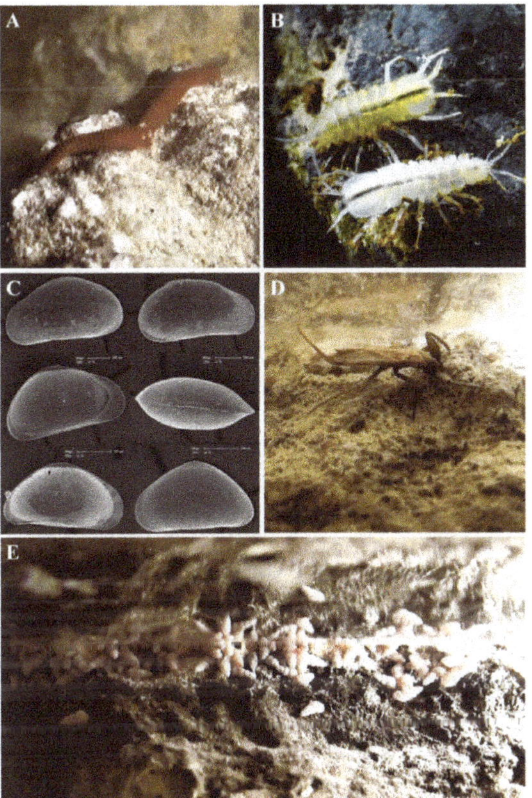

Figure 3. Aquatic invertebrates present in Movile Cave; (**A**). the leech (*Haemopis caeca*); (**B**). isopod crustaceans (*Asellus aquaticus infernus*); (**C**). ostracods (*Pseudocandona* sp. nov.), the scale length is 200 µm; (**D**). water scorpions (*Nepa anophthalma*); (**E**). gastropods (*Heleobia dobrogica*).

2.2. Terrestrial Fauna

The terrestrial fauna is more complex. It is composed of four species of isopods, six spider species, four pseudoscorpions, one acarian species, three chilopods, two millipedes, three springtails, two dipluran species, and five beetles (Tables 2 and 3). The largest invertebrate species and top predator in Movile Cave ecosystem is *Cryptops speleorex* (Figure 4B) [31]. These voracious centipedes are 8-10 cm long, and they roam continuously in search for prey, which is not scarce, and it ranges from the smallest collembolan or coleoptera species to the stout isopods *Trachelipus troglobius* (Figure 4D). The geophilid centipedes *Geophilus* sp. and *Clinopodes carinthiacus* are also among the predators in this ecosystem. They chase smaller prey, such as collembola, smaller isopods or the offspring of the larger isopods, and pseudoscorpions.

Figure 4. Terrestrial invertebrates present in Movile Cave; (**A**). the spider (*Agraecina cristiani*); (**B**). top predator chilopod (*Cryptops speleorex*); (**C**). millipede (*Archiboreoiulus serbansarbui*); (**D**). isopod (*Trachelipus troglobius*).

Numerous collembola (springtails) are present on the water surface, on top of the floating microbial biofilms, as well as on the biofilms that cover the cave walls in the remote Air-Bells [32,33]. These are very small hexapods, and therefore, they do not count much in the total biomass available for higher trophic levels. Instead, the three collembola species are significant for very high numbers. They achieve large densities because of the availability of luxurious resources, both in terms of type and amount of food. Collembola are everywhere in populated places in Movile Cave, such as in Lake Room or Air-Bells. Two of the three species jump continuously in all directions; therefore, they can easily become part of the menu of other cave inhabitants. Many of the small-size predators rely heavily on springtails as their primary food source.

The largest terrestrial species diversity and density in Movile Cave is present in Lake Room (Figure 2). Here, one can observe slender millipedes (*Archiboreoiulus serbansarbui*, in Figure 4C, and *Strongylosoma jaqueti*), the tiny *Haplophthalmus movilae* or the large and hunchbacked isopods *Trachelipus troglobius* (Figure 4D) approaching the lake banks for drinking, or more likely for grazing on microorganisms present along the lake shore, with the risk of being preyed upon by water scorpions (*Nepa anophthalma*). The millipedes and isopods are mostly present in the Lake Room and Air-Bell 1, most likely because O_2 is still abundant in these chambers. On the contrary, other isopods (i.e., *Caucasonethes vandeli pygmaeus* and *Armadillidium tabacarui*) are by far more abundant in the remote Air-Bells (2 and 3).

Table 2. List of troglobionts and stygobionts from Movile Cave.

	Aquatic/Terrestrial	Species	Taxonomic Affiliation	References
1	Aquatic	*Dendrocoelum obstinatum* *; Stocchino et al., 2017	Platyhelminthes, Dendrocoelidae	[30]
2	Aquatic	*Panagrolaimus* cf. *thienemani* *	Nematoda, Panagrolaimidae	[34]
3	Aquatic	*Chronogaster troglodytes* *; Poinar and Sarbu, 1994	Nematoda, Chronogasteridae	[35]
4	Aquatic	*Haemopis caeca* *,#; Manoleli et al., 1998	Annelida, Hirudinea, Haemopidae	[36]
5	Aquatic	*Helodrilus* sp. nov. *	Annelida, Clitellata, Lumbricidae	Martin, P., pers. comm.
6	Aquatic	*Heleobia dobrogica* *; Grossu and Negrea, 1989	Gastropoda, Moitessieriidae	[29]
7	Aquatic	*Pseudocandona* sp. nov. *	Crustacea, Ostracoda, Cyprididae	Danielopol, D., pers. comm.
8	Aquatic	*Eucyclops graeteri scythicus* *; Plesa, 1989	Crustacea, Copepoda, Cyclopidae	[37]
9	Aquatic	*Parapseudoleptomesochra italica*; Pesce and Petkovski, 1980	Crustacea, Copepoda, Harpacticoida	Rouch, pers. comm.
10	Aquatic	*Niphargus racovitzai* *; Dancau, 1970	Crustacea, Amphipoda, Niphargidae	[38]
11	Aquatic	*Niphargus dancaui* *,#; Brad et al., 2015	Crustacea, Amphipoda, Niphargidae	[39]
12	Aquatic	*Asellus aquaticus infernus* *,#; Turk-Prevorčnik and Blejec, 1998	Crustacea, Isopoda, Asellidae	[40]
13	Terrestrial	*Caucasonethes vandeli pygmaeus* *; Giurginca, 2020	Crustacea, Isopoda, Trichoniscidae	[41]
14	Terrestrial	*Haplophthalmus movilae* *; Gruia and Giurginca, 1998	Crustacea, Isopoda, Trichoniscidae	[42]
15	Terrestrial	*Trachelipus troglobius* *; Tabacaru and Boghean, 1989	Crustacea, Isopoda, Trachelipodidae	[43]
16	Terrestrial	*Armadillidium tabacarui* *; Gruia et al., 1994	Crustacea, Isopoda, Armadillidiidae	[44]
17	Terrestrial	*Chthonius monicae* *; Boghean, 1989	Arachnida, Pseudoscorpiones, Chthoniidae	[45]
18	Terrestrial	*Chthonius borissketi* *; Curčić et al., 2014	Arachnida, Pseudoscorpiones, Chthoniidae	[46]
19	Terrestrial	*Roncus dragobete* *; Curčić et al., 1993	Arachnida, Pseudoscorpiones, Neobisiidae	[47]
20	Terrestrial	*Roncus ciobanmos* *; Curčić et al., 1993	Arachnida, Pseudoscorpiones, Neobisiidae	[47]

Table 2. *Cont.*

	Aquatic/Terrestrial	Species	Taxonomic Affiliation	References
21	Terrestrial	*Palliduphantes constantinescui* *; Georgescu, 1989	Arachnida, Araneae, Linyphiidae	[48]
22	Terrestrial	*Agraecina cristiani* *,#; Georgescu, 1989	Arachnida, Araneae, Liocranidae	[48]
23	Terrestrial	*Kryptonesticus georgescuae* *; Nae, Sarbu, and Weiss, 2018	Arachnida, Araneae, Nesticidae	[49]
24	Terrestrial	*Hahnia caeca* *; Georgescu and Sarbu, 1992	Arachnida, Araneae, Hahniidae	[50]
25	Terrestrial	*Labidostomma motasi* *; Iavorschi, 1992	Arachnida, Acarina, Labidostommatidae	[51]
26	Terrestrial	*Geophilus* sp. nov. *	Chilopoda, Geophilidae	Baba, St., pers. comm.
27	Terrestrial	*Cryptops speleorex* *,#; Vahtera et al., 2020	Chilopoda, Cryptopidae	[31]
28	Terrestrial	*Archiboreoiulus serbansarbui* *,#; Giurginca et al., 2020	Diplopoda, Julida, Julidae	[52]
29	Terrestrial	*Onychiurus movilae* *; Gruia, 1989	Collembola, Onychiuridae	[53]
30	Terrestrial	*Oncopodura vioreli* *; Gruia, 1989	Collembola, Oncopoduridae	[53]
31	Terrestrial	*Plusiocampa isterina* *; Condé, 1993	Diplura, Campodeidae	[54]
32	Terrestrial	*Plusiocampa euxina* *; Condé, 1996	Diplura, Campodeidae	[55]
33	Terrestrial	*Medon dobrogicus* *; Decu and Georgescu, 1994	Coleoptera, Staphylinidae	[56]
34	Terrestrial	*Tychobythinus sulphydricus* *; Poggi and Sarbu, 2013	Coleoptera, Staphylinidae	[57]
35	Terrestrial	*Decumarellus sarbui* *; Poggi, 1994	Coleoptera, Staphylinidae	[58]
36	Terrestrial	*Bryaxis dolosus* *; Poggi and Sarbu, 2013	Coleoptera, Staphylinidae	[57]
37	Terrestrial	*Clivina subterranea* *; Decu et al., 1994	Coleoptera, Clivinidae	[59]
38	Aquatic	*Nepa anophthalma* *; Dedu et al., 1994	Hemiptera, Nepidae	[60]

*—species endemic to Movile Cave; #—species found in nearby springs and wells.

Table 3. List of troglophiles and stygophiles from Movile Cave.

	Aquatic/Terrestrial	Species	Taxonomic Affiliation	References
1	Aquatic	*Udonchus tenuicaudatus*; Cobb, 1913	Nematoda, Rhabdolaimidae	[34]
2	Aquatic	*Poikilolaimus* sp.	Nematoda, Rhabditidae	[34]
3	Aquatic	*Monhystrella* sp.	Nematoda, Monhysteridae	[34]
4	Aquatic	*Habrotrocha rosa*; Donner, 1949	Rotatoria, Habrotrochidae	Ricci, C., pers. comm.
5	Aquatic	*Habrotrocha bidens*; Gosse, 1851	Rotatoria, Habrotrochidae	Ricci, C., pers. comm.
6	Aquatic	*Aelosoma hyalinum*; Bunke, 1967	Annelida, Aeolosomatidae	Dumnicka, E., pers. comm.
7	Aquatic	*Aelosoma italica*; Bunke, 1967	Annelida, Aeolosomatidae	Dumnicka, E., pers. comm.
8	Aquatic	*Tropocyclops prasinus*; Fischer, 1860	Crustacea, Copepoda, Cyclopidae	[37]
9	Terrestrial	*Carniella brignolii*; Thaler and Steinberger, 1988	Arachnida, Araneae, Theridiiae	[61]
10	Terrestrial	*Dysdera hungarica*; Kulczynski, 1897	Arachnida, Araneae, Dysderidae	Weiss, L., pers. comm.
11	Terrestrial	*Clinopodes carinthiacus*; Latzel, 1880	Chilopoda, Geophilidae	Zapparoli, M., pers. comm.
12	Terrestrial	*Strongylosoma jaqueti*; Verhoeff, 1898	Diplopoda, Paradoxosomatidae	Tajovsky K., pers comm.
13	Terrestrial	*Pygmarrhopalites pygmaeus*; Wankel, 1860	Collembola, Arrhopalitidae	[62]

Armadillidium tabacarui form here large populations of up to 200 individuals per square meter. The density of *Caucasonethes vandeli pygmaeus* is practically impossible to estimate in Air-Bells 2 and 3 especially as the researcher must continue to breathe through a SCUBA regulator, and to wear a diving mask. *Caucasonethes vandeli pygmaeus* is an extremely small isopod, less than 2 mm long, it is translucid, and moves very fast.

Larger spider species, such as *Agraecina cristiani* (Figure 4A) and *Dysdera hungarica*, predate mainly on collembola, but also on smaller isopods (*Caucasonethes vandeli pygmaeus*, *Haplophthalmus movilae*, and *Armadillidium tabacarui*), which are present on the cave floor and on the walls in relatively high numbers, along with the smaller spiders and pseudoscorpions. *Palliduphantes constantinescui* and *Kryptonesticus georgescuae* are web-weaving spiders that catch small prey such as collembola or small isopods. Even smaller spiders (*Hahnia caeca* and *Carniella brignolii*) are content with springtails and mites.

Five Coleoptera species are present in Movile Cave. Of these, *Medon dobrogicus* and *Clivina subterranea* are frequently present in Air-Bells 2 and 3 where they form large populations. They run continuously on the walls in these sections of the cave, and predate upon collembola, juvenile individuals of the isopods *Armadillidium tabacarui*, *Caucasonethes vandeli pygmaeus*, and *Haplophthalmus movilae*, or they can even chase the small *Chthonius monicae* pseudoscorpions. Due to its very small size, the latter can only feed on springtails and mites, or the small *Caucasonethes vandeli pygmaeus* isopods. Smaller, but also predatory staphylinids beetles (*Tychobythinus sulphydricus*, *Decumarellus sarbui*, and *Bryaxis dolosus*), can only hunt and feed on collembola.

Movile Cave is a unique habitat that needs unquestionable protection. Pollution of the groundwater aquifers by the intensive farming in South-Eastern Dobrogea, or spilling of various contaminants in the environment as a result of the expansion of the residential areas around Mangalia, can lead to severe disturbances of this exceptional ecosystem. The cave is already part of Natura 2000 sites (Code ROSCI0114), it is accessible only for scientific research, and it can only be entered based on authorized permissions. New technological advances in research methods allow for better understanding of how life can prosper even in such extreme environments, like Movile Cave, in total darkness, low pH, hypoxia and anoxia, high sulfide-, CH_4, and CO_2 concentrations. Microbiological research has evolved significantly from characterization of enzymes produced by microbes and cultivation of sulfide oxidizers [63], to the first molecular characterization of microbial communities by basic fingerprinting techniques and generation of clone libraries [10,22], to the nowadays Next-Generation Sequencing approaches that allow the examination of tens of thousands of sequences or complete genomes [25,64]. Regarding the invertebrates, new and undescribed species are no longer a great surprise, but they have to be identified and studied before their possible disappearance. It also allows to document the extent and nature of evolutionary convergence across distinct lineages of stygobiontic crustaceans and to determine to what extent natural selection was the driver of the extreme modifications observed in certain species thriving in the cave. The structure of the food web in this special environment is being studied by direct observations of the feeding behaviors, stable isotope analysis, and metagenomic investigations of the gut content of the species that inhabit the cave. Information on food webs is important and tells ultimately on how organisms can find ways to survive, and even to thrive, in ecosystems that do not depend on Sun-derived energy. Here, the food is produced in situ by using natural and inexhaustible energy sources, such as the H_2S, CH_4 and NH_4^+ from the deep subterranean aquifer. Finally, research on symbiosis between crustaceans and bacteria with biotechnological applications, discovery of new species of cave bacteria with possible antibiotic resistance, and experiments on their possible use for human health, has also a great potential to be explored in more detail in the future. Therefore, Movile Cave still has a lot more to offer.

Author Contributions: All three coauthors (T.B., S.I. and S.M.S.) have equal contribution in the conceptualization, data curation, writing and reviewing of this manuscript. All authors have read and agreed to the published version of the manuscript.

Funding: T. Brad was supported by a grant of the Ministry of Research and Innovation in Romania, project number PN-III-P4-ID-PCCF-2016-0016 (DARKFOOD), and by EEA Grants 2014-2021, under Project contract no. 4/2019 (GROUNDWATERISK). S. Iepure and S. Sarbu were supported by grants of Ministry of Research and Innovation (UEFISCDI) projects number PN-III-P4-ID-PCE-2020-2843 (EVO-DEVO-CAVE) and PN-III-P4-ID-PCCF-2016-0016 (DARKFOOD).

Institutional Review Board Statement: Not applicable.

Informed Consent Statement: Not applicable.

Data Availability Statement: No new data were created or analyzed in this study. Data sharing is not applicable to this article.

Acknowledgments: The authors acknowledge the support of the Group for Underwater and Speleological Exploration (GESS) for logistics, access to the cave and laboratory house in Mangalia, where preliminary processing of samples occurred. We are grateful to our collaborators which have identified the various species present in Table 2 (personal communications).

Conflicts of Interest: The authors declare no conflict of interest.

References

1. Sarbu, S.; Lascu, C.; Brad, T. *Dobrogea: Movile Cave*; Springer International Publishing: Berlin/Heidelberg, Germany, 2019; pp. 429–436. ISBN 978-3-319-90745-1.
2. Lascu, C.; Popa, R.; Sarbu, S.M. Le Karst de Movile (Dobrogea de Sud). *Rev. Roum. Geogr.* **1994**, *38*, 85–94.
3. Sarbu, S.M.; Lascu, C. Condensation Corrosion in Movile Cave, Romania. *J. Cave Karst Stud.* **1997**, *59*, 99–102.

4. Sarbu, S.M.; Popa, R. A unique chemoautotrophically based cave ecosystem. In *The Natural History of Biospeleology*; Camacho, A.I., Ed.; Mus. Nat. de Hist. Naturales: Madrid, Spain, 1992; pp. 637–666.

5. Sarbu, S.M. Movile Cave: A chemoautotrophically based groundwater ecosystem. In *Subterranean Ecosystems*; Wilken, H., Culver, D.C., Humphreys, W.F., Eds.; Elsevier: Amsterdam, The Netherlands, 2000; pp. 319–343.

6. Riess, W.; Giere, O.; Kohls, O.; Sarbu, S. Anoxic Thermomineral Cave Waters and Bacterial Mats as Habitat for Freshwater Nematodes. *Aquat. Microb. Ecol.* **1999**, *18*, 157–164. [CrossRef]

7. Sarbu, S.M.; Kane, T.C.; Kinkle, B.K. A Chemoautotrophically Based Cave Ecosystem. *Science* **1996**, *272*, 1953–1955. [CrossRef] [PubMed]

8. Kumaresan, D.; Wischer, D.; Stephenson, J.; Hillebrand-Voiculescu, A.; Murrell, J.C. Microbiology of Movile Cave—A Chemolithoautotrophic Ecosystem. *Geomicrobiol. J.* **2014**, *31*, 186–193. [CrossRef]

9. Rohwerder, T.; Sand, W.; Lascu, C. Preliminary Evidence for a Sulphur Cycle in Movile Cave, Romania. *Acta Biotechnol.* **2003**, *23*, 101–107. [CrossRef]

10. Chen, Y.; Wu, L.; Boden, R.; Hillebrand, A.; Kumaresan, D.; Moussard, H.; Baciu, M.; Lu, Y.; Colin Murrell, J. Life without Light: Microbial Diversity and Evidence of Sulfur- and Ammonium-Based Chemolithotrophy in Movile Cave. *ISME J.* **2009**, *3*, 1093–1104. [CrossRef] [PubMed]

11. Flot, J.-F.; Bauermeister, J.; Brad, T.; Hillebrand-Voiculescu, A.; Sarbu, S.M.; Dattagupta, S. *Niphargus-Thiothrix* Associations May Be Widespread in Sulphidic Groundwater Ecosystems: Evidence from Southeastern Romania. *Mol. Ecol.* **2014**, *23*, 1405–1417. [CrossRef] [PubMed]

12. Sarbu, S.M.; Galdenzi, S.; Menichetti, M.; Gentile, G. Geology and Biology of Grotte Di Frasassi (Frasassi Caves) in Central Italy, an Ecological Multi-Disciplinary Study of a Hypogenic Underground Karst System. *Subterr. Ecosyst. Ecosyst. World* **2000**, *30*, 359–378.

13. Galassi, D.M.P.; Fiasca, B.; Di Lorenzo, T.; Montanari, A.; Porfirio, S.; Fattorini, S. Groundwater Biodiversity in a Chemoautotrophic Cave Ecosystem: How Geochemistry Regulates Microcrustacean Community Structure. *Aquat. Ecol.* **2017**, *51*, 75–90. [CrossRef]

14. Peterson, D.; Finger, K.; Iepure, S.; Mariani, S.; Montanari, A.; Namiotko, T. Ostracod Assemblages in the Frasassi Caves and Adjacent Sulfidic Spring and Sentino River in the Northeastern Apennines of Italy. *J. Cave Karst Stud. Natl. Speleol. Soc. Bull.* **2013**, *75*. [CrossRef]

15. Popa, I.; Brad, T.; Vaxevanopoulos, M.; Giurginca, A.; Baba, S.C.; Iepure, S.; Plaiasu, R.; Sarbu, S. Rich and Diverse Subterranean Invertebrate Communities Inhabiting Melissotrypa Cave in Central Greece. *Trav. Inst. Spéol. «Émile Racovitza»* **2019**, *58*, 65–78.

16. Por, F.D.; Dimentman, C.; Frumkin, A.; Naaman, I. Animal Life in the Chemoautotrophic Ecosystem of the Hypogenic Groundwater Cave of Ayyalon (Israel): A Summing Up. *Nat. Sci.* **2013**, *5*, 7–13. [CrossRef]

17. Frumkin, A.; Dimentman, C.; Naaman, I. Biogeography of Living Fossils as a Key for Geological Reconstruction of the East Mediterranean: Ayyalon-Nesher Ramla System, Israel. *Quat. Int.* **2020**. [CrossRef]

18. Fatemi, Y.; Malek Hosseini, M.-J.; Falniowski, A.; Hofman, S.; Kuntner, M.; Grego, J. Description of a New Genus and Species as the First Gastropod Species from Caves in Iran. *J. Cave Karst Stud.* **2019**, 233–243. [CrossRef]

19. Mousavi-Sabet, H.; Vatandoust, S.; Fatemi, Y.; Eagderi, S. Tashan Cave a New Cave Fish Locality for Iran; and *Garra tashanensis*, a New Blind Species from the Tigris River Drainage (Teleostei: Cyprinidae). *Fishtaxa* **2016**, *1*, 133–148.

20. Khalaji-Pirbalouty, V.; Fatemi, Y.; Malek-Hosseini, M.J.; Kuntner, M. A New Species of *Stenasellus* Dollfus, 1897 from Iran, with a Key to the Western Asian Species (Crustacea, Isopoda, Stenasellidae). *ZooKeys* **2018**, *766*, 39–50. [CrossRef] [PubMed]

21. Emerson, B.C.; Kolm, N. Species Diversity Can Drive Speciation. *Nature* **2005**, *434*, 1015–1017. [CrossRef]

22. Hutchens, E.; Radajewski, S.; Dumont, M.G.; McDonald, I.R.; Murrell, J.C. Analysis of Methanotrophic Bacteria in Movile Cave by Stable Isotope Probing. *Environ. Microbiol.* **2004**, *6*, 111–120. [CrossRef] [PubMed]

23. Schirmack, J.; Mangelsdorf, K.; Ganzert, L.; Sand, W.; Hillebrand-Voiculescu, A.; Wagner, D. *Methanobacterium movilense* Sp. Nov., a Hydrogenotrophic, Secondary-Alcohol-Utilizing Methanogen from the Anoxic Sediment of a Subsurface Lake. *Int. J. Syst. Evol. Microbiol.* **2014**, *64*, 522–527. [CrossRef]

24. Ganzert, L.; Schirmack, J.; Alawi, M.; Mangelsdorf, K.; Sand, W.; Hillebrand-Voiculescu, A.; Wagner, D. *Methanosarcina spelaei* Sp. Nov., a Methanogenic Archaeon Isolated from a Floating Biofilm of a Subsurface Sulphurous Lake. *Int. J. Syst. Evol. Microbiol.* **2014**. [CrossRef] [PubMed]

25. Bizic, M.; Brad, T.; Barbu-Tudoran, L.; Aerts, J.; Ionescu, D.; Popa, R.; Ody, J.; Flot, J.-F.; Tighe, S.; Vellone, D.; et al. Genomic and Morphologic Characterization of a Planktonic *Thiovulum* (Campylobacterota) Dominating the Surface Waters of the Sulfidic Movile Cave, Romania. *bioRxiv* **2020**. [CrossRef]

26. Nováková, A.; Hubka, V.; Valinová, Š.; Kolařík, M.; Hillebrand-Voiculescu, A.M. Cultivable Microscopic Fungi from an Underground Chemosynthesis-Based Ecosystem: A Preliminary Study. *Folia Microbiol. (Praha)* **2018**, *63*, 43–55. [CrossRef]

27. Reboul, G.; Moreira, D.; Bertolino, P.; Hillebrand-Voiculescu, A.M.; López-García, P. Microbial Eukaryotes in the Suboxic Chemosynthetic Ecosystem of Movile Cave, Romania. *Environ. Microbiol. Rep.* **2019**, *11*, 464–473. [CrossRef]

28. Muschiol, D.; Marković, M.; Threis, I.; Traunspurger, W. Predatory Copepods Can Control Nematode Populations: A Functional-Response Experiment with *Eucyclops subterraneus* and Bacterivorous Nematodes. *Fundam. Appl. Limnol. Arch. Für Hydrobiol.* **2008**, *172*, 317–324. [CrossRef]

29. Falniowski, A.; Szarowska, M.; Sîrbu, I.; Hillebrand-Voiculescu, A.; Baciu, M. *Heleobia dobrogica* (Grossu & Negrea, 1989)(Gastropoda: Rissooidea: Cochliopidae) and the Estimated Time of Its Isolation in a Continental Analogue of Hydrothermal Vents. *Molluscan Res.* **2008**, *28*, 165.

30. Stocchino, G.; Sluys, R.; Kawakatsu, M.; Sarbu, S.; Manconi, R. A New Species of Freshwater Flatworm (Platyhelminthes, Tricladida, Dendrocoelidae) Inhabiting a Chemoautotrophic Groundwater Ecosystem in Romania. *Eur. J. Taxon.* **2017**, *2017*. [CrossRef]

31. Vahtera, V.; Stoev, P.; Akkari, N. Five Million Years in the Darkness: A New Troglomorphic Species of *Cryptops* Leach, 1814 (Chilopoda, Scolopendromorpha) from Movile Cave, Romania. *ZooKeys* **2020**, *1004*, 1–26. [CrossRef]

32. Gruia, M. Quelques Considérations Sur La Faune de Collemboles de La Grotte de Movile, Roumanie. *Mém. Biospéol.* **1996**, *23*, 105–109.

33. Gruia, M. Sur La Faune de Collemboles de l'écosystème Exokarstique et Karstique de Movilé (Dobrogea Du Sud, Mangalia, Romania). *Mém. Biospéol.* **1998**, *25*, 45–52.

34. Muschiol, D.; Giere, O.; Traunspurger, W. Population Dynamics of a Cavernicolous Nematode Community in a Chemoautotrophic Groundwater System: Cavernicolous Nematode Community. *Limnol. Oceanogr.* **2015**, *60*. [CrossRef]

35. Poinar, G.; Sarbu, S.M. *Chronogaster troglodytes* Sp. n.(Nemata: Chronogasteridae) from Movile Cave, with a Review of Cavernicolous Nematodes. *Appl. Nematol.* **1994**, *17*, 231–237.

36. Manoleli, D.G.; Klemm, D.J.; Sarbu, S.M. *Haemopis caeca* (Annelida: Hirudinea: Arhynchobdellida: Haemopidae), A New Species Of Troglobitic Leech From A Chemoautotrophically Based Groundwater Ecosystem In Romania. *Proc. Biol. Soc. Wash.* **1998**, *111*, 222–229.

37. Plesa, C. Étude Préliminaire Des Cyclopides (Crustacea, Copepoda) de La Grotte "Peştera de La Movile", Mangalia (Roumanie). *Misc. Speologica Romanica* **1989**, *1*, 39–45.

38. Dancau, D. Sur un nouvel amphipode souterrain de Roumanie, *Pontoniphargus racovitzai* n.g., n.sp. In *Livre du Centenaire. Emile G. Racovitza 1868–1968*; Académie de la République Socialiste de Roumanie: Bucharest, Romania, 1970; pp. 275–285.

39. Brad, T.; Fišer, C.; Flot, J.-F.; Sarbu, S. *Niphargus dancaui* Sp. Nov. (Amphipoda, Niphargidae) - A New Species Thriving in Sulfidic Groundwaters in Southeastern Romania. *Eur. J. Taxon.* **2015**, *2015*. [CrossRef]

40. Turk-Prevorčnik, S.; Blejec, A. *Asellus aquaticus infernus*, New Subspecies (Isopoda: Asellota: Asellidae), From Romanian Hypogean Waters. *J. Crustac. Biol.* **1998**, *18*, 763–773. [CrossRef]

41. Giurginca, A.; Sarbu, S.M. *Caucasonethes vandeli pygmaeus* n.Ssp. (Crustacea, Isopoda, Oniscidea) from Movile Cave (Southern Dobrogea, Romania). *Trav. Inst. Spéol. «Émile Racovitza»* **2020**, *59*, 33–41.

42. Gruia, M.; Giurginca, A. *Haplophthalmus movilae* (Isopoda, Trichoniscidae), a New Troglobitic Species from Movile Cave, Dobrogea, Romania. *Mitteilungen Hambg. Zool. Mus. Inst.* **1998**, *95*, 133–142.

43. Tabacaru, I.; Boghean, V. Découverte En Dobrogea (Roumanie) D´ Une Espèce Troglobie Du Genre Trachelipus (Isopoda Oniscoidea Trachelipidae). *Misc. Speologica Romanica* **1989**, *1*, 53–75.

44. Gruia, M.; Iavorschi, V.; Sarbu, S. *Armadillidium tabacarui (Isopoda, Oniscidea, Armadillidiidae), A New Troglobitic Species From A Sulfurous Cave In Romania*; Biological Society of Washington: Washington, DC, USA, 1994.

45. Boghean, V. Sur Un Pseudoscorpion Cavernicole Nouveau, *Chthonius (C.) monicae* n.Sp. (Arachnida Pseudoscorpionida Chthonidae). *Misc. Speologica Romanica* **1989**, *1*, 77–83.

46. Curcic, B.P.M.; Sarbu, S.; Dimitrijevic, R.; Ćurčić, S. A New Cave Pseudoscorpion from the Region of Mangalia (Romania): *Chthonius (Ephippiochthonius) borissketi* n. Sp. (Chthoniidae, Pseudoscorpiones). *Arch. Biol. Sci.* **2014**, *66*, 955–961. [CrossRef]

47. Ćurčić, B.; Poinar, G.; Sarbu, S. New and Little-Known Species of Chthoniidae and Neobisiidae (Pseudoscorpiones, Arachnida) from the Movile Cave in Southern Dobrogea, Romania. *Bijdr. Tot Dierkd.* **1993**, *63*, 221–241. [CrossRef]

48. Georgescu, M. Sur Trois Taxa Nouveaux d'Araneides Troglobies de Dobrogea (Roumanie). *Misc. Speol. Romanica* **1989**, *1*, 85–102.

49. Nae, A.; Sarbu, S.; Weiss, I. *Kryptonesticus georgescuae* Spec. Nov. from Movile Cave, Romania (Araneae: Nesticidae). *Arachnol. Mitteilungen* **2018**, *55*, 22–24. [CrossRef]

50. Georgescu, M.; Sarbu, S.M. Description d'un Noveau Taxon: *Iberina caeca* de La Grotte: "Peştera de La Movile" (Araneae - Hahniidae). *Mém. Biospéol.* **1992**, *19*, 139–141.

51. Iavorschi, V. *Labidostoma motasi* n.Sp. (Nicoletiellidae) a New Species of the Mite of Romania. *Trav. Inst. Spéol. «Émile Racovitza»* **1992**, *31*, 47–51.

52. Giurginca, A.; Vanoaica, L.; Sustr, V.; Tajovský, K. A New Species of the Genus *Archiboreoiulus* Brolemann, 1921 (Diplopoda, Julida) from Movile Cave (Southern Dobrogea, Romania). *Zootaxa* **2020**, *4802*, 463–476. [CrossRef] [PubMed]

53. Gruia, M. Nouvelles Espèces Troglobiontes Des Collemboles de Roumanie. *Misc. Speol. Romanica* **1989**, *1*, 103–111.

54. Conde, B. Une lignée danubienne du genre *Plusiocampa* (Diploures Campodéidés); A Danubian lineage in the genus Plusiocampa (Diplura Campodeidae). *Rev. Suisse Zool.* **1993**, *100*, 735–745. [CrossRef]

55. Condé, B. Diploures Campodéidés de La Pestera de La Movile (Movile Cave), Dobroudja Méridionale (Roumanie). *Rev. Suisse Zool.* **1996**, *103*, 101–114. [CrossRef]

56. Decu, V.; Georgescu, M. Deux Espèces Nouvelles de *Medon* (*M. dobrogicus* et *M. paradobrogicus*) (Coleoptera, Staphylinidae) de La Grotte "Peştera de La Movile", Dobrogea Meridionale, Roumanie. *Mém. Biospéol.* **1994**, *21*, 47–51.

57. Poggi, R.; Sarbu, S. Two New Pselaphine Beetles from Movile Cave (Romania) (Coleoptera, Staphylinidae, Pselaphinae). *Ann. Mus. Civ. Storia Nat. Giacomo Doria* **2013**, *105*, 115–121.

58. Poggi, R. Descrizione Di Un Nuovo Pselafide Rumeno, Primo Rappresentante Cavernicolo Della Tribù Tyrini (Coleoptera Pselaphidae). *Boll. Della Soc. Entomol. Ital.* **1994**, *125*, 221–228.
59. Decu, V.; Nitu, E.; Juberthie, C. *Clivina subterranea* (Caraboidea, Scaritidae) Nouvelle Espèce de La Grotte "Pestera de La Movile", Dobrogea Meridionale, Roumanie. *Mém. Biospéol.* **1994**, *21*, 41–45.
60. Decu, V.; Gruia, M.; Keffer, S.L.; Sarbu, S.M. Stygobiotic Waterscorpion, *Nepa anophthalma*, n. Sp. (Heteroptera: Nepidae), from a Sulfurous Cave in Romania. *Ann. Entomol. Soc. Am.* **1994**, *87*, 755–761. [CrossRef]
61. Thaler-Knoflach, B.; Hänggi, A.; Kielhorn K-H.; von Broen, B. Revisiting the Taxonomy of the Rare and Tiny Comb-footed Spider *Carniella brignolii* (Araneae, Theridiidae). *Arachnologische Mitteilungen* **2014**, *47*, 7–13. [CrossRef]
62. Gruia, M. Collembola from Romanian caves. *Trav. Mus. D'Hist. Nat. "Grigore Antipa"* **2003**, *35*, 139–158.
63. Sarbu, S.M.; Vlasceanu, L.; Popa, R.; Sheridan, P.; Kinkle, B.K.; Kane, T.C. Microbial Mats in a Thermomineral Sulfurous Cave. In *Proceedings of the Microbial Mats*; Stal, L.J., Caumette, P., Eds.; Springer: Berlin/Heidelberg, Germany, 1994; pp. 45–50.
64. Kumaresan, D.; Stephenson, J.; Doxey, A.; Bandukwala, H.; Brooks, E.; Hillebrand-Voiculescu, A.; Whiteley, A.; Murrell, J. Aerobic Proteobacterial Methylotrophs in Movile Cave: Genomic and Metagenomic Analyses. *Microbiome* **2018**, *6*, 1. [CrossRef] [PubMed]

Correction

Correction: Brad et al. The Chemoautotrophically Based Movile Cave Groundwater Ecosystem, a Hotspot of Subterranean Biodiversity. *Diversity* 2021, *13*, 128

Traian Brad [1,2], Sanda Iepure [1,2] and Serban M. Sarbu [3,4,*]

1 "Emil Racoviță" Institute of Speleology, Str. Clinicilor nr. 5-7, 400006 Cluj-Napoca, Romania; traian.brad@academia-cj.ro (T.B.); sanda.iepure@academia-cj.ro (S.I.)
2 Institutul Român de Știință și Tehnologie, Str. Virgil Fulicea nr. 3, 400022 Cluj-Napoca, Romania
3 "Emil Racoviță" Institute of Speleology, Str. Frumoasa nr. 31, 010986 București, Romania
4 Department of Biological Sciences, California State University, Chico, CA 95929, USA
* Correspondence: serban.sarbu@yahoo.com

Citation: Brad, T.; Iepure, S.; Sarbu, S.M. Correction: Brad et al. The Chemoautotrophically Based Movile Cave Groundwater Ecosystem, a Hotspot of Subterranean Biodiversity. *Diversity* 2021, *13*, 128. *Diversity* 2021, *13*, 461. https://doi.org/10.3390/d13100461

Received: 21 May 2021
Accepted: 4 September 2021
Published: 24 September 2021

Publisher's Note: MDPI stays neutral with regard to jurisdictional claims in published maps and institutional affiliations.

The authors wish to make the following corrections to this paper [1]. The authors apologize for any inconvenience caused and state that the scientific conclusions are unaffected.

1. Replacing a sentence in Section 2.2 Terrestrial fauna, on page 5:

The geophilid centipedes *Geophilus* sp. and *Clinopodes carynthiacus* are also among the predators in this ecosystem.
With
The geophilid centipedes *Geophilus* sp. and *Clinopodes carinthiacus* are also among the predators in this ecosystem.

2. Replacing a sentence in Section 2.2 Terrestrial fauna, on page 6:

The three species jump continuously in all directions, and therefore they can easily become part of the menu of other cave inhabitants.
With
Two of the three species jump continuously in all directions; therefore, they can easily become part of the menu of other cave inhabitants.

3. Splitting of Table 2 (List of aquatic and terrestrial invertebrate species encountered and described in Movile Cave ecosystem) into troglobionts/stygobionts (Table 2) and troglophiles/stygophiles (Table 3), as following:

4. Replacing a sentence in Section 2.2 Terrestrial fauna, on page 9:

Caucasonethes vandeli pygmaeus is an extremely small isopod, about 1 mm long, it is translucid, and moves very fast.
With
Caucasonethes vandeli pygmaeus is an extremely small isopod, less than 2 mm long, it is translucid, and moves very fast.

5. Replacing a sentence in the Section: Funding on page 10:

S. Iepure and S. Sarbu were supported by grant PN-III-P4-ID-PCE-2020-2843 (EVO-DEVO-CAVE).
Should be replaced with
S. Iepure and S. Sarbu were supported by grants of Ministry of Research and Innovation (UEFISCDI) projects number PN-III-P4-ID-PCE-2020-2843 (EVO-DEVO-CAVE) and PN-III-P4-ID-PCCF-2016-0016 (DARKFOOD).

Table 2. List of troglobionts and stygobionts from Movile Cave.

	Aquatic/Terrestrial	Species	Taxonomic Affiliation	References
1	Aquatic	*Dendrocoelum obstinatum* *; Stocchino et al., 2017	Platyhelminthes, Dendrocoelidae	[30]
2	Aquatic	*Panagrolaimus* cf. *thienemani* *	Nematoda, Panagrolaimidae	[34]
3	Aquatic	*Chronogaster troglodytes* *; Poinar and Sarbu, 1994	Nematoda, Chronogasteridae	[35]
4	Aquatic	*Haemopis caeca* *,#; Manoleli et al., 1998	Annelida, Hirudinea, Haemopidae	[36]
5	Aquatic	*Helodrilus* sp. nov. *	Annelida, Clitellata, Lumbricidae	Martin, P., pers. comm.
6	Aquatic	*Heleobia dobrogica* *; Grossu and Negrea, 1989	Gastropoda, Moitessieriidae	[29]
7	Aquatic	*Pseudocandona* sp. nov. *	Crustacea, Ostracoda, Cyprididae	Danielopol, D., pers. comm.
8	Aquatic	*Eucyclops graeteri scythicus* *; Plesa, 1989	Crustacea, Copepoda, Cyclopidae	[37]
9	Aquatic	*Parapseudoleptomesochra italica*; Pesce and Petkovski, 1980	Crustacea, Copepoda, Harpacticoida	Rouch, pers. comm.
10	Aquatic	*Niphargus racovitzai* *; Dancau, 1970	Crustacea, Amphipoda, Niphargidae	[38]
11	Aquatic	*Niphargus dancaui* *,#; Brad et al., 2015	Crustacea, Amphipoda, Niphargidae	[39]
12	Aquatic	*Asellus aquaticus infernus* *,#; Turk-Prevorčnik and Blejec, 1998	Crustacea, Isopoda, Asellidae	[40]
13	Terrestrial	*Caucasonethes vandeli pygmaeus* *; Giurginca, 2020	Crustacea, Isopoda, Trichoniscidae	[41]
14	Terrestrial	*Haplophthalmus movilae* *; Gruia and Giurginca, 1998	Crustacea, Isopoda, Trichoniscidae	[42]
15	Terrestrial	*Trachelipus troglobius* *; Tabacaru and Boghean, 1989	Crustacea, Isopoda, Trachelipodidae	[43]
16	Terrestrial	*Armadillidium tabacarui* *; Gruia et al., 1994	Crustacea, Isopoda, Armadillidiidae	[44]
17	Terrestrial	*Chthonius monicae* *; Boghean, 1989	Arachnida, Pseudoscorpiones, Chthoniidae	[45]
18	Terrestrial	*Chthonius borissketi* *; Curčić et al., 2014	Arachnida, Pseudoscorpiones, Chthoniidae	[46]
19	Terrestrial	*Roncus dragobete* *; Curčić et al., 1993	Arachnida, Pseudoscorpiones, Neobisiidae	[47]
20	Terrestrial	*Roncus ciobanmos* *; Curčić et al., 1993	Arachnida, Pseudoscorpiones, Neobisiidae	[47]
21	Terrestrial	*Palliduphantes constantinescui* *; Georgescu, 1989	Arachnida, Araneae, Linyphiidae	[48]
22	Terrestrial	*Agraecina cristiani* *,#; Georgescu, 1989	Arachnida, Araneae, Liocranidae	[48]
23	Terrestrial	*Kryptonesticus georgescuae* *; Nae, Sarbu, and Weiss, 2018	Arachnida, Araneae, Nesticidae	[49]
24	Terrestrial	*Hahnia caeca* *; Georgescu and Sarbu, 1992	Arachnida, Araneae, Hahniidae	[50]
25	Terrestrial	*Labidostomma motasi* *; Iavorschi, 1992	Arachnida, Acarina, Labidostommatidae	[51]
26	Terrestrial	*Geophilus* sp. nov. *	Chilopoda, Geophilidae	Baba, St., pers. comm.
27	Terrestrial	*Cryptops speleorex* *,#; Vahtera et al., 2020	Chilopoda, Cryptopidae	[31]
28	Terrestrial	*Archiboreoiulus serbansarbui* *,#; Giurginca et al., 2020	Diplopoda, Julida, Julidae	[52]
29	Terrestrial	*Onychiurus movilae* *; Gruia, 1989	Collembola, Onychiuridae	[53]
30	Terrestrial	*Oncopodura vioreli* *; Gruia, 1989	Collembola, Oncopoduridae	[53]
31	Terrestrial	*Plusiocampa isterina* *; Condé, 1993	Diplura, Campodeidae	[54]
32	Terrestrial	*Plusiocampa euxina* *; Condé, 1996	Diplura, Campodeidae	[55]
33	Terrestrial	*Medon dobrogicus* *; Decu and Georgescu, 1994	Coleoptera, Staphylinidae	[56]
34	Terrestrial	*Tychobythinus sulphydricus* *; Poggi and Sarbu, 2013	Coleoptera, Staphylinidae	[57]
35	Terrestrial	*Decumarellus sarbui* *; Poggi, 1994	Coleoptera, Staphylinidae	[58]
36	Terrestrial	*Bryaxis dolosus* *; Poggi and Sarbu, 2013	Coleoptera, Staphylinidae	[57]
37	Terrestrial	*Clivina subterranea* *; Decu et al., 1994	Coleoptera, Clivinidae	[59]
38	Aquatic	*Nepa anophthalma* *; Dedu et al., 1994	Hemiptera, Nepidae	[60]

*—species endemic to Movile Cave; #—species found in nearby springs and wells.

Table 3. List of troglophiles and stygophiles from Movile Cave.

	Aquatic/Terrestrial	Species	Taxonomic Affiliation	References
1	Aquatic	*Udonchus tenuicaudatus*; Cobb, 1913	Nematoda, Rhabdolaimidae	[34]
2	Aquatic	*Poikilolaimus* sp.	Nematoda, Rhabditidae	[34]
3	Aquatic	*Monhystrella* sp.	Nematoda, Monhysteridae	[34]
4	Aquatic	*Habrotrocha rosa*; Donner, 1949	Rotatoria, Habrotrochidae	Ricci, C., pers. comm.
5	Aquatic	*Habrotrocha bidens*; Gosse, 1851	Rotatoria, Habrotrochidae	Ricci, C., pers. comm.
6	Aquatic	*Aelosoma hyalinum*; Bunke, 1967	Annelida, Aeolosomatidae	Dumnicka, E., pers. comm.
7	Aquatic	*Aelosoma italica*; Bunke, 1967	Annelida, Aeolosomatidae	Dumnicka, E., pers. comm.
8	Aquatic	*Tropocyclops prasinus*; Fischer, 1860	Crustacea, Copepoda, Cyclopidae	[37]
9	Terrestrial	*Carniella brignolii*; Thaler and Steinberger, 1988	Arachnida, Araneae, Theridiiae	[48]
10	Terrestrial	*Dysdera hungarica*; Kulczynski, 1897	Arachnida, Araneae, Dysderidae	Weiss, L., pers. comm.
11	Terrestrial	*Clinopodes carinthiacus*; Latzel, 1880	Chilopoda, Geophilidae	Zapparoli, M., pers. comm.
12	Terrestrial	*Strongylosoma jaqueti*; Verhoeff, 1898	Diplopoda, Paradoxosomatidae	Tajovsky K., pers comm.
13	Terrestrial	*Pygmarrhopalites pygmaeus*; Wankel, 1860	Collembola, Arrhopalitidae	[55]

Reference

1. Brad, T.; Iepure, S.; Sarbu, S.M. The Chemoautotrophically Based Movile Cave Groundwater Ecosystem, a Hotspot of Subterranean Biodiversity. *Diversity* **2021**, *13*, 128. [CrossRef]

Article

The Subterranean Fauna of Križna Jama, Slovenia

Slavko Polak [1],* and Tanja Pipan [2]

[1] Notranjska Museum Postojna, Zavod Znanje Postojna, Kolodvorska Cesta 3, SI-6230 Postojna, Slovenia
[2] Research Centre of the Slovenian Academy of Science and Arts, Karst Research Institute, Novi trg 2, SI-1000 Ljubljana, Slovenia; tanja.pipan@zrc-sazu.si
* Correspondence: slavko.polak@notranjski-muzej.si

Abstract: The karstic cave Križna jama in the South Western part of Slovenia is one of the largest, well known and most beautiful Slovene water caves. The cave consists of more than 8 km of corridors with impressive halls, colossal dripstone formations, a subterranean river and numerous lakes. Considering the subterranean fauna, Križna jama has been identified amongst the richest caves in the world. So far, 60 troglobionts, the obligate subterranean species among them 32 aquatic and 28 terrestrial taxa have been recorded and documented. Križna jama has scientific importance, as well as ten subterranean taxa, which have been described based on specimens from this cave. Despite Križna jama is relatively well-studied, new recent unexpected findings are promising. Thus, further discoveries of specialized subterranean species in the cave are expected.

Keywords: caves; Križna jama; biospeleology; subterranean biodiversity; hotspots; troglobiont; checklist

1. Introduction

Križna jama (registered in Slovene Cave Cadaster, Cad. No. 65) is developed under the Križna gora (857 m) hill near the village Lož in the South Western part of Slovenia. It is one of the largest, well known and most beautiful Slovene water caves. Due to its natural beauty, it has been promoted in the Slovene edition of National Geographic magazine [1]. In the first global subterranean biodiversity comparative study [2], Križna jama was listed among the richest caves of the Dinaric range. The cave consists of 8273 m of corridors with impressive halls, colossal dripstone formations, a subterranean river and many lakes (Figure 1). The main passage, called Jezerski or Glavni rov, located approximately in the middle of the cave, at a spot named "Kalvarija", bifurcates into two separate corridors: Pisani rov and Blatni rov or Blata (Figure 2). Cavers recently discovered some new unknown and still unmeasured passages (Troha A. per. comm.). The cave is mainly horizontal with a difference of only 32 m between the highest and the lowest point of the cave. The lowest drypoint is at the water level of the lakes in Kittl shafts or Kittlova brezna, where the subterranean river sinks into the passage that cannot be accessed.

The unique features of Križna jama are its breathtaking clean and emerald green subterranean lakes (Figure 1). In two passages, Jezerski and Pisani rov, there are 22 lakes. Summing these to the lakes in Blatni rov, there are 45 lakes in Križna jama altogether. During the rainy seasons, the running water on its way through the karstic massif dissolves limestones and saturates with calcium carbonate. On its course through the cave, streams deposit carbonates on the riverbeds. Especially on its stream rapids, the permanently deposited sinter creates rimstone barriers that dam and accumulate numerous lakes. An average rate of sinter deposition in Križna jama is estimated to be approx. 0.256 mm per year [3]. From a geological and hydrological point of view, Križna jama is located in a syncline between the Bloke plateau and the Idrija fault zone, where the karstic Ljubljanica river flows through a series of karst poljes. It lies within the aquifer of the triangle formed by the Bloke plateau (720 m), the Cerknica polje (550 m) and the Loško polje (575 m). The oldest rocks of the syncline are the Upper Triassic dolomites, while the youngest are the Upper Jurassic limestones. Between them, there are a series of limestone strata, with lenses or nests of dolomites. The syncline shows a relatively weak faulting but local faults, together with tectonised bedding planes, can guide the formation of some Križna jama passages [4,5].

During periods of low water levels, Križna jama is mainly characterised by the flow of percolated water from the nearby hilly karst area. When water levels are high, Križna jama drains a small extent of allogenic water from the Bloke plateau, mainly from the Bloščica and the Farovščica streams too [5]. Underground water courses and connections have been elucidated by a series of tracer tests [6,7]. The water flowing through Križna jama sinks near the cave entrance in the deep Kittl shaft or abyss which ends with a syphon and reapers downstream in the Štebrščica spring at the edge of the Cerknica polje [6]. Speleo-divers tried to follow the stream in the Kittl shaft in an attempt to find further subterranean connections. Passing the flooded syphons at the depth of 70 m however, appears to be technically too difficult. Instead, cavers in the attempt to find easier access to the subterranean cave passages, tried to find some hidden entrance on the surface. In a long-term effort after 200 h of digging, in 1991 cavers finally enlarged a small hole (known as Dihalnik v Grdem dolu). Until that time, only the edible dormice (*Glis glis*) have regularly used this small entrance. This new small entrance led to the discovery of additional 1415 m long virgin cave, popularly called "Križna jama 2". Due to the fragile dripstone formations, this cave is extremely sensitive to visit. After proper investigation and measuring, this cave has been closed and strictly protected, with less than 30 cavers ever visited it. Although it is undoubtedly part of the same cave system, Križna jama 2 (registered in Slovene Cave Cadaster, Cad. No. 6286, as Dihalnik v Grdem dolu) is considered as a separate cave object from Križna jama.

Figure 1. The unique features of Križna jama are its breathtaking clean and emerald green subterranean lakes.

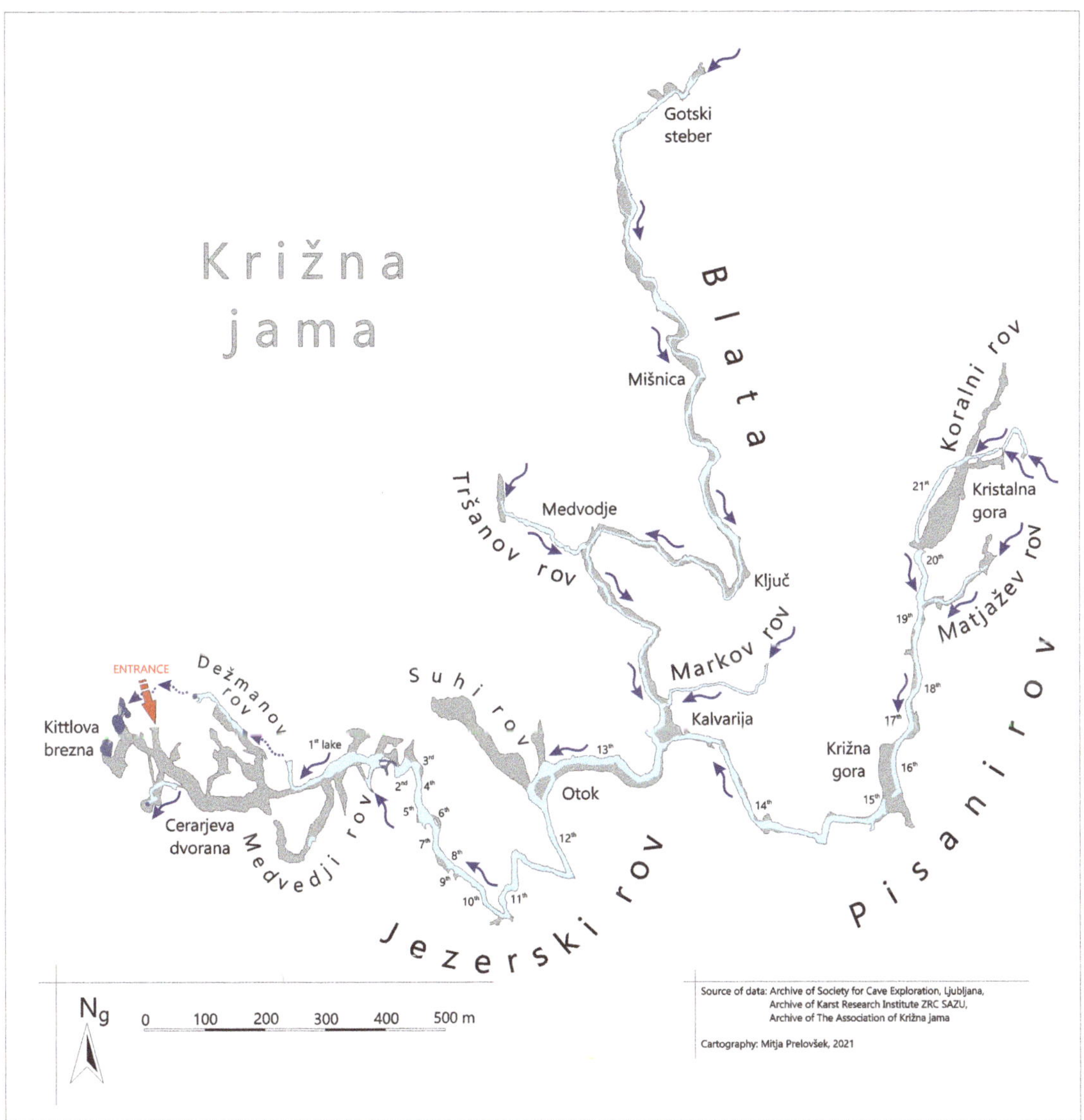

Figure 2. Plan of Križna jama. Archive of Karst Research Institute ZRC SAZU, cartography Mitja Prelovšek.

Križna jama crystal clean lakes and dripstone formations attracted numerous explorers, cavers and visitors early in the 20th century. Uncontrolled visits in the early time resulted in some damage of dripstone formations and local pollution of the cave. The local cavers soon closed the entrance and within a couple of volunteer actions successfully cleaned up the cave. Križna jama nowadays accepts about 10,000 guided visitors per year, but only a small part of the visitors are lucky enough to enjoy the beautiful lakes in the inner parts of the cave by boat. Due to the fragility of the sinter dripstones and the sensitivity of the cave, the number of guided visits is limited and strictly controlled.

2. Materials and Methods

To complete an updated list of troglobiotic fauna in Križna jama, we consulted the available literature. Species recognized as troglophiles and trogloxenes are discussed in the text but not listed in Table 1. In searching for subterranean fauna reports in dispersed publications within the last, almost 200 years, we encountered difficulties concerning the different names used for Križna jama. In our review we recorded many different names, all referring to Križna jama: Mrzla jama am Fuss des Kreuzgeberges [8], Berühmte Grotte von Podlaz [9], Mrzla Jama [10], Krizna jama, ou Kreuzöhle sur le versant D. du Kreuzberg [11], Mrzla jama, á Blazka Poliza, distr. De Logatesc and Krizna jama/Kreuzhöhle/sur le Kreuzberg, prés de Lass [12] and Kreuzberghöhle (1700 m N Laas and N Hang des Kreuzberges) [13]. Križna jama [14] cave should not be mistaken with another cave with a similar name—Mrzla jama pri Ložu (registered in Slovene Cave Cadaster, Cad. No. 79), located 1.7 km heading south. A third "Mrzla jama" cave (1.5 km heading north) is a small cave with a water stream called Mrzla jama pri Bločicah (registered in Slovene Cave Cadaster, Cad. No. 1176). The fauna of all these three caves is however similar or probably equal.

Table 1. An updated list of stygobiotic and troglobiotic species encountered in the Križna jama. Tb.—terrestrial (troglobiont), Stb.—aquatic (stygobiont), Stb. pop.—stygobiotic population, Tl.—type locality in Križna jama, new—new, unpublished data.

Taxonomic Group	Familia	Genus/Species/Subspecies	Status	References
Ciliata	Lagenophryidae	*Lagenophrys monolistrae* Stammer, 1935	Stb.	[14,15]
Turbellaria	Dendrocoelidae	*Dendrocoelum* cf. *spelaeum* (Kenk, 1924)	Stb.	[14]
Turbellaria	Scutariellidae	*Stygodyticola hadzii* Matjašič, 1958	Stb.	[14,16]
Gastropoda	Hydrobiidae	*Belgrandiella superior* Kuščer, 1932	Stb.	[14,17,18]
Gastropoda	Hydrobiidae	*Belgrandiella crucis* (Kuščer, 1928)	Stb., Tl.	[13,14,19]
Gastropoda	Hydrobiidae	*Belgrandiella schleschi* (Kuščer, 1932)	Stb., Tl.	[13,14,19]
Gastropoda	Hydrobiidae	*Hauffenia michleri* Kuščer, 1932	Stb.	[13,14,19]
Gastropoda	Moitessieriidae	*Phaladilhiopsis* sp.	Stb.	[14]
Gastropoda	Ellobiidae	*Zospeum exiguum* Kuščer, 1932	Stb., Tl.	[14,18–20]
Gastropoda	Ellobiidae	*Zospeum kusceri* A. J. Wagner, 1912	Tb.	[14,18,20]
Gastropoda	Ellobiidae	*Zospeum isselianum* Pollonera, 1887	Tb.	[14,18,20]
Oligochaeta	Haplotaxidae	*Delaya bureschi* (Michaelsen, 1925)	Stb.	[14]
Oligochaeta	Lumbriculidae	*Trichodrilus strandi* Hrabe, 1936	Stb.	[14]
Oligochaeta	Lumbriculidae	*Trichodrilus ptujensis* Hrabe, 1963	Stb.	[14]
Oligochaeta	Lumbriculidae	*Trichodrillus pragensis* Vejdovsky, 1876	Stb.	[14]
Oligochaeta	Naididae	*Rhyacodrilus omodeoi* Martinez-Ansemil, Sambugar & Giani, 1997	Stb., Tl.	[14]
Oligochaeta	Naididae	*Rhyacodrilus sketi* Karaman, 1974	Stb.	[14]
Oligochaeta	Naididae	*Tubifex pescei* (Dumnicka, 1980)	Stb.	[14]
Ostracoda	Entocytheridae	*Sphaeromicola* sp.	Stb.	[14]
Copepoda	Canthocamptidae	*Elaphoidella jeanneli* (Chappuis, 1928)	Stb.	[14]
Copepoda	Canthocamptidae	*Elaphoidella stammeri* Chappuis, 1936	Stb.	[14]
Copepoda	Canthocamptidae	*Lessinocamptus* n.sp. Stoch unpubl.	Stb., Tl.	[14,21]
Copepoda	Canthocamptidae	*Bryocamptus (Rheocamptus) balcanicus* s.l. (Kiefer, 1933)	Stb.	[14]
Copepoda	Cyclopidae	*Megacyclops viridis* s.l. (Jurine, 1820)	Stb.	[14]
Copepoda	Cyclopidae	*Acanthocyclops kieferi* (Chappuis, 1925)	Stb.	[14]
Copepoda	Cyclopidae	*Acanthocyclops troglophilus* (Kiefer, 1932)	Stb.	[14]
Copepoda	Cyclopidae	*Diacyclops languidoides goticus* Kiefer, 1931	Stb.	[14]
Copepoda	Cyclopidae	*Diacyclops charon* (Kiefer, 1931)	Stb.	[14]
Copepoda	Cyclopidae	*Speocyclops infernus* (Kiefer, 1930)	Stb.	[14]
Amphipoda	Niphargidae	*Niphargus orcinus* Joseph, 1869	Stb., Tl.	[13,14,22]
Amphipoda	Niphargidae	*Niphargus wolfi* Schellenberg, 1933	Stb.	[14]
Amphipoda	Niphargidae	*Niphargus stygius* (Schiödte, 1847)	Stb.	[14]
Amphipoda	Crangonyctidae	*Synurella ambulans* (F. Müller, 1846)	Stb. pop.	[14]

Table 1. *Cont.*

Taxonomic Group	Familia	Genus/Species/Subspecies	Status	References
Isopoda	Trichoniscidae	*Titanethes albus* (C. Koch, 1841)	Tb.	[14]
Isopoda	Trichoniscidae	*Androniscus stygius tschameri* Strouhal, 1935	Tb.	[14,23]
Isopoda	Sphaeromatidae	*Monolistra racovitzai* Strouhal, 1928	Stb., Tl.	[13,14,24]
Araneae	Dysderidae	*Stalita taenaria* Schiödte, 1847	Tb.	[10,13,14,25–28]
Araneae	Dysderidae	*Parastalita stygia* (Joseph, 1882)	Tb.	[10,13,14,26,28]
Araneae	Dysderidae	*Mesostalita nocturna* (Roewer, 1931)	Tb.	new
Araneae	Linyphiidae	*Troglohyphantes excavatus* Fage, 1919	Tb.	[10,13,14,29,30]
Pseudoscorpiones	Neobissidae	*Neobisium spelaeum* (Schiödte, 1847)	Tb.	[14,31]
Pseudoscorpiones	Chthoniidae	*Chthonius (Globochthonius) speleophylus* Hadži, 1930	Tb.	[14,32]
Opiliones	Nemastomatidae	*Hadzinia ferrani* Novak & Kozel, 2014	Tb.	[33]
Diplopoda	Attemsiidae	*Attemsia falcifera* Verhoeff, 1899	Tb.	[13,14,34]
Diplopoda	Anthogonidae	*Haasia largescutata paligera* (Strasser, 1940)	Tb.	[35]
Diplopoda	Polydemidae	*Brachydesmus inferus concavus* Attems, 1898	Tb.	[35,36]
Diplura	Campodeidae	*Plusiocampa (Stygiocampa) nivea* (Joseph, 1882)	Tb.	[13,14,25]
Collembola	Arrhopalitidae	*Arrhopalites/Pygmarrhopalites* sp.	Tb.	[10]
Collembola	Paronellidae	*Troglopedetes pallidus* Absolon, 1907	Tb.	new
Collembola	Oncopoduridae	*Oncopodura cavernarum* Stach 1934	Tb.	new
Collembola	Onychiuridae	*Absolonia gigantea* (Absolon, 1901)	Tb.	new
Collembola	Onychiuridae	*Onychiurus/Onychiurides* sp.	Tb.	new
Collembola	Tomoceridae	*Tritomurus scutellatus* Frauenfeld, 1854	Tb.	[10,13,14]
Coleoptera	Carabidae	*Typhlotrechus bilimekii frigens* Jeannel 1928	Tb., Tl.	[13,14,37,38]
Coleoptera	Carabidae	*Anophthalmus heteromorphus* (G. Müller 1923)	Tb., Tl.	[13,14,37–40]
Coleoptera	Staphylinidae	*Machaerites ravasinii* G. Müller, 1922	Tb.	[14,41–44]
Coleoptera	Leiodidae	*Bathyscimorphus (Drovenikia) trifurcatus* Jeannel, 1924	Tb., Tl.	[12–14,45,46]
Coleoptera	Leiodidae	*Bathysciotes khevenhuelleri* (Miller, 1852)	Tb.	[12–14]
Coleoptera	Leiodidae	*Aphaobius milleri* (Schmidt, 1855)	Tb.	[12–14,47]
Coleoptera	Leiodidae	*Leptodirus hochenwartii* Schmidt, 1832	Tb.	[8,11–14]

In the stygobiotic and troglobiotic faunal checklist (Table 1), we report only valid names of taxa. Those were checked using online world taxa databases as WoRMS for aquatic taxa and MilliBase, Pseudoscorpions of the World, MolluscaBase, etc. For some taxa, we updated recently synonymized names. Species listed in the former checklist that are not accepted as valid taxa were excluded from the list. In the older literature, some dubious records are persistently appearing. For the precautionary reason, we excluded such taxa from the list if the specimens were not collected or reported recently. In some cases, taxonomic experts explicitly stated that specimens found in Križna jama belong to a new, yet undescribed species.

As the last published checklist of Križna jama troglobionts [14] was mostly focused on aquatic fauna and there was a lack of terrestrial fauna studies, in the autumn of 2020 and early spring of 2021, we sampled the terrestrial fauna. In a few excursions, we investigated the cave up to the point called Križna gora (Figure 2). We searched for fauna mostly by eye on speleothems, cave walls, among rocks and especially near the bat guano deposits. The use of rubber boats, provided by the managers of Križna jama, was essential to investigate deeper parts of the cave. To collect additional terrestrial fauna we deployed a smaller number of plastic pit-fall traps along the transect in the surveyed cave corridor. The traps were baited with cheese and rotten meat. As the traps were without a fixative, we examined them after five days. We took macrophotographs of most of the captured animal species "in situ" in the cave. Specimens needed for future research were stored in alcohol and are deposited in the Zoological Collection of the Notranjska Museum Postojna. Collembolan fauna was preliminarily determined by Marko Lukić (Zagreb) but a detailed study of the collected specimens is still pending. Especially at the cave entrance, we sifted the soil and gravel with entomological sieves but using this method we did not find any significant troglobiotic fauna. During our recent field trips, we also sampled aquatic fauna. Larger crustaceans were sampled with aquatic nets in the water of the cave river, pools and lakes.

Some of the sampled aquatic fauna was later photographed in the laboratory. A large sample of gravel and silt was collected from the bottom of the lakes, sieved and examined in the laboratory to study smaller invertebrate species, especially molluscs and annelids. Microscopic slides with informative details of the species were made for the identification of some invertebrate species. In some cases, taxa are identified to genus level since no detailed and final study on the collected specimens is made.

3. Results

3.1. A Historical Overview

Local people have been visiting Križna jama for millennia. Evidence of that are fragments of pottery dating from the Bronze Age found within the cave. The oldest signature found on a cave wall is dated back to 1557. The first explorations of the cave are documented by the evidence that Jožef (Josip) Cerar (in German Johann Zörrer) visited Križna jama in November 1824 and in July 1825 with a group of local people from Lož and Cerknica. His report, the cave description and its first map, unfortunately, did not appear until 1838 [48]. He was the first to report that some cave passages contain cave bear (*Ursus speleus*) bones. In his report, he used the German name of the cave "Heiligen-kreutz", despite the locals knew the cave as "Mrzla jama" (cold cave). The first known printed account of this cave comes from an Englishman John James Tobin diary who accompanied Sir Humphry Davy douring his cave visit, in 1832. At that time, only the first 500 m or so of the cave were accessible and known. Tobin's report of proteus (*Proteus anguinus*) seen in the cave is very interesting [49]. Soon after that, Aleksander Škofic in 1847 and Adolf Schmidl in 1854 wrote detailed descriptions of the cave. Their reports about plenty of cave bear bones and teeth made Križna cave famous. These reports encouraged the Austrian geologist Ferdinand von Hochstetter to start digging bones in Križna jama in 1878 and 1879. Hochstetter and his workers collected out an impressive number of 4600 cave bear bones from more than 100 individuals [50]. Two complete cave bear skeleton reconstructions, displayed in the Museum of Natural History in Vienna, originated from this time [51]. The first to report about the presence of the cave slender-neck beetle *Leptodirus hochenwartii* in Križna jama was Heinrich Müller in 1857 [8]. The German naturalist Gustav Joseph visited several Slovenian caves in the period from 1853 to 1881. Despite the fact that his collection is not preserved, he published basic data about the Križna jama fauna in one of his publications [10]. Additional faunal reports from Križna jama can be found in Hamann's catalogue [25]. In the 19th century, mostly cave beetle collectors visited Križna jama. There were Eduard Knirsch, Joseph Müller, Anton Haucke, Josip Sever, Alfonz Gspan, Ivan Dolar and Egon Pretner. Roman Kenk and Albin Seliškar visited Križna jama in 1928 and first collected aquatic fauna as well. Ljudevit Kuščer studied and published descriptions of some Križna jama new subterranean snails [17,19]. The first list of Križna jama fauna reported and described taxa by various European leading taxa specialists [12,24,52] was published in Wolf's catalogue [13]. After World War II, Egon Pretner made many visits to Križna jama, mostly leading foreign and other Slovene biologists who sampled the fauna. Jože Bole investigated the gastropods of Križna jama and its surrounding springs [18,53]. From the 70's onwards, Boris Sket and his colleagues visited Križna jama on several excursions and gathered samples, mainly focusing on aquatic fauna [14]. Nevertheless, no detailed ecological investigations have been done in the cave. Based on collected samples, significant progress has been made about the presence of some less investigated animal groups, such as Oligochaeta [54,55]. Sket published a second paper in Slovene about the Križna jama fauna in 1986 [56], and a scientific overview with Fabio Stoch in 2014 [14]. Recently, the authors carried out an additional survey, mainly focused on terrestrial fauna. In this paper, new and so far unpublished data on the presence of some terrestrial troglobiotic fauna is given.

3.2. The Subterranean fauna of Križna Jama

The entrance hall of Križna jama is voluminous and due to the artificially enlarged entrance, daylight penetrates deep inside the cave. Within the first part of the cave next to the entrance, numerous troglophilic and trogloxene animals can be regularly seen. Especially in winter geometrid moths *Triphosa dubitata* and noctuids *Scoliopteryx libatryx*, as well as cave crickets (*Troglophilus* spp.), are common as parietal fauna on the walls. Different species of harvestmen (Opiliones), spiders (Araneae), caddisflies (Trichoptera) and various flies and mosquito (Diptera) considered as troglophiles and trogloxenes are common here. In the bottom of the entrance hall rich with humus, some other soil fauna, not considered as troglobiotic, is present. Stone martens (*Martes foina*) and edible dormice (*Glis glis*) regularly penetrate deeper into the cave corridors. On the contrary, the tawny owl (*Stryx aluco*) is usually found dormant only at the cave entrance.

So far, seven species of bats have been recorded in Križna jama [57,58]. The most common is the lesser horseshoe bat (*Rhinolophus hipposideros*). It overwinters in both Križna jama and in Križna jama 2 in significant numbers. Regularly about 900 individuals overwinter here and this represents the second-highest concentration of bats compared to other Slovene caves [58]. Other species like the great mouse-eared bat (*Myotis myotis*), Daubenton's bat (*Myotis daubentonii*), the serotine (*Eptesicus serotinus*) and the barbastelle bats (*Barbastella barbastellus*) are occasionally found in smaller numbers. Schreibers' bat (*Miniopterus schreibersii*) have so far been spotted in Križna jama only once [58]. There are no bat breeding colonies in Križna jama. Bat parasites such as the bat thick (*Ixodes vespertilionis*) can be regularly found on walls next to the hibernating bats. Numerous troglophilic and troglobiotic collembolans, mites, diplopods and beetles are attracted by owl pellets and bat guano as well as by marten scats and dormice droppings all over the cave.

Among the troglobionts cave beetles are relatively numerous. The most famous of all, the slenderneck beetle *Leptodirus hochenwartii*, is rare and mostly found in the deeper cave spaces far from the entrance. In the entrance hall, as well all along the immense cave tunnels of Križna jama, the small bathyscioid leptodirine beetle *Bathyscimorphus trifurcatus* (Figure 3b) is the commonest troglobiotic beetle [11,45,46]. The leptodirine beetle *Bathysciotes khevenhuelleri* is regularly trapped in pitfall traps in the entrance hall between the stones and the soil in the habitat that resembles the MSS, while the leptodirine *Aphaobius milleri* have been collected only sporadically [47]. *Bathysciotes khevenhuelleri* is relatively common in Snežnik-Javorniki wider area in mesovoid shallow substratum (or "Milieu souterrain superficiel"—MSS) [59], while *Aphaobius milleri* prefers to some extent colder cave microclimates in the area which can be found right after the Križna jama entrance. Two species of troglobiotic ground beetles are known from Križna jama and for both taxa Križna jama is their type locality. The subspecies of the trechine *Typhlotrechus bilimeki frigens* (Figure 3a) is relatively common [37,38] in many places, while the endemic *Anophthalmus heteromorphus* is extremely rare and thus much sought by collectors [38–40]. The troglobiotic pselaphine beetle *Machaerites ravsinii* is occasionally found among the rocks in the entrance hall of Križna jama [41–44]. Some other subterranean beetle species occasionally reported for Križna jama are doubtful and need further recent confirmation.

The Collembolan fauna of Križna jama has not been studied in details until recently. Joseph [10] described two species of *Sminthurus* from Križna jama, but they are not considered valid in later works. In Wolf's catalogue [13] there are records of five species of collembolans, which were questioned as doubtful in the last Križna jama list of troglobionts [14]. In recent samplings, the collembolan species *Troglopedetes pallidus* (Figure 3b), *Oncopodura cavernarum*, *Absolonia gigantea* and some undefined species of Onychuiridae and Arrhopalitidae are documented. Small, eyeless and strongly troglomorphic collembolan specimens of taxonomically not yet determined species of the complex *Arrhopalites/Pygmarrhopalites* (Figure 3e) can be found on bat guano regularly. Troglobiotic collembolans are all common on the bat guano and similar organic debris, especially stone marten scats in deeper parts of the cave. *Tritomurus scutellatus* is common only near the entrance, but still within the cave's dark zone. The dipluran species *Plusiocampa (Stygiocampa) nivea* is rare but widespread in Križna jama.

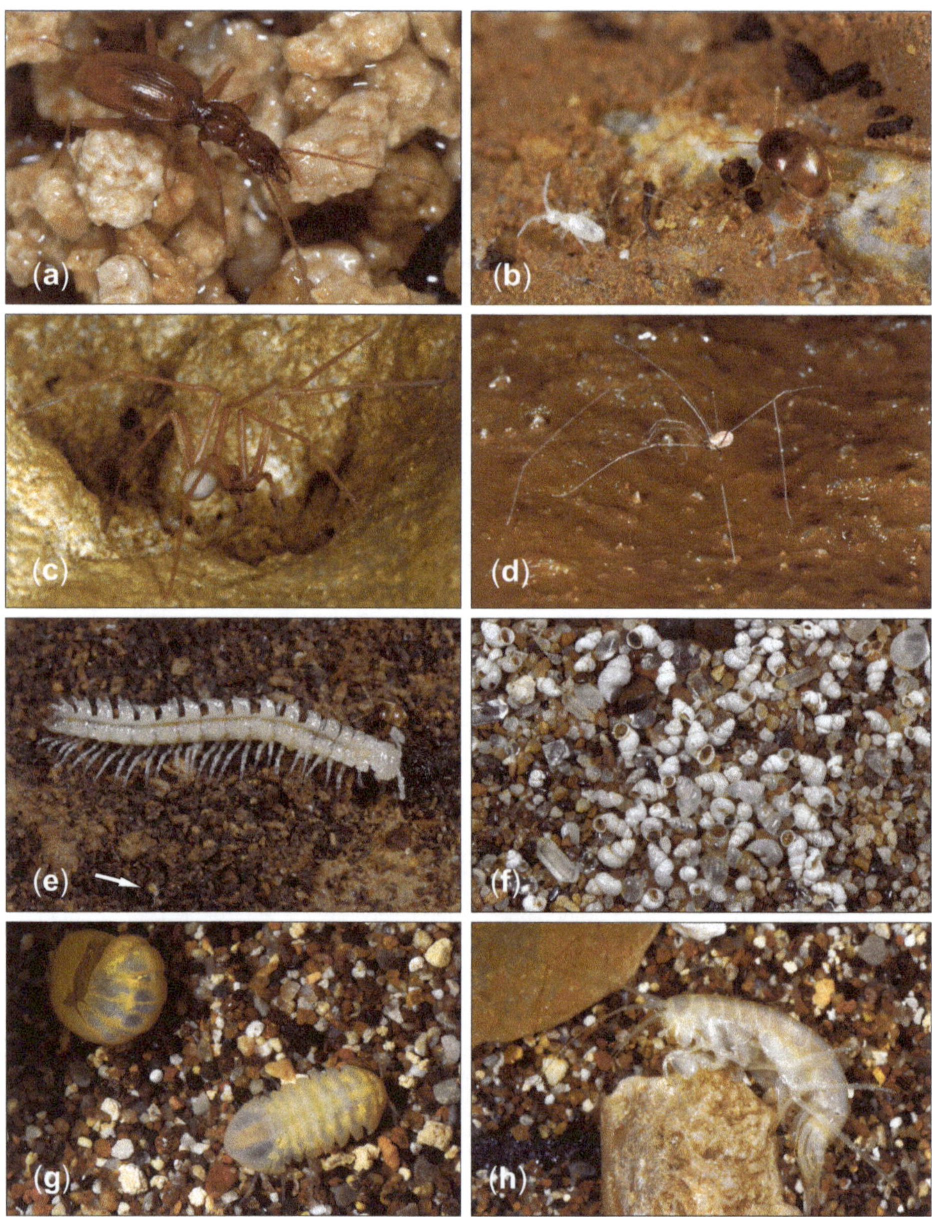

Figure 3. Troglobiotic and stygobiotic fauna from Križna jama: (**a**) Ground beetle (*Typhlotrechus bilimekii frigens*). (**b**) Leptodirine beetle (*Bathyscimorphus (Drovenikia) trifurcatus*), on the right and Collembolan (*Troglopedetes pallidus*), on the left. (**c**) Cave spider (*Parastalita stygia*). (**d**) Cave Harvestman (*Hadzinia ferrani*). (**e**) Cave Millipede (*Brachydesmus inferus concavus*) and Collembolan (*Pygmarrhopalites* sp.), indicated with white arrow. (**f**) Various species of aquatic snails from family Hydrobiidae and quartz crystals. (**g**) Sphaeromatid isopod (*Monolistra racovitzai*). (**h**) Cave amphipod (*Niphargus orcinus*).

The troglobiotic arachnid fauna of Križna jama is rich as well. So far, only one of the lyniphiid *Troglohyphantes* species is reported from Križna jama [10,14,27,29,30]. The troglobiotic dysderid spider species *Stalita taenaria* and *Parastalita stygia* (Figure 3c) live in Križna jama sympatrically [26–28]. The coexistence of these two related species in the same cave is rare [60]. Surprisingly enough, a third distinctly smaller dysderid species *Mesostalita nocturna* was found deep inside Križna jama during one of the last excursions (Delić T. per. comm.). In one of the recent ecological studies [60], it was demonstrated that the shift in the trophic niche amongst related species minimizes the interspecific competition and enables related dysderid spider species coexistence. The fact that the Križna jama is a tourist cave and is regularly visited, the recent discovery of fully troglomorphic harvestmen was an unexpected surprise. A detailed study of the only known single male and female specimens showed that they belong to a recently described species *Hadzinia ferrani* (Figure 3d) from the cave Ferranova buža, situated 40 km northwest from Križna jama [33,61]. Amongst the pseudoscoripons in Križna jama two troglobiotic species are known [31,32]. The relatively giant *Neobysium speleum* is rarely encountered on wet walls and stalactites of the deeper parts of the cave. The second smaller species *Chthonius (Globochthonius) spelaeophilus* is reported recently from the cave too [32].

Among the Myriapod, a only troglobiotic millipedes have been recorded in Križna jama so far, although the cave centipede *Lithobius stygius* is expected to be found also as it was recently confirmed in nearby Mrzla jama pri Ložu [62]. The most abundant among the millipedes are the polydesmid *Brachydesmus inferus concavus* (Figure 3e) [35,36,63], which was reported as *B. subterraneus* or *B.* sp. in former works [13,14]. On any organic material, even in the deeper parts of the cave, all development stages (larvae and mating adults) of this millipede are commonly seen. Two other millipede species, *Attemsia fulcifera* [13,14,34], reported as *A. pretneri* in older works and *Hassia largescutata paligera*, reported as *Acherosoma cornuatum paligerum*, endemic species of the region, are living in Križna jama too [35,63–65].

Two species of terrestrial isopod crustaceans are regularly found in Križna jama. The giant cave woodlouse *Titanethes albus* lives on wet spots in deeper parts of the cave, while the smaller species *Androniscus stygius tschammeri* [23] is usually found on rotting wood or organic debris near the entrance of the cave. Visitors to Križna jama can regularly encounter the aquatic sphaeromatid isopod *Monolistra racovitzai* (Figure 3g) [13,14,24]. In some permanent cave lakes, these crustaceans slowly crawl on rocks and flooded rimstones. In the same cave lakes, the monolistrians are regularly accompanied by the relatively rarer predatory amphipod crustacean *Niphargus orcinus* (Figure 3h) [14,22]. This *Niphargus* species can reach up to 3 cm in length and it is a relatively giant invertebrate. Studying crustaceans specimens of the genus *Monolistra* and *Niphargus* under the microscope, the presence of ostracodan *Sphaeromicola stammeri* and turbellarian *Stygodyticola hadzii* as epizoans or exterior parasites have been found [14,16]. Two smaller amphipod species *Niphargus wolfi* and *Niphargus stygius* live in Križna jama too. While the former is usually found in loam and rimstone pools fed by percolated water dripping from the cave ceilings, *N. stygius* is found in cave lakes too [14]. In such pools, significant fauna of copepod crustaceans like *Elaphoidella* spp., *Speocyclops infernus* and *Acanthocyclops kieferi* have been sampled [14,21,66,67]. These pools can be occasionally flooded by the subterranean river during the floods. Larger waters, the lakes and mainstream rivers in both corridors of the Blata and Pisani rov, are oligotrophic and the fauna here is relatively scarce. Besides *Monolistra* and *Niphargus*, smaller copepod fauna such as *Acanthocyclops troglophilus* and *Diacyclops charon* have been recorded [14]. Despite most of the river corridor bottoms being covered in sinter deposits in some places, especially in the corridor Blata, the river bottom consists of very loose silt and loam mixed with sand. Here the large and colourful troglobiotic oligochaete *Delaya bureschi* is common. Six other troglobiotic oligochaetes have been recorded for Križna jama so far [14,55]. In the subterranean waters of Križna jama two species of troglobiotic turbellarians are reported as well.

In the lakes, river and streams of Križna jama, six aquatic, including four hydrobiid, snail species are found [17–19,53,68]. Based on specimens from Križna jama, two species

Belgrandiella crucis and *B. schleschi*, were described. *Belgrandiella superior* and *Hauffenia michleri*, have been described from cave waters from a surrounding Ljubljanica river catchment area. Specimens of the aquatic snail from the genus *Paladilhilopsis* are still to be determined [14]. Aquatic snails live all along permanent water streams, where specimens can be found mainly on the bottom among the rocks, gravel and silt. In the lakes, where small amounts of organic debris drop from the surface (such as particles of fallen leaves), the populations of snails are more abundant. Besides hydrobiid snails, three species of primarily terrestrial troglobiotic snails live in Križna jama. The species *Zospeum kusceri* and *Z. isselianum* can be usually found on wet walls and stalactites. A third species *Zospeum exiguum*, described from Križna jama, is considered as at least a partly aquatic. Living specimens of *Z. exiguum* were collected mainly on the submerged rocks in lakes and streams so far [19,20]. However, the easiest way to collect empty snail shells is to collect sand on certain spots on the lake edges, where empty shells are deposited in significant numbers as a tanatocenoses (Figure 3f).

Sifting gravel and sand samples from the mainstream pools, in search of snails, recently revealed unexpected cave sedimentary deposits. In previous Križna jama fluvial sediment studies [69,70] allochthone oolitic bauxite ooids and quartz pebbles amongst the autochthonous fragments of broken sinter were noted. Searching for snails, we found tiny quartz crystals of a special biterminated shape as well (Figure 3f). Most of the quartz crystals in Križna jama are notably eroded, but some brilliant specimens can be found as well. Such biterminated quartz crystals, popularly called "cerknica diamonds", are unique and in the region known from the narrow strip of land characteristic by porous sucrose dolomites on the Slivnica Mountain on the eastern side of Cerknica [71]. Accompanied by bauxite ooids, these fluvial sediments clearly show their origin and a fluvial transport from Slivnica—Bloke plateau and demonstrate the river catchment area of at least part of Križna jama running waters. Identification of water catchment area that drains into the Križna jama water system is essential for understanding zoogeographical reasons for species composition as well as for nature conservation reasons. For some rare events of periodic water pollution in the Križna jama, the source stream Farovščica in the Bloke plateau has already been identified, using water tracing methods [6].

The largest aquatic cave animal in Križna jama is to be the cave salamander or proteus (*Proteus anguinus*). Its presence in Križna jama was reported by Tobin in 1832 [49] later by Joseph to [10] and cited by Wolf [13]. Since then, no one has ever seen or mentioned proteus in Križna jama. Sket [72] declared this report as a "most probably erroneous" and therefore proteus has been later omitted from the fauna list of Križna jama. Proteus has a wide holo-dinaric distribution and it is present along all the Ljubljanica river drainage area. The closest confirmed localities of proteus are some wells in Loško polje, only a few kilometres away from Križna jama. Its absence in this cave is for this reason of great scientific interest. Similarly to the absence of proteus, the cave shrimp *Troglocaris* has never been reported in the Križna jama, not even in the wider Lož area [73]. The reason for this deficiency might be the same as for the absence of any Thiaridae, Neritidae and Unionidae Mollusca in the neighbouring Cerknica lake [14,18,53,68].

4. Discussion

In one of the first reports on the Križna jama subterranean biodiversity [2] there are listed 45 troglobionts and stygobionts from Križna jama, among them 29 aquatic and 16 terrestrial species what listed this cave on the list of the world's richest caves [2,74]. In a more recent and updated list [14], the total number of troglobionts and stygobionts reached 50. Among those, 32 species are aquatic and 18 species as terrestrial. In this paper, the number of stygobionts remains the same but we omitted some doubtful taxa reported in older works and added significant new terrestrial troglobiotic taxa that were documented recently. Therefore, we present a list of 60 troglobiotic species, among them 32 aquatic and 28 terrestrial (Table 1). No new taxa have been recently described on basis of the specimens from Križna jama and thus the number of taxa with the type locality in this cave remains 10 species and subspecies. The fauna of Križna jama is similar to the fauna of other biodiverse caves in the Notranjska region like the Postojna-Planina cave system. In the Križna jama, there are some additional faunistic elements among the terrestrial fauna, which are related to the Dolenjska karst and are not present in Notranjska karst caves. The spider *Parstalita stygia* and millipede *Brachydesmus inferus* are such examples. Aquatic fauna is the most characteristic of the Ljubljanica river drainage system [68]. Geographically located in the temperate climatic zone and in the Dinaric mountain range, which is well known for its rich subterranean biodiversity the high number of troglobionts is not surprising. However, most hot spots of high subterranean biodiversity tend to have high primary productivity or rich organic input from the surface [2]. Contrary to some caves as the Postojna-Planina cave system, Logarček and some other caves situated in the region popularly called Notranjska triangle and well known as a prominent subterranean fauna hotspot [68], Križna jama is to be classified as oligotrophic.

There is no sinking river from the surface passing through the cave that brings significant organic material as for example Pivka River in the Postojna–Planina cave system. Waters in Križna jama are thus clean and of good quality. This is probably the main reason why there are so few non-troglobiotic elements in Križna jama waters. Fauna in Križna jama is due to lack of food, not abundant, but it is well diversified. The only exception could be attributed to Tršanov rov or the so-called Stransko jezero, where occasionally we can find a small amount of organic debris coming from the surface. The aquatic fauna is the richest there. In this part, small stonefly larvae enter the cave and are a presumably welcome food source for cave animals such as *Niphargus orcinus*.

The entrance to the Križna jama is situated on the foothill of Križna gora Mountain. The cave is horizontal and extends directly into the mountain. The thickness of the limestone ceiling rapidly increases and measures 50, 100 and 150 m above the first lake close to the entrance, in the middle of the cave and in the deepest part of the cave, respectively. This immense, to some extent cracked and partly saturated habitat called epikarst, is from a biological point of view mostly unexplored and undersampled [75–77]. Epikarst tends to be a hotspot within a hotspot because the heterogeneous nature of epikarst allows for high species richness. Both the aquatic and terrestrial fauna of epikarst tends to be abundant, with epikarst copepods being the most diverse and best-studied group in some other biodiverse caves [76,77]. To further improve the knowledge of the faunal composition of Križna jama, sampling of the epikarst water seems to be the most promising, both direct sampling of dripping water as well as a detailed sampling of pools and lakes filled with percolating water. The thickness of rock deposits above the cave prevents penetration of organic particles and also incidental soil fauna elements. So food sources scarcity was noted throughout the inner parts of Križna jama too. These are the reasons that in Križna jama the terrestrial troglobiotic fauna is not abundant in numbers but it consists of numerous troglobiotic taxa without the presence of not-troglobiotic representatives. The cave is relatively well studied, however the recent finding of some, even big-sized arachnid species, so far overlooked is promising. Thus, further discoveries of specialized subterranean species in the cave are expected.

Author Contributions: Conceptualization, investigation, writing—original draft preparation, S.P.; writing—review and editing, T.P. Both authors have read and agreed to the published version of the manuscript.

Funding: This research received no external funding.

Institutional Review Board Statement: Not applicable.

Informed Consent Statement: Not applicable.

Data Availability Statement: Not applicable.

Acknowledgments: We are deeply thankful to the local caretakers of Križna jama, namely Gašper Modic and Alojz Troha, for their help during our Križna jama visits. Tone Novak (Slovenj Gradec), Peter Kozel (Maribor) and Teo Delić (Ljubljana) accompanied us and participated in some recent cave surveys. We wish to thank Marko Lukić (Zagreb) for the preliminary determination of Collembola. Franjo Drole and Mitja Prelovšek (both IZRK ZRC SAZU Postojna) provided us with detailed speleological data and map of Križna jama.

Conflicts of Interest: The authors declare no conflict of interest.

References

1. Simić, M.; Gedei, P. Križna jama in jama Dihalnik v Grdem dolu, skrivnostni lepotici pod Križno goro v zaledju Cerkniškega polja. *Natl. Geogr. Slov. Ed.* **2018**, 26–43.
2. Culver, D.C.; Sket, B. Hotspots of subterranean biodiversity in caves and wells. *J. Cave Karst Stud.* **2000**, *62*, 11–17.
3. Prelovšek, M. Speleogenesis and Flowstone Deposition in Križna jama. In *Križna jama: Palaeontology, Zoology and Geology of Križna jama in Slovenia*; Pacher, M., Pohar, V., Rabeder, G., Eds.; Verlag der Österreichischen Akademie der Wissenschaften: Wien, Austria, 2014; pp. 21–26.
4. Knez, M.; Prelovšek, M. The Geological Settings of Križna jama. In *Križna jama: Palaeontology, Zoology and Geology of Križna jama in Slovenia*; Verlag der Österreichischen Akademie der Wissenschaften: Wien, Austria, 2014; pp. 15–20.
5. Prelovšek, M. The Hydrogeological Setting of Križna jama. In *Križna jama: Palaeontology, Zoology and Geology of Križna jama in Slovenia*; Pacher, M., Pohar, V., Rabeder, G., Eds.; Verlag der Österreichischen Akademie der Wissenschaften: Wien, Austria, 2014; pp. 27–33.
6. Kogovšek, J.; Prelovšek, M.; Petrič, M. Underground water flow between Bloke plateau and Cerknica polje and hydrologic function of Križna jama, Slovenia. *Acta Carsol.* **2008**, *37*, 213–225. [CrossRef]
7. Novak, D. O barvanju potoka v Križni jami. *Geogr. Vestn.* **1969**, *41*, 75–79.
8. Müller, H. Uber Lebensswiese der Augenlosen Käfer in der Krainer Höhlen. *Stettiner Entomol. Zeitung* **1857**, *18*, 65–74.
9. Hauffen, H. Beiträge zur Grottenkunde Krain's. *Jahresh. des Vereines des Krainischen Landes-Mus.* **1858**, *2*, 40–53.
10. Joseph, G. Erfahrungen im wissenschaftlichen Sammeln und Beobachten der den Krainer Tropfsteigrotten eigenen Arthropoden. *Separat-Abdruck aus der Berliner Entomol. Zeitschrift* **1882**, *26*, 1–100.
11. Jeannel, R. Revision des Bathysciinae. *Biospeol. XIX—Arch. Zool. Expérimentale Générale* **1911**, *5*, 1–641.
12. Jeannel, R. Monographie des Bathysciinae. *Arch. Zool. Expérimentale Générale* **1924**, *64*, 1–436.
13. Wolf, B. *Animalium Cavernarum Catalogus*; Dr. W. Junk's Gravenhage: Berlin, Germany, 1937.
14. Sket, B.; Stoch, F. Recent Fauna of the Cave Križna jama in Slovenia. In *Križna jama: Palaeontology, Zoology and Geology of Križna jama in Slovenia*; Pacher, M., Pohar, V., Rabeder, G., Eds.; Verlag der Österreichischen Akademie der Wissenschaften: Wien, Austria, 2014; pp. 45–55.
15. Stammer, H.J. Zwei neue troglobionte Protozoen: Speleophrya troglocaridis n. g., n. sp. von den Antennen der Hohlengarnele Troglocaris schmidti Dorm. und Lagenophrys monolistrae n. sp. vod den Kiemen (Pleopoden) der Hohlenasselgattung Monolistra. *Arch. Für Protistenkd.* **1935**, *84*, 518–527.
16. Matjašič, J. Verlaufige mitteilungen uber Europaische Temnocephalen. *Biološki Vestn.* **1958**, *6*, 60–65.
17. Kuščer, L. Drei neue Höhlenschnecken. *Glas. Muzejskega Društva Slov.* **1928**, *7–8*, 50–51.
18. Bole, J. Mehkužci Cerkniškega jezera in okolice. *Acta Carsol.* **1979**, *8*, 204–236.
19. Kuščer, L. Höhlen- und Quellenschnecken aus dem Flussgebiet Ljubljanica. *Arch. Für Molluskenkd.* **1932**, *64*, 48–62.
20. Bole, J. Rod Zospeum Bourguignat 1856 (Gastropoda, Ellobiidae) v Jugoslaviji. *Razpr. SAZU* **1974**, *17*, 249–291.
21. Stoch, F. A new genus and two new species of Canthocamptidae (Copepoda, Harpactioida) from caves in northern Italy. *Hydrobiologia* **1997**, *350*, 49–61. [CrossRef]
22. Joseph, G. Über die Grotten in der Krainer Gebirge und deren Tierwelt. *Jahresbericht der Schlesischen Gesellschaft für Vaterländische Kult. Breslau* **1869**, *46*, 48–52.
23. Strouhal, H. Landasseln aus Balkanhöhlen, gesammelt von Prof. Dr. Karl Absolon. 10. Mitteilung. In *Studien aus dem Gebiete der allgemeinen Karstforschung, der wissenschaftlichen Höhlenkunde, der Eiszeitforschung und den Nachbargebieten*; Barvič & Novotný: Brno, Czech Republic, 1939; pp. 1–37.
24. Strouhal, H. Eine neue Höhlen-Sphaeromatide (Isop.). *Zool. Anz.* **1928**, *77*, 84–92.
25. Hamman, O. *Europäische Höhlenfauna. Eine Darstellung der in den Höhlen Europas lebenden Tierwelt mit besonderer Berücksichtigung der Höhlenfauna Krains*; Hermann Costenoble: Jena, Germany, 1896.
26. Kratochvil, J. Cavernicole Dysderae. *Prirodoved. Pr. Ust. Československe Akad. ved v Brne* **1934**, *2*, 165–226.
27. Kratochvil, J. Liste générale des araignées cavernicoles en Yougoslavie. *Prirodosl. Razpr.* **1934**, *2*, 165–226.
28. Deeleman-Reinhold, C.L. Beitrag zur Kenntnis höhlenbewohnender Dysderidae (Araneida) aus Jugoslawien. *Razprave-Dissertationes SAZU, Ljubljana* **1971**, *14*, 93–120.
29. Deeleman-Reinhold, C.L. Revision of the cave-dwelling and related spiders of the genus Troglohyphantes Joseph (Linyphiidae), with special reference to the Yugoslav species. *Razprave-Dissertationes SAZU, Ljubljana* **1978**, *23*, 1–221.
30. Fage, L. Etudes sur les araignées cavernicoles. III. Sur le genre Troglohyphantes Biospelogica XL. *Arch. Zool. Expérimentale Générale* **1919**, *55*, 55–148.

31. Beier, M. Die Höhlenpseudoscorpione der Balkanhalbinsel. *Biospeol. Balc.* **1939**, *4*, 1–83.
32. Gardini, G. Gli Chthonius (Globochthonius) D'Italia e di Slovenia (Pseudoscoripines Chthonidae). *Gortania, Bot. Zoolgia* **2009**, *31*, 53–64.
33. Kozel, P.; Delić, T.; Novak, T. Nemaspela borkoae sp. nov. (Opiliones: Nemastomatidae), the second species of the genus from the Dinaric Karst. *Eur. J. Taxon.* **2020**, *717*, 90–107. [CrossRef]
34. Mršić, N. Attemsiidae (Diplopoda) of Yugoslavia. *Razpr. SAZU* **1987**, *27*, 101–168.
35. Strasser, K. Diplopoden des jugoslavischen Draubanats. *Prirodosl. Razpr.* **1940**, *4*, 13–85.
36. Mršić, N. Polydesmida (Diplopoda) of Yugoslavia. *Razpr. SAZU* **1988**, *29*, 69–112.
37. Jeannel, R. *Monographie des Trechinae (3)*; Les Trechini Cavernicoles; L'Abeille: Paris, France, 1928.
38. Drovenik, B.; Peks, H. Catalogus faunae Carbidae del Balkanländer, Coleoptera, Carabidae. *Shwanfelder Colopterologische Mitteilungen* **1994**, *1*, 1–103.
39. Daffner, H. Revision der Anophthalmus-Arten und -Rassen mit lang und dicht behaarter Körperoberseite. *Mitteilungen der Münchner Entomol. Gesellschaft* **1996**, *86*, 33–78.
40. Müller, J.G. Über einige Krainer Anophthalmen. *Wiener Entomol. Zeitung* **1921**, *38*, 91–99.
41. Nonveiller, G.; Pavićević, D. Description d'une sous-espèce nouvelle et de six espèces nouvelles du genre Machaerites Miller, 1855 de Slovénie et de Croatie (Coleoptera, Pselaphinae, Bythinini). *Nouv. Rev. Entomol.* **2001**, *18*, 317–333.
42. Jeannel, R. Les Psélaphides troglobies de la Slovénie. *Notes Biospéologiques* **1954**, *9*, 7–15.
43. Poggi, R. Forme nuove o poco note di Pselaphidae cavernicoli del Friuli-Venezia Giulia e della Jugoslavia. *Mem. della Soc. Entomol. Ital.* **1991**, *70*, 201–224.
44. Müller, G. I Pselafidi cavernicoli del Carso Adriatico settentrionale (Venezia Giulia e Carniola). *Boll. della Soc. Adriat. di Sci. Nat. Trieste* **1947**, *43*, 133–146.
45. Pretner, E. Coleoptera, Catopidae, Bathysciinae. *Cat. Faunae Jugoslaviae, SAZU, Ljubljana* **1968**, *3*, 1–60.
46. Bognolo, M. Il Genere Bathyscimorphus (Coleoptera: Cholevidae). *Coleoptera* **2002**, *6*, 1–33.
47. Bognolo, M.; Vailati, D. Revision of the genus Aphaobius Abeille de Perrin, 1878 (Coleoptera, Cholevidae, Leptodirinae). *Scopolia* **2010**, *68*, 1–75.
48. Zörrer, J. Beschreibung einer Berghöhle bei Heiligen Kreuz unweit Laas. *Beiträge zur Naturgeschichte, Landwirtschaft und Topogr. Herz. Krain* **1838**, *1*, 76–88.
49. Shaw, T. *Foreign Travellers in the Slovene Karst 1486–1900*; ZRC Publishing: Ljubljana, Slovenia, 2008.
50. Hohstetter, F. Die Kreuzberghöhle bei Laas in Krain. *Denkschriften der Kais. Akad. der Wissenschaften, Math. Cl.* **1881**, *43*, 1–18.
51. Pohar, V. Križna jama: Description and History of Research. In *Križna jama: Palaeontology, Zoology and Geology of Križna jama in Slovenia*; Pachner, M., Pohar, V., Rabeder, G., Eds.; Verlag der Österreichischen Akademie der Wissenschaften: Wien, Austria, 2014; pp. 1–6.
52. Müller, J.G. Vier neue Anophthalmen aus Krain (Col. Carab.). *Wiener Entomol. Zeitung* **1923**, *40*, 101–106. [CrossRef]
53. Bole, J. Taksonomska, ekološka in zoogeografska problematika družine Hydrobiidae (Gastropoda) iz porečja Ljubljanice. *Razpr. (Dissetationes), Cl. IV; SAZU* **1967**, *10*, 57–108.
54. Martinez-Ansemil, E.; Sambugar, B.; Giani, N. Groundwater Oligochaetes from southern-Europe. I. A new genus and three new species of Rhyacodrilinae (Tubificidae) with a redescription of Tubifex pescei (Dumnicka) comb. n. *Ann. Limnol.* **1997**, *33*, 33–44. [CrossRef]
55. Karaman, S. *Taksonomska, Zoogeografska i Ekološka Studija Oligochaeta u Področju Planine, Cerknice i Postojne*; Univerza v Ljubljani: Ljubljana, Slovenija, 1978.
56. Sket, B. O favni v Križni jami. In *Notranjski List III*; Kulturna skupnost Cerknica: Cerknica, Slovenia, 1986; pp. 25–28.
57. Presetnik, P.; Koselj, K.; Zagmajster, M. *Atlas netopirjev (Chiroptera) Slovenije. Atlas of the bats of Slovenia*; Presetnik, P., Koselj, K., Zagmajster, M., Eds.; Center za kartografijo favne in flore: Miklavž na Dravskem polju, Slovenia, 2009.
58. Presetnik, P.; Troha, A. Prezimujoči netopirji. In *Križna Jama*; Kržič, M., Ed.; Društvo ljubiteljev Križne jame: Grahovo, Slovenia, 2010; pp. 24–32.
59. Pipan, T.; López, H.; Oromí, P.; Polak, S.; Culver, D.C. Temperature variation and the presence of troglobionts in terrestrial shallow subterranean habitats. *J. Nat. Hist.* **2010**, *45*, 253–273. [CrossRef]
60. Pavlek, M.; Mammola, S. Niche-based processes explaining the distributions of closely related subterranean spiders. *J. Biogeogr.* **2021**, *48*, 118–133. [CrossRef]
61. Novak, T.; Kozel, P. Hadzinia ferrani, sp. n. (Opiliones: Nemastomatidae), a highly specialized troglobiotic harvestman from Slovenia. *Zootaxa* **2014**, *3841*, 135–145. [CrossRef] [PubMed]
62. Kos, A. *Genetska Raznolikost Strig Kompleksa vrsta Lithobius Stygius v Severnih Dinaridih, Magistrasko Delo*; Univerza v Ljubljani: Ljubljana, Slovenia, 2021.
63. Strasser, K. Die Diplopoden Sloweniens. *Acta Carsol.* **1966**, *4*, 159–220.
64. Strasser, K. Neue Acherosomen (Diplopoda Ascospermophora). *Prirodosl. Razpr.* **1935**, *2*, 231–244.
65. Strasser, K. Uber Diplopoden Jugoslawiens. *Senckkenbergiana Biol.* **1971**, *52*, 313–345.
66. Chappius, P.A. Copépodes (première serie) avec l'énumeration de tous les Copépodes cavernicoles connus en 1931, Biospeologica 59. *Arch. Zool. Expérimentale Générale* **1933**, *76*, 1–57.

67. Petkovski, T.K. Neue höhlenbewonende Harpacticoida (Crustacea, Copepoda) aus Slovenien. *Acta Musei Maced. Sci. Nat.* **1983**, *16*, 177–205.
68. Sket, B. The cave Fauna in the triangle Cerknica-Postojna-Planina (Slovenia, Yugoslavia), its conservation importance. *Varst. Narave* **1979**, *13*, 45–59.
69. Kralj, P. Sedimentary Deposits in Kittl's Bear Gallery in Križna Jama Cave (Slovenija). In *Križna jama: Palaeontology, Zoology and Geology of Križna jama in Slovenia*; Pacher, M., Pohar, V., Rabeder, G., Eds.; Verlag der Österreichischen Akademie der Wissenschaften: Wien, Austria, 2014; pp. 35–43.
70. Gospodarič, R. Fluvialni sedimenti v Križni jami (Fluvial sediments in Križna jama). *Acta Carsol.* **1974**, *6*, 327–366.
71. Jeršek, M.; Žorž, M. Kremen iz okolice Cerknice. In Mineralna bogastva Slovenije. *Scopolia* **2006**, *3*, 410–417.
72. Sket, B. Distribution of Proteus (Amphibia: Urodela: Proteidae) and its possible explanation. *J. Biogeogr.* **1997**, *24*, 263–280. [CrossRef]
73. Sket, B.; Zakšek, V. European cave shrimp species (Decapoda: Caridea: Atyidae), redefined after a phylogenetic study; redefinition of some taxa, a new genus and four new Troglocaris species. *Zool. J. Linn. Soc.* **2009**, *155*, 786–818. [CrossRef]
74. Culver, D.C.; Pipan, T. Subterranean Ecosystems. In *Encyclopedia of Biodiversity*; Levin, A.S., Ed.; Academic Press: Wlatham, MA, USA, 2013; pp. 49–62.
75. Pipan, T.; Brancelj, A. Ratio of copepods (Crustacea: Copepoda) in fauna of percolation water in six karst caves in Slovenia = Delež ceponožcev (Crustacea: Copepoda) v favni preniklih voda v šestih kraških jamah v Sloveniji. *Acta Carsol.* **2001**, *30*, 257–265.
76. Culver, D.; Pipan, T. *Shallow Subterranean Habitats. Ecology, Evolution, and Conservation*; Oxford University Press: Oxford, UK, 2014.
77. Pipan, T. *Epikarst—A Promising Habitat: Copepod fauna, Its Diversity and Ecology: A Case Study from Slovenia (Europe)*; ZRC SAZU Publishing: Ljubljana, Slovenia, 2005.

Article

Biodiversity in the Cueva del Viento Lava Tube System (Tenerife, Canary Islands)

Pedro Oromí [1,*] and Sergio Socorro [2]

1   Departamento de Biología Animal, Facultad de Biología, Universidad de La Laguna, 38205 La Laguna, Spain
2   Museo de Ciencias Naturales de Tenerife, Fuente Morales s/n, 38003 Santa Cruz, Spain;
    socher@telefonica.net
*   Correspondence: poromi@ull.edu.es

Abstract: Cueva del Viento and Cueva de Felipe Reventón are lava tubes located in Tenerife, Canary Islands, and are considered the volcanic caves with the greatest cave-dwelling diversity in the world. Geological aspects of the island relevant to the formation of these caves are discussed, and their most outstanding internal geomorphological structures are described. An analysis of the environmental parameters relevant to animal communities is made, and an updated list of the cave-adapted species and their way of life into the caves is provided. Some paleontological data and comments on the conservation status of these tubes are included.

Keywords: lava tubes; geology; fauna; troglobionts; paleontology; conservation; Canary Islands

check for **updates**

Citation: Oromí, P.; Socorro, S. Biodiversity in the Cueva del Viento Lava Tube System (Tenerife, Canary Islands). *Diversity* **2021**, *13*, 226. https://doi.org/10.3390/d13060226

Academic Editors: Michael Wink, Tanja Pipan, David C. Culver and Louis Deharveng

Received: 29 April 2021
Accepted: 19 May 2021
Published: 23 May 2021

Publisher's Note: MDPI stays neutral with regard to jurisdictional claims in published maps and institutional affiliations.

1. Introduction

Caves are mostly associated with sedimentary terrains subjected to karst processes, mainly in carbonate rocks but also in gypsum, rock salt, conglomerate, etc. [1]. These solution caves are the result of a chemical erosive process that progresses slowly (millions of years) and can result in important cavity dimensions, such as the 667 km of Mammoth Cave (USA) or the 371 km of the Sac Actun/Dos Ojos System (Mexico) [2]. Volcanic terrains typically lack limestone and other sedimentary rocks, at least with sufficient thickness and age to originate karst caves, with a few exceptions when the reef limestone has been uplifted and later karstified [3]. However, other kinds of caves are frequent in volcanic terrains, formed in a completely different way and different type of rocks, usually fluid basalts but sometimes also phonolites [4]. Volcanic caves, most of them lava tubes, have a constructive rather than erosive origin, are exactly the same age as their host rock, and become destroyed in a very short geological time span compared to karst ones [5]. This, together with the fact that the whole volcanic areas have much smaller extension than karstifiable sedimentary rocks, means that volcanic caves are much less abundant worldwide and, in general, of considerably smaller dimensions than karst ones [6]. It had been claimed that volcanic caves did not have any adapted cave-dwelling fauna [7]. However, this was later disproved for Hawaiian lava tubes and later for many other tropical and non-tropical volcanic areas [5,8–11]. Conversely, some lava tubes are particularly rich in cave-adapted fauna and feature prominently in global cave rankings. Thus, among the 27 caves richest in cave-adapted species worldwide in 2019, 23 were karst caves, and only four were lava tubes: one in Australia (Bayliss Cave) and three on the Canary Islands [12]. However, the lava tubes are surprisingly well ranked: Jameos del Agua lava tube for groundwater animals, and Cueva de Felipe Reventón and Cueva del Viento for underground terrestrial richness. Cueva de Felipe Reventón and Cueva del Viento are lava tubes located in Tenerife, Canary Islands, and have been the most attractive caves for scientific studies from this archipelago, together with the anchialine part of Cueva de la Corona lava tube on the island of Lanzarote [13–17].

2. Location and General Characteristics

The Canary archipelago is located in the eastern Atlantic between 27°37′ to 29°25′ N and 13°20′ to 18°10′ W, off the coast of south Morocco. It comprises eight main inhabited islands of volcanic origin, with geological ages as emerged land decreasing from 21 Ma for the eastern Fuerteventura to 1 Ma for the westernmost El Hierro. A remarkable variety of volcanic rocks, type of eruptions, local climates and landscapes is found across this archipelago. Tenerife is in the center of the island chain, with an intermediate age of 12 Ma [18]. It is the largest (2034 km^2), the highest (3714 m at Teide Peak) and the most diverse island both in terms of ecosystems and of plant and animal species [19]. The abundance of subrecent and recent basaltic lavas allows the existence of many lava tube caves and other subterranean environments that harbor a rich cave-adapted fauna [20]. The highest density of lava tubes is found around Icod de Los Vinos, in the northwest of the island, an area covered by the lavas of Teide and Pico Viejo stratovolcanos from the central-western part of the island [21] (Figure 1). In these lava flows is the Cueva del Viento System, which includes Cueva de Felipe Reventón and Cueva del Viento, as well as other minor lava tubes. The two main caves are situated close to each other with entrances between 572 and 847 m a.s.l.; but the extreme lowest and highest internal galleries of Cueva del Viento are at 395 and 880 m, respectively.

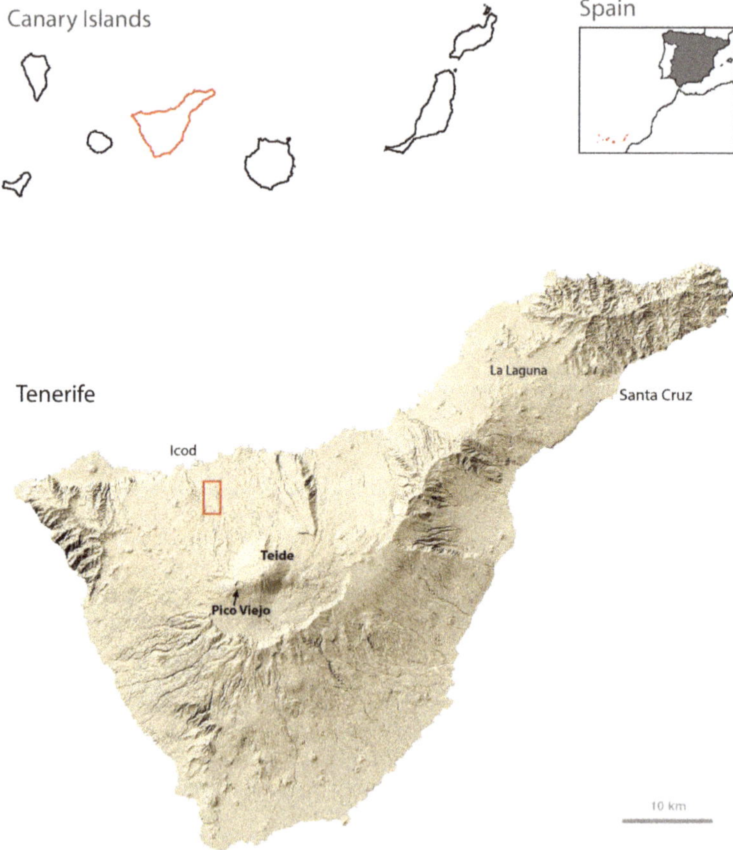

Figure 1. Maps of the Canary Islands and of Tenerife island, with indication of the area (red rectangle) where the Cueva del Viento System lies, and Pico Viejo volcano whose lavas produced the caves. By S. Socorro.

Both caves originated in the same lava flow and probably simultaneously, but they are not connected and have traditionally been considered as independent caves. The maps of their surveys show that Cueva de Felipe Reventón is roughly aligned with one branch of Cueva del Viento, as if it were an extension of the latter (Figure 2). The corresponding ends facing each other are blocked by solidified lava siphons, leaving an unknown 90 m long stretch that had to connect the two caves during their formation. The access between caves is impossible for humans, but not for small animals through the cracks of the same lava flow. From the biospeleological point of view, they would be part of the same Cueva del Viento System.

Viento cave complex

Surveyed: 17.032 m + 1.845 m (Felipe Reventón)

Figure 2. Survey of Cueva del Viento System, with different colors for the main sections, and indication of all accessible entrances. Assemblage partially based on other previously published maps [22,23], with permission. © S. Socorro.

The climate of this area is mild (annual mean 15.1 °C) and moderately humid by the island standards. However, the potential vegetation is not a laurel forest as in other places at the same altitude and orientation, but a mixed pine forest dominated by the Canary pine (*Pinus canariensis*) with big heathers (*Erica arborea*) and some broadleaf trees such as wax myrtle (*Morella faya*), laurel (*Laurus novocanariensis*) and Canary holly (*Ilex canariensis*). This area is moderately inhabited, with many scattered houses and cultivation fields but never with a compact urban structure, and only the upper part of the cave system (Cueva del Sobrado) develops under a fairly well-preserved pine forest.

Cueva del Viento has five easily practicable entrances and two small skylights, all originated by roof collapses. Two of these openings are within a public property belonging to the Cabildo de Tenerife (the local government of the island), the cave sector in between being a show cave; the other five openings are in private land. Four different names are used for the main entries and their corresponding cave sectors: Cueva Piquetes, Cueva de Breveritas, Galería Belén and Cueva del Sobrado, each with their own entrance though Sobrado has four (Figure 2). Piquetes and Breveritas are artificially separated by a cellar of a house; Belén is a short sector of Breveritas isolated from the rest by two stone collapses; and the formerly considered as independent Cueva del Viento and Cueva del Sobrado are actually connected by a narrow natural passage, and now considered as a single cave. For study purposes the whole cave has been divided into the following sections, from the lowest to the highest and using entrances or inner collapses as limits: Piquetes, Breveritas Inferior, Breveritas Superior, Belén, Ingleses, Sobrado Inferior and Sobrado Superior. Breveritas Inferior, Breveritas Superior and Ingleses would be popularly known as Cueva del Viento proper, with a single entrance called Breveritas. All the main galleries of Cueva del Viento have reasonable large dimensions and are easily accessible (Figure 3).

Figure 3. Passage of Cueva del Viento (Breveritas) with the original ceiling evidenced by the abundant small stalactites of lava (red arrows) and with dangling roots (white arrows). © S. Socorro.

The mapped extent of Cueva de Felipe Reventón is 1845 m so far measured, but the complete survey has never been finished and it is probably something over 3000 m in length [24]. The only entrance to Felipe Reventón is in a public property also belonging to the Cabildo de Tenerife, but most of the upper part of the cave lies under private land with a few houses. This entrance is at 628 m a.s.l. just at the lower end of the cave, and no different sector names are used for partial galleries. It is a rather labyrinthine system of interconnected galleries with a relatively short distance (438 m) between upper and lower extremes (Figure 2), and the internal dimensions are generally somewhat smaller than the main galleries of Cueva del Viento.

3. Discovery, Exploration and Scientific Study

The first detailed description of a volcanic cave in the Canary Islands was in 1776 by Bethencourt de Castro brothers on Cueva de San Marcos, in the same Pico Viejo lava flow as the Viento System but next to the coast [21]. The objective of the expedition was to describe and survey the cave, and "to continue to the summit . . . , where some say it has communication with another cave called del Viento", then already known by this name.

The first modern expedition was carried out in 1969 by the local Sección de Exploraciones Vulcanoespeleológicas de La Guancha (SEVG). A year later they completed the topography of what they considered the entire cave, but the map was not published. Members of the Club Montañés de Barcelona contacted SEVG and published a geological study and the first topography [13]. Shortly afterwards, the American vulcanospeleologist W.R. Halliday explored the cave and, after analyzing the survey of SEVG, stated that the known part was 6021 m long and listed it as the longest volcanic cavity in the world [25].

In 1973 and 1974 the British C. Wood and his team explored the cave and published an extensive geological study of Cueva de San Marcos and Cueva del Viento proper, with a new topography of the latter including a new 2340 m branch that runs through a level lower than the rest of the cave [14]. The cave reached 10,002 m, but by then it could no longer be claimed as the longest known volcanic cavity in the world as both Leviathan Cave (Kenya) and Kazumura Cave (Hawaii) exceeded 11 km [26].

Between 1986 and 1988, the GIET (Grupo de Investigaciones Espeleológicas de Tenerife) of the University of La Laguna found several unpublished sections of the deep branch discovered by Wood and Mills (Ingleses gallery), the total length rising to 10,964 m [22]. The same team carried out the first biological study of the complete Cueva del Viento System [27].

In 1988, members of the Grupo de Espeleología de Tenerife Benisahare (henceforth Benisahare) discovered a connection with the well-known Cueva del Sobrado whose topography, commissioned by the Cabildo of Tenerife in 1989 and made by J.L. Martín, added another 3570 m in one of the most intricate areas of Cueva del Viento.

During the exhibition "Labyrinths of lava" of the Museum of Natural Sciences of Tenerife, Sergio Socorro and collaborators presented in 1990 a three-dimensional topographic assembly of the whole Cueva del Viento, in which the "axis" was taken as in the Wood and Mills topography given the greater accuracy of the orientation of the sections.

The removal of rubble from the lower entrance of Cueva del Sobrado in 1994, in a work directed by the later J.J. Hernández Pacheco for the Cabildo of Tenerife, uncovered a 17 m deep pit connecting with a new lower branch 2346 m long [23]. The published topographies total 17,032 m of galleries, but searching, exploration and surveying has been continued by the Benisahare group, with a provisional figure of 18,249 m, although still unpublished. Cueva del Viento is not the first but the seventh longest among volcanic caves [6], but many specialists agree that its complexity of passageways and its geomorphology are the most outstanding among this type of caves [14].

The first topographic and biological survey of Cueva de Felipe Reventón was performed by the GIET of La Laguna University [28], whose members together with Benisahare have continued the task over the years [16].

4. Geology

The Teide volcano represents the culmination of the evolution of an intraplate oceanic island. The Canary Islands are located on a thick oceanic crust close to a continent. This results in subsidence (sinking of the crust due to the weight of an island) being minimal, which means that these islands can remain emerged for more than 20 million years. In turn, this allows long periods of magmatic evolution and successive eruptive cycles, producing stratovolcanos like Teide. In Tenerife, during the last three million years, two previous stratovolcanos have already grown up and eventually slid down to the sea. At present, after 200,000 years of the third cycle of construction of a stratovolcano, the current Teide is still active in the center of the island [29].

When an island is located right on the hot spot in its initial formation, basaltic magmas coming directly from the earth's mantle predominate. This is called a basaltic shield island; most of the Hawaiian Islands do not go beyond that phase (while active). Conversely, Tenerife is already in a mature state, away from the hot spot, such that in its entrails various magmatic chambers can interact with each other [29].

All the caves of the Cueva del Viento System are volcanic cavities framed in this context. They were formed 27,030 ± 430 years ago in lavas from the first eruptive phase of the Pico Viejo volcano, located on the slopes of Teide [29] (see Figure 1).

The composition of lava in the Cueva del Viento System does not correspond to a basalt in the strict sense, which would be the fluid lavas that normally develop as a "pahoehoe" flow (a more fluid basaltic lava with a smooth to ropy surface), where lava tubes are usually formed. The so-called plagioclase basalts of Pico Viejo, in which the caves lie, constitute somewhat evolved lavas in their geochemistry. These lavas had already begun magmatic evolution, so that their composition is between tephrites and tephryphonolites, with the particularity of containing a high proportion of plagioclase crystals [4]. These crystals were already formed in the magma before the eruption and increased the viscosity of the lava. In other words, the fluidity of the lava is diminished by abundant solid particles in suspension that the plagioclases make up. This lack of fluidity could be compensated by the steep slope of the area and decreased eruptive flow or distributed among several emission centers, making them compatible with the pahoehoe flow mechanism that usually forms the lava tube caves.

On the surface, the area where the different entrances of the Cueva del Viento System are distributed is covered by subsequent flows from Teide-Pico Viejo volcanic building. These more evolved lavas (tephryphonolites and phonolites) as well as the more recent phonolites of Roques Blancos (1800 BP) [29], formed as huge channels of viscous lava now covering most of the flow that originated the caves, so that its outcrops are very scarce in the area concerned.

This succession of flows determines, in part, the surface vegetation. As a deep soil has not developed, most of the recent flows are occupied by a mixed forest dominated by *Pinus canariensis*, a tree capable of growing under these conditions, unlike the laurel forest species that might be expected with this altitude and climate.

These caves follow the downward slope of the island dropping 485 m at an average inclination of 11°. They have a unique geomorphological complexity among the known volcanic tubes, characterized by the convergence and bifurcation of their numerous galleries [14]. This type of tube formation is the most difficult to understand since a great complexity is observed in their structure and arrangement of the branches. Both Cueva del Viento and Cueva de Felipe Reventón are good models of this type of lava tubes. The eruption probably lasted for several months producing flows hundreds of meters wide, leading to the formation of this complex cave with branches that run in three superimposed floors, corresponding to various episodes of the initial activity of Pico Viejo volcano [14].

In these types of tubes, the way in which the pahoehoe spreads greatly influences how the spatial arrangement of the underground network of ducts can be complicated. Furthermore, after a network of ducts is established, a new flow can overlap on top of the former, creating a new system that can become intercommunicating with the underlying passages. In Breveritas Superior and Sobrado galleries there are two vertical pots through which cascades of lava flowed from a higher tube to a lower one.

5. Geomorphology

The complex network of passages in Cueva del Viento proper is distributed in three main levels (Figure 4):

- An upper one (Level 3) that connects the main tubes through narrow shallower passages;
- An intermediate one (Level 2) that includes Sobrado Superior (3.57 km) and most of Cueva del Viento (7.82 km);
- A deep Level 1 through which Ingleses (3.14 km) and Sobrado Inferior (2.35 km) galleries lie.

The axis Sala de la Cruz–Pájaros Gallery–Belén–Breveritas Entrance–Piquetes (see Figure 2) probably corresponds to a lava channel that was initially open to the sky in the pahoehoe flow [14]. In some sectors of this axis, closure sutures are observed on the cave ceiling and bulges on the outer surface of the terrain (Figure 5).

In general, the internal surfaces within the cavity consist of cornices, terraces, benches, and tiers due to variations and stabilizations that the level of the lava flow undergoes within the already formed tube. The different levels of flow stabilization are marked symmetrically on both walls, sometimes only by flux striations.

Changes in viscosity are also marked. The smooth surfaces correspond to phases in which the lava circulated with great fluidity (Figure 6); instead, a rough floor would be the last residual flow that was solidifying (Figure 7).

Figure 4. Complex system of tubes at different levels, the highest ones captured by upwards re-excavation and collapses of the ceiling. Ingleses Gallery. © S. Socorro.

Figure 5. Ceiling with a closure suture of a formerly open lava channel, in Breveritas Gallery. © S. Socorro.

Figure 6. Network of conduits with smooth surface formed by very fluid lava in Cueva de Felipe Reventón. Foreground: last blocks fallen from the roof after the eruption. © S. Socorro.

Figure 7. Vitreous black stripe on the walls of Ingleses Gallery. Re-excavation of the ceiling showing the pahoehoe structure. The very last flow stopped and formed a rough step. © S. Socorro.

In the Sobrado Superior sector, the geomorphological beauty of its galleries and its didactic interest are notable. There are side terraces, multiple levels marked by lava when changing height, effects of centrifugal force on lateral benches, etc. The area called the Labyrinth is an intricate network of interconnected tubes of almost circular section and smooth walls, which can rival the other existing labyrinth in Cueva de Felipe Reventón (Figure 6).

Unlike the upper galleries, that of Ingleses (Level 1) evidently did not receive solidifying lava streams and the flow was never stabilized for long enough to form side terraces and benches, which are scarce and not very prominent here. Due to the distance from the surface, the lava is "cleaner" with few subsequent mineral coatings deposited by infiltrated water, as it is more likely to settle at higher levels. In addition, several lava flows from later eruptions have covered the ground surface above these galleries, increasing their distance from the exterior.

The spectacular black band that can be seen along the main gallery of Ingleses must have been formed by a pulse of very fluid lava that flooded the conduit to a certain height (Figures 7 and 8). The rapid fall in the lava level left a thin layer that quickly cooled and vitrified in contact with the walls.

Figure 8. Large re-excavation of the ceiling and walls in the Ingleses Gallery, showing the parallel layers of the pahoehoe flow. Most of the fallen blocks that fell from the roof during the tube formation were swept away by the lava, while the latest to fall stayed in position on the floor. © S. Socorro.

In various passages of Ingleses one can see how the hot flows through the tube "corroded from the inside" the solidified lava in the pahoehoe flow on the surface that had formed the roof of the tube. In other words, while the flow and configuration of the ducts last, the stream drags along loose pieces detached from the "puff pastry" of the cavity ceiling, which constitutes the pahoehoe flow itself, and that today we see sectioned in walls and roofs (Figure 8). These detached blocks are also seen welded to the ground in various places throughout the cavity.

Sometimes, the magnitude of this "re-excavation towards the ceiling" is so great that other higher ducts are intercepted from below (Figure 7). Such conduits have not yet been explored, and they could correspond to Level 2 at which the main galleries of Breveritas Superior formed above Ingleses Gallery. These geological cuts should be like those observed by Ollier and Brown [30], which inspired their now outdated theory to explain the formation of tubes, based on what they called "layered lava" and related it to a supposed laminar flow; however, it corresponds to the complex movement of pahoehoe flows with various lobes, embryonic tubes, lava channels, and sometimes continuous mantles of lava.

The cylindrical tube configuration only occurs when the current reaches the ceiling persistently, to consolidate a continuous inner lining. Otherwise, as in many sections, the ceiling is irregular and may show pahoehoe "puff pastry" structures, although more or less blurred by mineral deposits in the upper levels of Breveritas.

The main branches of Cueva de Felipe Reventón and the connecting galleries correspond largely to embryonic lava tube systems that develop in the complex advance of a pahoehoe stream. For this reason, these areas are made up of complicated labyrinths and tubes at different heights, where an authentic tangle of lava outbreaks is interwoven that, at the time, constituted the advanced front of the pahoehoe flow (Figure 6).

6. The Cave Environment

Lava tube caves originate very quickly in geology and remain very close to the surface, usually a few meters deep throughout their length [31]. This is the start point for their ecological succession, erosion, and final destruction due to collapse or to clay silting; this happens in a much shorter time span than karstic caves: a maximum of 500,000 years for Hawaiian caves [5]. Some Canary tubes have persisted for longer, especially when they are gently inclined and develop in dry climates, as in the 992,000 ± 21,000 years old Cueva del Llano formed by drainage of a lava lake in the subdesert Fuerteventura island, or the exceptional case of Cueva de Aslobas in Gran Canaria with more than 13 Ma [32,33]. The environment inside very recent lava tubes is highly influenced by surface weather, due to the abundant interconnected cracks and voids of the lava and to the absence of covering soil. Therefore, this habitat is not yet suitable for cave-adapted fauna. Ecological succession on the surface will provide a layer of soil and vegetation on the middle-aged lavas over a period dependent on local climate. A parallel, slightly delayed ecological succession will take place inside the tube until it reaches a humid and stable cave environment [34]. The process is sometimes accelerated by deposits of small-sized pyroclasts (lapilli and cinders) that almost immediately cover the surface providing sufficient isolation for cave conditions, as happens in Cueva de Don Justo on El Hierro island [34].

In limestone caves ecological succession advances downwards creating new habitats ever deeper, but in lava tubes the situation is reversed, and succession progresses upwards as protective soil and vegetation develop on the surface [35]. Thus, deeper galleries of the cave (or deeper parts of a big gallery) mature before shallower ones. Usually, in the Canary Islands lava tubes reach this mature ecological stage between many hundreds and some thousand years depending on the local climate, and they become senile and ecologically decaying after a few hundred thousand years. Therefore, Cueva del Viento and Cueva de Felipe Reventón are now in their best stage to host a rich adapted fauna due to their age (ca. 27,000 ybp) [21]. Their situation in a vast area of moderately recent lavas still with an inner network of cracks and voids allows obligate subterranean fauna to reach the caves from other underground habitats. The moderate altitude of the area is also favorable given the local climate on the north slope of the island, mesic at middle altitudes between 500 and 1200 m a.s.l. due to humid trade winds, but drier both at lower and higher altitudes. Other lava tubes located in the dry lowlands usually harbor a much poorer cave-adapted fauna [20,34].

Stable temperatures and high humidity prevail throughout the year in the most suitable caves, a necessary requirement for cave adapted fauna. The deeper a gallery the greater the environmental stability, but with lower availability of trophic resources. The absence of water streams inside the tubes slows down the advance of organic matter in depth. Thus, the deepest passages of lower levels in Cueva del Viento proper differ from those in shallower ones, the former with scarcer food resources, more constant temperature and humidity, and sometimes higher CO_2 concentration [36]. In addition to trophic scarcity, these parameters are unsuitable for poorly adapted species and lead to the dominance of troglobionts (i.e., obligate inhabitants of subterranean habitats, usually but not necessarily with the morphological syndrome of lack of pigment, eyelessness and elongated appendages) [1]. This fact was already verified in lava tubes of Australia and Hawaii [10] and is noticeable in Sobrado Inferior and especially in Ingleses, both at the third lower level at least 15 m below surface with a scarce, almost exclusively troglobiotic fauna [36]. In Cueva de Felipe Reventón there are no different levels of galleries and environmental features are very similar along the cave. The main contrast of ecological significance between its galleries is their substrate, sometimes bare lava but often with remarkable earth accumulation derived from the overlying soil through very small roof collapses or thick cracks.

Most lava tubes are shallow, unless they have been covered by later flows belonging to more recent eruptions. This situation close to the surface and the abundant cracks in the lava allow the plant roots to reach the cave, often dangling from the ceiling (Figures 3 and 9). This is an important resource for the animal community, especially for root-feeding and sap-sucking species, the latter being markedly abundant in individuals [31]. These root-dependent species are common in lava tubes around the world, especially from volcanic archipelagos but also in limestone caves, although the latter almost exclusively in the tropics [37]. As many as 18 cave-adapted planthopper species are found in the non-tropical volcanic Macaronesian Islands (Canary Is., Madeira and Azores) but only four species are known from karst caves in the rest of West Palearctic [38–40]. In some sections of the whole Viento cave system, especially in Sobrado Superior, sap-sucking insects are by far the most abundant cave-dwelling animals, represented by the planthopper *Tachycixius lavatubus* Remane and Hoch (Figure 10). However, the roots do not reach all passages of the cave, especially those at the deepest galleries or those in the cave sections underlying thick overlapping modern lavas. Nevertheless, once the organic matter has reached the shallower parts of the lava tube, it can diffuse horizontally (and more slowly vertically) towards the rest of the cave and be integrated in the food chain of the ecosystem.

Figure 9. The tender roots hanging from the ceiling are an important input of organic matter. © S. Socorro.

Besides the main entrances, there are along both caves several partial collapses of the original compact ceiling, facilitating the connection with the overlying soil and sometimes even with the surface. This allows the entry by many trogloxenes (animals not linked to the caves) and subtroglophiles (only linked for part of its life cycle) [41], which form an important resource for predatory species in many parts of the lava tubes, sometimes far from any large entrance. Abundant flying insects can be found in such spots, like the subtroglophile *Megaselia* scuttle flies that probably are common prey for a diversity of troglobiotic web-spiders, which could not survive only feeding on other troglobionts. In the twilight zone close to entrances, these flying insects are preyed on by the abundant spider *Meta bourneti* Simon, a typical eutroglophile with permanent populations completing its life cycle inside and rarely outside the caves [41].

Figure 10. The planthopper *Tachycixius lavatubus* is the most abundant troglobiont in these caves. © S. Socorro.

Relative humidity is high and ceiling drip very frequent and even abundant in many parts, but running water is rarely present except just after heavy rains. Almost all water quickly percolates through the abundant cracks of the lava floor connected with a network of voids in the host rock. Thus, no significant water ponds are present, and no groundwater species have ever been found, given that the water table lies much deeper than these caves. Environmental features in Cueva del Viento and Cueva de Felipe Reventón are very similar since they are in the same lava flow, roughly at the same altitude and at an identical stage of maturity, and they share 32 of their troglobiotic species. There are two notable differences between the two caves: (i) the greater depth of the lowest galleries in Cueva del Viento (e.g., Ingleses), with the consequent greater shortage of organic matter input and of troglobiont individuals, and (ii) the greater accumulation of soil in some passages of Cueva de Felipe Reventón, providing a different habitat for the occurrence of some species never found in the rest of the cave system.

7. Animal Ecology and Diversity

The ecological driver for the high diversity in Cueva del Viento System is evidently the high-productivity due to abundant roots because of being very close to the surface [42]. As many as 79 different troglobionts are known so far from the island of Tenerife. In the northeastern geologically old Anaga peninsula there are nine troglobionts, two occurring in the only existing cave (a volcanic pit), and the rest in the mesocavernous shallow substratum (MSS), a colluvial subterranean habitat from the mountains of old areas [1]. Eight of these species are exclusive to this particular area and vicariant with respect to closely related ones from the rest of the island. This is probably due to an ancient ecological isolation, given the high proportion of local Anaga endemics, both underground and surface species [15,43,44]. Among the 72 troglobionts occurring in the rest of the island (where the Cueva del Viento System is located) 14 are known only from the MSS and 58 can be found in MSS and/or in caves, 42 of which are present in Cueva de Felipe Reventón and/or Cueva del Viento (38 and 36 respectively) (see Table 1). Thus, the fauna of these two caves is an important reflection of the troglobiotic fauna of the island. Only 9 troglobionts are exclusive to the Cueva del Viento System, the remaining 33 species being also found in some or many other subterranean places in Tenerife, either caves or MSS. This reflects the importance of MSS and other subsurface habitats of volcanic terrains as a means of dispersal for troglobionts. Some of them can be found in rather distant caves that are unconnected since they occur in lava flows geographically and chronologically separated. Such distant populations of a single species usually show morphological, behavioral or genetic differences that reflect the slow but finally successful dispersal of troglobionts through underground spaces, even without proper caves in between.

Troglobiotic lifestyle does not always imply apparent morphological adaptations (eye and pigment regression, lengthening of appendages). There are different life forms of troglobionts: troglomorphic, with these characters highly accentuated and large body size; euedaphomorphic, with eye and pigment regression but small body and short appendages; and hypogeomorphic, with variable regressive characters but unchanged appendages and body size compared to surface species [41]. Some eyeless and depigmented species with an euedaphomorphic or hypogeomorphic rather than troglomorphic body pattern are more frequent in Cueva de Felipe Reventón than in Cueva del Viento, especially small carabid and histerid beetles [45,46].

Primary consumers are basically those feeding on roots. Two species feed on fresh root tissues: the moth *Schrankia costaestrigalis* (Stephens) and the scarce weevil *Oromia hephaestos* (Figure 11). The eutroglophile *Schrankia costaestrigalis* is very abundant, especially in Felipe Reventón, where the caterpillars feed on tender roots and pupate in cocoons hanging from the ceiling and covered with dry root pieces. The adult moths reproduce and live permanently in the caves, but unlike *S. howarthi* Davis and Medeiros from Hawaiian caves [47], the females have no troglomorphic adaptations and can fly perfectly. *Oromia hephaestos* belongs to an eyeless endemic genus with species occurring either in caves or in the MSS from the islands of La Gomera, Tenerife, and Gran Canaria, with a notable radiation in the latter [48]. The only sapsucker species is the planthopper *Tachycixius lavatubus*, by far the most abundant troglobiont in both caves. This highly troglomorphic species belongs to a genus with other two hypogeomorphic species occurring in colluvial MSS, and a scarce surface-dwelling species. As in other planthoppers, males emit a specific vibration transmitted through the roots, and then answered by females [49]. *Tachycixius lavatubus* can be found in almost any cave around the island, although several different allopatric species probably exist, according to unpublished population genetic and bioacoustic studies [50].

Table 1. Troglobiotic species present in Cueva de Felipe Reventón and Cueva del Viento. *: Canarian endemic genera. EUED: euedaphomorphic; TROG: troglomorphic; HYPO: hypogeomorphic. +: presence; ++: species exclusive to these caves. After Oromí [15] and unpublished own data.

Order	Family	Species	Life Form	Felipe Reventón	Viento
Pseudoscorpiones	Syarinidae	*Microcreagrina subterranea* Mahnert, 1993	EUED	+	+
	Chthoniidae	*Lagynochthonius curvidigitatus* Mahnert, 1997	TROG	++	
		Occidenchthonius oromii Zaragoza, 2017	EUED	++	
		Paraliochthonius setiger (Mahnert, 1997)	TROG	+	+
		Paraliochthonius superstes (Mahnert, 1986)	TROG	+	+
Araneae	Dysderidae	*Dysdera ambulotenta* Ribera, Ferr. and Blasco, 1985	TROG	+	+
		Dysdera esquiveli Ribera and Blasco, 1986	TROG	+	+
		Dysdera labradaensis Wunderlich, 1992	TROG	+	+
		Dysdera sibyllina Arnedo, 2007	TROG	++	++
		Dysdera unguimmanis Ribera, Ferr. and Blasco, 1985	TROG	+	+
	Pholcidae	*Spermophorides reventoni* Wunderlich, 1992	HYPO	++	++
	Liocranidae	*Agraecina canariensis* Wunderlich, 1992	TROG	+	+
	Linyphiidae	*Metopobactrus cavernicola* Wunderlich, 1992	TROG	++	++
		Troglohyphantes oromii (Ribera and Blasco, 1986)	TROG	+	+
		Walckenaeria cavernicola Wunderlich, 1992	TROG		+
	Nesticidae	* *Canarionesticus quadridentatus* Wunderlich, 1992	TROG	+	+
Glomerida	Glomeridae	*Glomeris speobia* Golovatch and Enghoff, 2003	TROG	+	+
Julida	Julidae	*Dolichoiulus labradae* Enghoff, 1992	TROG	+	+
		Dolichoiulus ypsilon Enghoff, 1992	TROG		+
Scolopendromorpha	Cryptopidae	*Cryptops vulcanicus* Zapparoli, 1990	TROG	+	
Lithobiomorpha	Lithobiidae	*Lithobius speleovulcanus* Serra, 1984	TROG	+	+
Isopoda	Trichoniscidae	*Trichoniscus* cf. *bassoti* Vandel, 1960	HYPO	+	+
	Armadillidae	*Venezillo tenerifensis* Dalens, 1984	TROG	+	+
	Porcellionidae	*Porcellio martini* Dalens, 1984	HYPO	+	+
Entomobryomorpha	Paronellidae	*Troglopedetes* cf. *cavernicolus* Delamare, 1944	HYPO	++	++
		Troglopedetes cf. *vandeli* Cassagnau and Delamare, 1951	HYPO	++	++
Blattodea	Blattellidae	*Loboptera subterranea* Martín and Oromí, 1987	TROG	+	+
		Loboptera troglobia Izquierdo and Martín, 1990	TROG	+	+
Hemiptera	Cixiidae	*Tachycixius lavatubus* Remane and Hoch, 1988	TROG	+	+
Coleoptera	Carabidae	* *Gietopus martini* (Machado, 1992)	HYPO	+	+
		Lymnastis subovatus Machado, 1992	EUED	+	+
		Lymnastis thoracicus Machado, 1992	EUED	+	+
		* *Spelaeovulcania canariensis* Machado, 1987	HYPO	+	+
		* *Wolltinerfia tenerifae* (Machado, 1984)	HYPO	+	+
	Staphylinidae	*Alevonota oromii* Assing, 2002	EUED		+
		Alevonota outereloi Gamarra and Hdez., 1989	TROG		+
		Domene alticola Oromí and Hernández, 1986	TROG	+	+
		Domene vulcanica Oromí and Hernández, 1986	TROG	+	+
		Micranops bifossicapitatus (Outerelo and Oromí, 1987)	EUED	+	
		Micranops spelaeus Frisch and Oromí, 2006	TROG	++	
	Histeridae	*Aeletes oromii* Yélamos, 1995	EUED	+	
	Curculionidae	* *Oromia hephaestos* A. Zarazaga, 1987	EUED	++	++
Total species		42		38	36

Figure 11. *Oromia hephaestos* is a scarce weevil known only from the Cueva del Viento System. © H. López.

Detritivores are more varied than root-feeding species, but due to the broad spectrum of feeding behavior necessary to survive in the cave, it is often difficult to know the preferences of each millipede, woodlouse, springtail or cockroach. The pill-millipede *Glomeris speobia* and the woodlouse *Venezillo tenerifensis* are among the most widespread troglobionts on the island and are regularly seen in both the Icod caves discussed here (Figures 12 and 13). The millipede genus *Dolichoiulus* is a paradigm of island radiation with 21 endemic species to Tenerife, three of them troglobionts [51] (Figure 14); *Dolichoiulus labradae* and *D. ypsilon* are becoming rarer in these caves, probably due to the increasing abundance of the alien, invasive troglophile *Blaniulus guttulatus* (F.). The woodlouse *Porcellio martini* is also scarce, found here and there in other caves of the island. Another woodlouse *Trichoniscus bassoti* and the springtails *Troglopedetes vandeli* and *T. cavernicolus* may not be correctly identified [52,53], since the presence of non-endemic troglobionts on an oceanic island is not conceivable due to their inability to disperse over distances across the sea.

Cockroaches of the genus *Loboptera* are among the most characteristic components of the troglobiotic fauna of the western Canary Islands, with eight species on Tenerife, seven allopatric to each other but the eighth (*L. troglobia*) widely distributed and sympatric to many others. They are among the biggest local troglobionts, eyeless, wingless, with reduced number of ovarioles and living either in caves or in the MSS [54]. Two species occur in both Cueva de Felipe Reventón and Cueva del Viento: the big and scarce *L. subterranea* which is exclusive to the Icod area, and the smaller and more troglomorphic and widespread *L. troglobia* (Figure 15).

Most predatory troglobionts have reduced populations, some with very few individuals, and many species act also as scavengers. Five different pseudoscorpions are found here, all endemic to the island except the eyeless *Microcreagrina subterranea* which also occurs on other islands. The remaining species are Chthoniidae, two of them especially troglomorphic: *Paraliochthonius superstes* and *P. setiger*; only *Lagynochthonius curvidigitatus* is exclusive to Cueva de Felipe Reventón [55]. The highly diversified spider genus *Dysdera* is represented by five species occurring in both caves, one of them (*D. unguimmanis*) highly troglomorphic and probably parthenogenetic, since no males have ever been found (Figure 16). Each one has a different epigean sister species, so they originated by five independent colonization events of the underground environment [56]. Among troglobiotic web spiders *Troglohyphantes oromii* is the most abundant (Figure 17), not only in these two caves but in the rest of the island except in Anaga; a similar situation is that of the widespread centipede *Lithobius speleovulcanus*.

Figure 12. The eyeless *Glomeris speobia* is the only troglobiotic millipede so far known in the Canary Islands. © P. Oromí.

Figure 13. The woodlouse *Venezillo tenerifensis* is widespread in almost all suitable underground environments on Tenerife. © P. Oromí.

Figure 14. *Dolichoiulus ypsilon*, one of the two troglobiotic millipede species of the genus occurring in the Icod caves. © P. Oromí.

Figure 15. Male of *Loboptera troglobia*, the most specialized cockroach occurring in many subterranean environments on the island. © P. Oromí.

All troglobiont ground beetles are eyeless but otherwise not especially troglomorphic (Figure 18); Pterostichinae are medium size beetles in two endemic genera (*Wolltinerfia* and *Gietopus*), while Trechinae are smaller and are represented by the monospecific endemic genus *Spelaeovulcania canariensis* and two euedaphomorphic species of *Lymnastis* (Figure 19). The diversity of subterranean rove beetles is remarkable, both in the Canary Islands (27 spp.) and in Icod caves (6 spp.) [57], and they are usually more troglomorphic than ground beetles, as shown by the marked elongation of body and appendages in *Domene alticola* and *D. vulcanica* (Figure 20).

Figure 16. Female of *Dysdera unguimmanis*, one of the five different troglobiotic species of the genus recorded in these caves. © P. Oromí.

Figure 17. Female of *Troglohyphantes oromii*, the most abundant spider in Cueva del Viento System. © P. Oromí.

Besides the subterranean obligate fauna, the shallowness of these caves facilitates the input of trogloxene and subtroglophile animals. Thus, the total number of arthropod species found in them roughly doubles the number of troglobionts [58], but the trogloxenes are almost absent in the deepest levels of Breveritas Superior and Sobrado galleries. Local eutroglophiles are usually widespread species, some introduced like the millipedes *Blaniulus guttulatus* and *Choneiulus subterraneus* (Silvestri), and some probably native species like the spider *Meta bourneti*, the psocid *Psyllipsocus ramburii* (Selys-Longchamps) and the moth *Schrankia costaestrigalis*. The abundant larvae of the subtroglophile winter crane fly *Trichocera maculipennis* Meigen capture their preys in sticky silk threads, being important predators in most of the cave communities of Tenerife.

Figure 18. The eyeless ground beetle *Gietopus martini* belongs to a monospecific genus endemic to Tenerife. © P. Oromí.

Figure 19. The small ground beetle *Spelaeovulcania canariensis* has no other related species on the Canary Islands. © H. López.

Figure 20. The highly troglomorphic *Domene vulcanica* is one of the six cave-adapted rove beetles known from these caves. © P. Oromí.

8. Paleontology of the Caves

Like in other volcanic archipelagos, Canary Islands lava tubes and pits are the best paleontological sites, especially for quaternary subfossils of vertebrates. The accumulation of bird bones is usually not far from cave entrances, which sometimes function as traps that allow the birds to come in by accident, and get injured and lost, and soon die. Lizards and small mammals can actively enter and explore the habitat, reaching longer and deeper distances due to their higher resistance to starvation. Actually, in Cueva del Viento it is rather frequent to see lost or recently dead lizards and young rabbits which penetrate through narrow, small roof collapses existing all along the lava tube due to its shallowness. The Cueva del Viento System is one of the most important paleontological sites in the archipelago, where many remains of several already extinct vertebrates have been found. Bones of the Tenerife Giant Lizard *Gallotia goliath* (Mertens), over one meter long, were collected abundantly in Piquetes galleries in the lower part of Cueva del Viento [59], underlying a surface habitat where these reptiles were probably common given the relatively warm climate and the absence of forest. The Tenerife giant rat *Canariomys bravoi* Crusafont and Petit was also present in this part of the cave, but particularly more abundant in galleries at higher altitude, especially in Sobrado Inferior passages (Figure 21).

In 1994 an important paleontological site was discovered at the end of a 600 m long Pájaros Gallery in Breveritas Superior (see Figure 2), with bone remains of more than 200 vertebrate individuals comprising 6% lizards, 46% mammals (mainly extinct giant rats and recent rabbits and rats) and 48% birds, among which there were some extant species and some others extinct on the island or completely extinct [60] (Table 2). Such accumulation of bones away from accessible entrances is explained by the presence of an old collapse that interrupts the gallery just before an entrance of the cave (Galería Belén, see Figure 2) by which most of the animals could have entered before this erosional event. Two extinct bird species were found at this spot: the Canary Quail (*Coturnix gomerae* Jaume, McMinn and Alcover) and the Long-legged Bunting (*Emberiza alcoveri* Rando and López), the latter being the only flightless bunting so far known [61].

Figure 21. Skull and other subfossil remains of the Tenerife giant rat found in Sobrado Inferior. © P. Oromí.

Table 2. Bone remains found in Pájaros Gallery (Breveritas Superior). *: Extinct species. **: Extant species no longer present in Tenerife [59–61].

Group	Scientific Name	Common Name
Reptiles	*Gallotia goliath* (Mertens) *	Tenerife giant lizard
	Gallotia galloti (Oudart)	Tenerife common lizard
Birds	*Coturnix gomerae* Jaume et al. *	Canary quail
	Clamydotis undulata (Jacquin) **	Houbara
	Pyrrhocorax pyrrhocorax (L.) **	Re-billed clough
	Emberiza alcoveri Rando and López *	Long-legged bunting
	Columba sp.	cf. Laurel pigeon
	Turdus sp.	cf. Blackbird
Mammals	*Canariomys bravoi* Crusafont and Petit *	Tenerife giant rat
	Rattus rattus (L.)	Black rat
	Oryctolagus cuniculus (L.)	Rabbit

The Tenerife giant rat was probably the only one of the abovementioned vertebrates that used the cave as part of its habitat, as the introduced Brown Rat does today [62]; the other species would be inside just by accident. Bones of the giant rat have also been found in Cueva de Felipe Reventón, though less abundantly. Some of these bones were dated with C_{14} with ca. 25,000 BP, the minimum age applicable to the whole cave system [63]. Extinct species disappeared soon after the arrival of the aboriginal people (ca. 2500 ybp) [64].

9. Conservation Status

The Cueva del Viento System is in an inhabited area, so it is threatened much depending on the caves and sectors. The main problem is that the whole municipality of Icod de los Vinos lacks a sewage system, and the solution for each house is to have a cesspool or, alternatively and much worse, to drill a hole until reaching the highly porous rock which very often is just a cave. The local people call this "to find the volcano", quite easy given the abundance of holes and caves in the underground of the area. Thus, the lower part of Cueva del Viento (Piquetes) is polluted by raw sewage and dirty mud, and is almost impassable; Breveritas Inferior is also contaminated with filtered dirty waters and even bad smell, but the central part (Breveritas Superior) is reasonably well preserved though continually endangered given the uncontrolled construction of houses; the relatively short Galería Belén is just below a single house, enough to cause a sewer smell; the access to Sobrado Inferior is through a controlled entrance to Sobrado show cave and the visitors are not allowed, but some traces of pollutants were detected in past analyses [16]; the upper part (Sobrado Superior) is in a mostly natural area under pine forest and is owned by the local government, double-gated and in very good condition. The only solution for this important contamination would be a sewage system in densely inhabited zones, and individual treatment plants for isolated houses. Unfortunately, the municipality has neither money nor special will to do it, and with time more and more illegal houses have been built. All entrances are gated except Piquetes, Breveritas Inferior and Belén, which are not visited due to unpleasant conditions.

A 150 m long stretch of Sobrado Superior is dedicated to public visits, but this does not cause a noticeable impact to the environment because it is between two large entrances. Ventilation is higher and humidity lower than in the rest of the cave, and 20 years ago, before its adaptation for visits, it had been verified that it lacked troglobionts. The visitors enter the cave in groups of maximum 10 people, each with a personal headlamp since the tube has no electric lighting to prevent the growth of green algae and mosses. The touristic use of a small portion devoid of adapted fauna compensates for the control which is exercised over the adjacent richer parts of Sobrado.

Cueva de Felipe Reventón has a solid gate in the only entrance in public land and only a few people can visit it. However, the upper part of the cave lies under private land and a house is nearby, with danger of sewage filtering. There are no apparent signs of contamination and the troglobiotic fauna is rich, but water analyses made in 2000 revealed the presence of nitrates and nitrites (unpublished own data).

Cavers and conservationists have been for many years arguing for the protection of the area, always without success. A LIFE-Nature project on the biodiversity and conservation of caves in Tenerife, La Palma and El Hierro islands was carried out in 1999–2001 and included a project for a sewage system in the Icod area, but the installation has never been performed [15]. A project by the Cabildo de Tenerife to declare a protected area around these caves is dragging on for many years without taking place, and it will be increasingly difficult due to pressures by landowners who want to build.

Author Contributions: P.O. has written the Introduction, Location and General Characteristics, Cave Environment, Animal Ecology, Paleontology and Conservation sections; S.S. has written the History, Geology and Geomorphology sections. Both authors have structured the article and selected the figures. All authors have read and agreed to the published version of the manuscript.

Funding: This research received no external funding.

Institutional Review Board Statement: Not applicable.

Informed Consent Statement: Not applicable.

Data Availability Statement: No new data were created or analyzed in this study. Data sharing is not applicable to this article.

Acknowledgments: We are indebted to Heriberto López for some pictures, and to Philip Ashmole for critical revision of the contents and the English version. Three anonymous reviewers have also contributed to the improvement of the text. Most of the information included herein results from the work of all GIET and Benisahare members over many years surveying lava tubes and studying their fauna.

Conflicts of Interest: The authors declare no conflict of interest.

References

1. Culver, D.C.; Pipan, T. *The Biology of Caves and Other Subterranean Habitats*; Oxford University Press: Oxford, UK, 2019; p. 301.
2. Gulden, B. World's Longest Caves. 2021. Available online: http://www.caverbob.com/wlong.htm (accessed on 10 March 2021).
3. Halliday, W.R. History and status of the Moiliili karst, Hawaii. *J. Cave Karst Stud.* **1998**, *60*, 141–145.
4. Carracedo, J.C. Claves para la identificación de las formas volcánicas (chapter 6). In *El Volcán Teide*; Carracedo, J.C., Ed.; Ediciones Promociones Saquiro: Santa Cruz de Tenerife, Spain, 2008; Volume 2, pp. 7–63.
5. Howarth, F.G. The cavernicolous fauna of Hawaiian lava tubes, 1. Introduction. *Pac. Insects* **1973**, *15*, 139–151.
6. Gulden, B. World's Longest Lava Tubes. 2021. Available online: http://www.caverbob.com/lava.htm (accessed on 10 March 2021).
7. Leleup, N. Origine et évolution des faunes troglobies terrestres holarctiques et intertropicales. *Stalactite* **1963**, *6*, 199–201.
8. Oromí, P.; Martín, J.L. *Apteranopsis canariensis* n.sp., un nuevo coleóptero cavernícola de Tenerife (Staphylinidae). *Nouv. Rev. Entomol. (NS)* **1984**, *1*, 41–48.
9. Peck, S.B. Biology of the Idaho lava tube beetle *Glacicavicola* (Coleoptera: Leiodidae). *Natl. Speleol. Soc. Bull.* **1974**, *36*, 1–3.
10. Howarth, F.G.; Stone, F.D. Elevated carbon dioxide levels in Bayliss Cave, Australia: Implications for the evolution of oblígate cave species. *Pac. Sci.* **1990**, *44*, 207–218.
11. Borges, P.A.V.; Oromí, P. Cave-dwelling ground beetles of the Azores (Col., Carabidae). *Mém. Biospéleol.* **1991**, *18*, 185–191.
12. Deharveng, L.; Bedos, A. Biodiversity in the tropics. In *Encyclopedia of Caves*, 3rd ed.; White, W.B., Culver, D.C., Pipan, T., Eds.; Academic Press: London, UK, 2019; pp. 146–162.
13. Montoriol, J.; de Mier, J. Estudio vulcanoespeleológico de la Cueva del Viento (Icod de los Vinos, Tenerife, Canarias). *Speleon* **1974**, *21*, 5–24.
14. Wood, C.; Mills, M.T. Geology of the lava tube caves around Icod de los Vinos, Tenerife. *Trans. Br. Cave Res. Assoc.* **1977**, *4*, 453–469.
15. Oromí, P. Researches in lava tubes. In *Cave Ecology*; Moldovan, O.T., Kováč, L., Halse, S., Eds.; Springer: Cham, Switzerland, 2018; pp. 369–381.
16. Arechavaleta, M.; Sala, L.; Oromí, P. La fauna invertebrada de la Cueva de Felipe Reventón (Icod de los Vinos, Tenerife, Islas Canarias). *Vieraea* **1999**, *27*, 229–244.
17. Martínez, A.; Gonzalez, B.C.; Núñez, J.; Wilkens, H.; Oromí, P.; Iliffe, T.M.; Worsaae, K. *Guide to the Anchialine Ecosystems of Jameos del Agua and Túnel de la Atlántida*; Cabildo de Lanzarote: Arrecife, Spain, 2016; pp. 1–380.
18. Carracedo, J.C.; Troll, V.R. North Atlantic islands: The Macaronesian archipelagos (Azores, Madeira, Canaries and Cape Verde). In *Encyclopedia of Geology*; Elsevier: Amsterdam, The Netherlands, 2021; pp. 671–699.
19. Martín, J.L. (Ed.) *Atlas de Biodiversidad de Canarias*; Gobierno de Canarias: Santa Cruz de Tenerife, Spain, 2010; pp. 1–287.
20. Oromí, P. Canary Islands: Biospeleology. In *Encyclopedia of Caves and Karst Science*; Gunn, J., Ed.; Fitzroy Dearborn: New York, NY, USA, 2004; pp. 179–181.
21. Carracedo, J.C.; Rodríguez-Badiola, E.; Paris, R.; Pérez-Torrado, F.J.; Rodríguez, A.; Socorro, J.S. Erupciones del Pico Viejo. In *El Volcán Teide*; Carracedo, J.C., Ed.; Ediciones Promociones Saquiro: Santa Cruz de Tenerife, Spain, 2008; Volume 3, pp. 126–145.
22. Hernández, J.J.; Medina, A.L.; Izquierdo, I.; Vera, A.; García, H. Topografía y Espeleometría. In *La Cueva del Viento*; Oromí, P., Ed.; Canary Is. Government: Santa Cruz de Tenerife, Spain, 1995; pp. 9–14.
23. Lainez, A. Galería "Hernández Pacheco" un nuevo descubrimiento en el complejo de la Cueva del Viento-Sobrado (Icod de Los Vinos, Tenerife). In *Proceedings of the 7th International Symposium on Vulcanospeleology, Santa Cruz La Palma, Spain, 4–11 November 1994*; Oromí, P., Ed.; Libros de la Frontera: Sant Cugat Vallés, Spain, 1996; pp. 69–73.
24. Hernández, J.J.; Oromí, P.; Lainez, A.; Ortega, G.; Pérez, A.E.; López, J.S.; Medina, A.L.; Izquierdo, I.; Sala, L.; Zurita, N.; et al. *Catálogo Espeleológico de Tenerife*; Cabildo de Tenerife: Santa Cruz de Tenerife, Spain, 1995; pp. 1–168.
25. Halliday, W.R. Vulcanospeleological abstract. *Cascade Caver* **1972**, *11*, 12.
26. Wood, C. Lava tubes: Their morphogenesis and role in flow formation. *Cascade Caver* **1979**, *18*, 15–17.
27. Martín, J.L.; Oromí, P.; Izquierdo, I.; Medina, A.L.; González, J.M. Biología. In *La Cueva del Viento*; Oromí, P., Ed.; Canary Is. Government: Santa Cruz de Tenerife, Spain, 1995; pp. 31–66.
28. Hernández, J.J.; Izquierdo, I.; Medina, A.L.; Oromí, P. Introducción al estudio de la Cueva de Felipe Reventón (Tenerife, Islas Canarias). In *Proceedings of the II Symposium Regional on Speleology, Burgos, Spain*; Federación Castellana Espeleología: Burgos, Spain, 1984; pp. 107–122.
29. Carracedo, J.C.; Rodríguez-Badiola, E.; Guillou, H.; Paterne, M.; Scaillet, S.; Pérez-Torrado, F.J.; Paris, R.; Fra-Paleo, U.; Hansen, A. Eruptive and structural history of Teide volcano and rift zones of Tenerife, Canary Islands. *GSA Bull.* **2007**, *119*, 1027–1051. [CrossRef]
30. Ollier, C.D.; Brown, M.C. Lava caves of Victoria. *Bull. Volcanol.* **1965**, *28*, 215–229. [CrossRef]
31. Stone, F.D.; Howarth, F.G.; Hoch, H.; Asche, M. Root communities in lava tubes. In *Encyclopedia of Caves*; White, W.B., Culver, D.C., Eds.; Academic Press: London, UK, 2012; pp. 658–664.
32. Carracedo, J.C.; Pérez Torrado, F.J.; Guillou, H. Estudio geológico y datación de la cueva. In *La Cueva del Llano. Centro de Interpretación*; Socorro, J.S., Ed.; Cabildo Insular de Fuerteventura: Puerto del Rosario, Spain, 2006; pp. 8–9.
33. Fernández, O.; Naranjo, M.; Martín, S. Cueva de Aslobas: Hallazgo del tubo volcánico más antiguo de las Islas Canarias. In *Proceedings of the 1st International Convention on Speleology, Barcelona, Spain*; Federació Catalana Espeleologia: Barcelona, Spain, 2015; pp. 75–82.

34. Ashmole, N.P.; Oromí, P.; Ashmole, M.J.; Martín, J.L. Primary faunal succession in volcanic terrain: Lava and cave studies in the Canary Islands. *Biol. J. Linn. Soc.* **1992**, *46*, 207–234. [CrossRef]

35. Howarth, F.G. A comparison of the ecology and evolution of cave-adapted faunas in volcanic and karstic caves. In *Proceedings of the 7th International Symposium on Vulcanospeleology, Santa Cruz de la Palma, Spain, 4–11 November 1994*; Oromí, P., Ed.; Libros de la Frontera: Santa Cugat Vallés, Spain, 1996; pp. 63–68.

36. Arechavaleta, M.; Oromí, P.; Sala, L.L.; Martín, C. Distribution of carbon dioxide concentration in Cueva del Viento (Tenerife, Canary Islands). In *Proceedings of the 7th International Symposium on Vulcanospeleology, Santa Cruz de la Palma, Spain, 4–11 November 1994*; Oromí, P., Ed.; Libros de la Frontera: Sant Cugat Vallés, Spain, 1996; pp. 11–14.

37. Hoch, H. Homoptera (Auchenorrhyncha Fulgoroidea). In *Encyclopaedia Biospeologica*; Juberthie, C., Decu, V., Eds.; Société de Biospéologie: Moulis, France, 1994; Volume 1, pp. 313–325.

38. Hoch, H.; Asche, M. Evolution and speciation of cave-dwelling Fulgoroidea in the Canary Islands (Homoptera: Cixiidae and Meenoplidae). *Zool. J. Linn. Soc.* **1993**, *109*, 53–101. [CrossRef]

39. Hoch, H. *Trirhacus helenae* sp. n.; a new cave-dwelling planthopper from Croatia (Hemiptera: Fulgoromorpha: Cixiidae). *Dtsch. Entomol. Z.* **2013**, *60*, 155–161.

40. Hoch, H.; Sendra, A.; Montagud, S.; Teruel, S.; Lopes Ferreira, R. First record of a cavernicolous Kinnaridae from the Old World (Hemiptera, Auchenorrhyncha, Fulgoromorpha, Kinnaridae, Adolendini) provides testimony of an ancient fauna. *Subterr. Biol.* **2021**, *37*, 1–26. [CrossRef]

41. Deharveng, L.; Bedos, A. Diversity of terrestrial invertebrates in subterranean hábitats. In *Cave Ecology*; Moldovan, O.T., Kováč, L., Halse, S., Eds.; Springer: Cham, Switzerland, 2018; pp. 107–172.

42. Culver, D.C.; Pipan, T. Subterranean ecosystems. In *Encyclopedia of Biodiversity*, 2nd ed.; Levin, S.A., Ed.; Academic Press: Waltham, MA, USA, 2013; Volume 7, pp. 49–62.

43. Oromí, P.; Martín, J.L. The Canary Islands. Subterranean fauna, characterization and composition. In *The Natural History of Biospeleology*; Camacho, A.I., Ed.; C.S.I.C.: Madrid, Spain, 1992; pp. 527–567.

44. Moya, O.; Contreras, H.G.; Oromí, P.; Juan, C. Genetic structure, phylogeography and demography of two ground-beetle species endemic to the Tenerife laurel forest (Canary Islands). *Mol. Ecol.* **2004**, *13*, 3153–3167. [CrossRef] [PubMed]

45. Machado, A. Nuevos Trechodinae y Trechinae de las Islas Canarias (Coleoptera, Carabidae). *Fragm. Entomol.* **1987**, *19*, 323–338.

46. Yélamos, T. Un nuevo *Aeletes* Horn, 1873 de las Islas Canarias (Coleoptera, Histeridae). *Vieraea* **1995**, *24*, 87–90.

47. Howarth, F.G.; Medeiros, F.J.; Stone, F. Hawaiian lava tube cave associated Lepidoptera from the collections of Francis G: Howarth and Fred Stone. *Bishop Mus. Occas. Pap.* **2020**, *129*, 37–54.

48. García, R.; Andújar, C.; Oromí, P.; López, H. *Oromia orahan* (Curculionidae, Molytinae), a new subterranean species for the Canarian underground biodiversity. *Subterr. Biol.* **2020**, *35*, 1–14. [CrossRef]

49. Hoch, H.; Wessel, A. Communication by substrate-born vibrations in cave planthoppers. In *Insect Sounds and Communication: Physiology, Behavior, Ecology and Evolution*; Taylor and Francis: Boca Ratón, FL, USA, 2006; Volume 13, pp. 187–199.

50. Arnedo, M.A.; (Universitat de Barcelona, Barselona, Spain); Hoch, H.; (Museum für Naturkunde, Berlin, Germany). Personal communication, 2021.

51. Enghoff, H. *Dolichoiulus*—A mostly Macaronesian multitude of millipedes (Diplopoda, Julida, Julidae). *Entomol. Scand. Suppl.* **1992**, *40*, 1–158.

52. Dalens, H. Isopodes terrestres rencontrés dans les cavités volcaniques de l'île de Tenerife. *Trav. Lab. Écobiol. Arthr. Édaph.* **1984**, *5*, 5–19.

53. da Gama, M.M.; Ferreira, C.S. Collembola from two caves of Tenerife, Canary Islands. *Mém Biospéol.* **2000**, *27*, 143–145.

54. Izquierdo, I.; Oromí, P.; Bellés, X. Number of ovarioles and degree of dependance with respect to the underground environment in the Canarian species of the genus *Loboptera* Brunner (Blattaria, Blattellidae). *Mém. Biospéol.* **1990**, *17*, 107–111.

55. Mahnert, V. New species and records of pseudoscorpions (Arachnida, Pseudoscorpiones) from the Canary Islands. *Rev. Suisse Zool.* **1997**, *104*, 559–585. [CrossRef]

56. Arnedo, M.A.; Oromí, P.; Múrria, C.; Macías, N.; Ribera, C. The dark side of an island radiation: Systematics and evolution of troglobitic spiders of the genus *Dysdera* Latreille (Araneae: Dysderidae) in the Canary Islands. *Invertebr. Syst.* **2007**, *21*, 623–660. [CrossRef]

57. Hlaváč, P.; Oromí, P.; Bordoni, A. Catalogue of troglobitic Staphylinidae (Pselaphinae excluded) of the world. *Subterr. Biol.* **2006**, *4*, 19–28.

58. Mesa, F.M.; Pérez, A.J.; Oromí, P. La Cueva del Viento (Icod de los Vinos). Recopilatorio de su catálogo faunístico y subfósil. In *El Karst y el Hombre: Las Cuevas como Patrimonio Mundial*; Andrea, B., Durán, J., Eds.; Asociación Cuevas Turísticas Españolas: Nerja, Spain, 2016; pp. 35–48.

59. Marrero, A.; García Cruz, C.M. Nuevo yacimiento de restos subfósiles de dos vertebrados extintos de la isla de Tenerife (Canarias), *Lacerta maxima* Bravo, 1953 y *Canariomys bravoi* Crusafont & Petit, 1964. *Vieraea* **1978**, *7*, 165–174.

60. Rando, J.C.; López, M. Un nuevo yacimiento de vertebrados fósiles en Tenerife (Islas Canarias). In *Proceedings of the 7th International Symposium on Vulcanospeleology, Santa Cruz de la Palma, Spain, 4–11 November 1994*; Oromí, P., Ed.; Libros de la Frontera: Sant Cugat Vallés, Spain, 1996; pp. 171–173.

61. Rando, J.C.; López, M. A new species of extinct flightless passerine (Emberizidae: *Emberiza*) from the Canary Islands. *Condor* **1999**, *101*, 1–13. [CrossRef]

62. Hutterer, R.; Oromí, P. La rata gigante de la Isla Santa Cruz, Galápagos: Algunos datos y problemas. *Res. Cient. Proy. Galápagos TFMC* **1993**, *4*, 63–76.
63. Michaux, J.; López-Martínez, N.; Hernández-Pacheco, J.J. A ^{14}C dating of *Canariomys bravoi* (Mammalia, Rodentia), the extinct giant rat from Tenerife (Canary Islands, Spain), and the recent history of the endemic mammals in the archipelago. *Vie Milieu* **1996**, *46*, 261–266.
64. Rando, J.C.; Alcover, J.A.; Navarro, J.F.; Michaux, J.; Hutterer, R. Poniendo fechas a una catástrofe: ^{14}C, cronologías y causas de la extinción de vertebrados en Canarias. *El Indiferente* **2011**, *21*, 6–15.

diversity

Article

Stygobiont Diversity in the San Marcos Artesian Well and Edwards Aquifer Groundwater Ecosystem, Texas, USA

Benjamin T. Hutchins [1,*], J. Randy Gibson [2], Peter H. Diaz [3] and Benjamin F. Schwartz [1,4]

[1] Edwards Aquifer Research and Data Center, Texas State University, 601 University Dr., San Marcos, TX 78666, USA; bs37@txstate.edu

[2] U.S. Fish and Wildlife Service, San Marcos Aquatic Resources Center, 500 E. McCarty Ln., San Marcos, TX 78666, USA; randy_gibson@fws.gov

[3] U.S. Fish and Wildlife Service, Texas Fish and Wildlife Conservation Office, 500 E. McCarty Ln., San Marcos, TX 78666, USA; pete_diaz@fws.gov

[4] Department of Biology, Texas State University, 601 University Dr., San Marcos, TX 78666, USA

* Correspondence: bh1333@txstate.edu

Abstract: The Edwards Aquifer and related Edwards-Trinity Aquifer of Central Texas, USA, is a global hotspot of stygobiont biodiversity. We summarize 125 years of biological investigation at the San Marcos Artesian Well (SMAW), the best studied and most biodiverse groundwater site (55 stygobiont taxa: 39 described and 16 undescribed) within the Edwards Aquifer Groundwater Ecosystem. Cluster analysis and redundancy analysis (RDA) incorporating temporally derived, distance-based Moran's Eigenvector Mapping (dbMem) illustrate temporal dynamics in community composition in 85 high-frequency samples from the SMAW. Although hydraulic variability related to precipitation and discharge partially explained changes in community composition at the SMAW, a large amount of temporal autocorrelation between samples remains unexplained. We summarize potential mechanisms by which hydraulic changes can affect community structure in deep, phreatic karst aquifers. We also compile information on 12 other Edwards and Edwards-Trinity Aquifer sites with 10 or more documented stygobionts and used distance-based RDA to assess the relative influences of distance and site type on three measures of β-diversity. Distance between sites was the most important predictor of total dissimilarity and replacement, although site type was also important. Species richness difference was not predicted by either distance or site type.

Keywords: phreatic karst aquifer; stygobite; species richness; temporal dynamics; beta-diversity

Citation: Hutchins, B.T.; Gibson, J.R.; Diaz, P.H.; Schwartz, B.F. Stygobiont Diversity in the San Marcos Artesian Well and Edwards Aquifer Groundwater Ecosystem, Texas, USA. *Diversity* **2021**, *13*, 234. https://doi.org/10.3390/d13060234

Academic Editors: Michael Wink, Tanja Pipan, David C. Culver and Louis Deharveng

Received: 22 April 2021
Accepted: 20 May 2021
Published: 26 May 2021

Publisher's Note: MDPI stays neutral with regard to jurisdictional claims in published maps and institutional affiliations.

1. Introduction

The karstic Edwards Aquifer of Central Texas (USA) supplies water for more than 2 million people [1] and is recognized for high stygobiont biodiversity [2]. The 10,500 km^2 Edwards Aquifer occurs in a broad arc of Cretaceous limestones that stretch approximately 400 km across Central Texas, USA, and is hydrologically connected to the 91,744 km^2 Edwards-Trinity Aquifer (Figure 1). Edwards-equivalent limestones also extend into northern Coahuila, Mexico [3]. Late Cretaceous through early Miocene uplift of the Edwards Plateau exposed Edwards limestones along the Balcones Fault Zone: a series of en-echelon, high-angle faults downthrown to the southeast [4,5]. Increased permeability along faults in exposed Edwards limestones allowed meteoric recharge and dissolution, forming complex west–east and southwest–northeast flowpaths within hydrologically connected segments of the aquifer [6]. Present-day flowpaths are overprinted on hypogenically-derived permeability [7]. To the south and east, freshwater in the aquifer is confined below non-karstic units and juxtaposed against a lower-permeability zone of sulfide-rich, saline water along a steep freshwater–saline water interface (FWSWI) [8]. Shallower flowpaths and less faulting dominate in the Edwards-Trinity system to the north and west of the Edwards Aquifer. This region also contains many active stream caves.

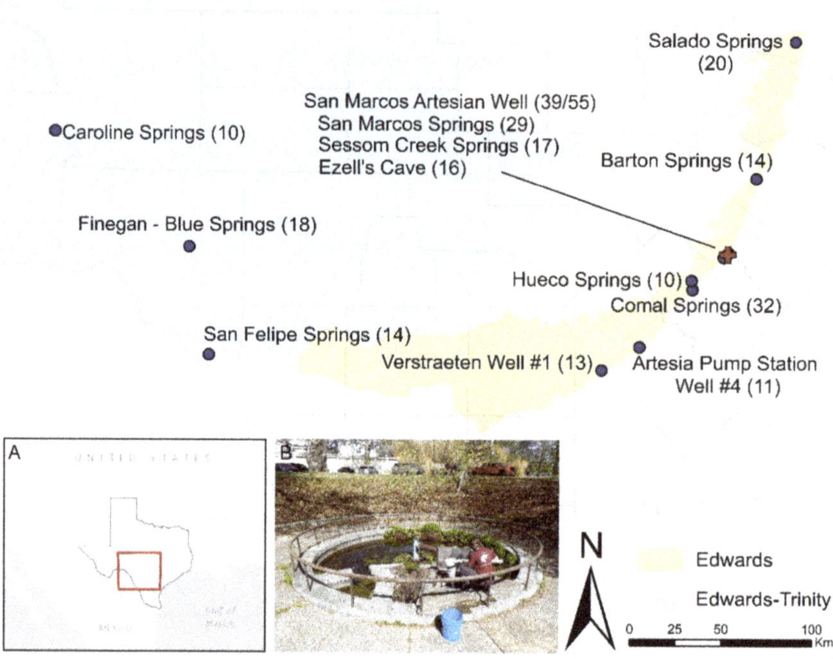

Figure 1. San Marcos Artesian Well and other diverse groundwater sites (>10 stygobionts) in the Edwards and Edwards-Trinity Aquifers. Species richness in parentheses. Inset A: Overview map. Inset B: San Marcos Artesian Well. Two numbers reported for the SMAW are published records (39) used in β-diversity analyses and published + unpublished, undetermined taxa (55).

The San Marcos Artesian Well (SMAW, state well number 6701828), on the Texas State University, San Marcos campus, is a flowing freshwater artesian well completed in late 1895 or early 1896 in the confined portion of the Edwards aquifer. The well intersects a 1.5 m tall phreatic conduit at a depth of −59.5 m [9]. Nearby saline wells illustrate proximity (i.e., <100 m) to the FWSWI. Dye tracing showed hydraulic connectivity between nearby Ezell's Cave (2.9 km to the southwest), the SMAW, and Deep Hole Spring (part of the San Marcos Springs Complex, 500 m northeast) [10]. From 2015 to 2020, discharge from the well averaged 16 L/s. During November 2013, 15-min continuous data documented average water properties of: temperature 22.3 °C (± 0.007), dissolved oxygen 5.3 mg/L (±0.01), and electrical conductivity 608 μS/cm (±0.5).

The SMAW and, to a lesser extent, springs, caves, and other wells in the Edwards and Edwards-Trinity Aquifers have been the focus of numerous studies ranging from taxonomic to ecological investigations. In the literature, the SMAW has been referred to as SWTSU Well, Texas State Artesian Well, and artesian well at San Marcos, and it is probably the source for most of the data for 'San Marcos Springs' in the first global list of subterranean biodiversity hotspots [11]. San Marcos Springs are hydrologically connected to the SMAW but inundated by a shallow reservoir, making them more difficult to sample. Consequently, fewer stygobionts are documented from San Marcos Springs relative to the SMAW.

The SMAW was the site of the first biospeleological investigations in Texas with the descriptions of the salamander *Eurycea rathbuni* (Stejneger, 1896), the shrimp *Palaemon antrorum* (Benedict, 1896), the isopod *Cirolanides texensis* (Benedict, 1896), and the amphipod *Stygobromus flagellatus* (Benedict, 1896). Soon after, Ulrich [12] described the isopod *Lirceolus smithii* (Ulrich, 1902) and two species of cyclopoid copepods. After those initial descriptions, taxonomic work in the Edwards Aquifer slowed until the late 1970s when Glenn Longley began a second phase of investigation with systematic sampling of the well via drift nets. Longley initiated important collaborations with taxonomists including John Holsinger (Amphipoda) and Robert Hershler (Gastropoda) and supported graduate students such as Henry Karnei that investigated Edwards Aquifer biodiversity. Through his collaboration and directorship of the Edwards Aquifer Research and Data Center, San Marcos, TX, USA (EARDC), created in 1979, the number of stygobionts recorded from the SMAW increased from eight to 25 between 1976 and 2000. Holsinger and Longley [9] and Longley [2] illustrated that the SMAW had a globally diverse stygobiont fauna, although previous studies had illustrated that faunal composition at the well was apparently distinct compared to other USA stygofauna sites [12–17]. Holsinger [15] and Holsinger and Longley [9] emphasized the presence of both marine and freshwater derived species at the site.

A third phase of biologic investigation began in the mid-2010s, when Benjamin Schwartz became director of the EARDC upon Longley's retirement. Schwartz also facilitated systematic sampling and taxonomic collaborations, most importantly with Okan Külköylüoğlu (Ostracoda) [18–23]. He also initiated morphometric and molecular analyses [23,24] to identify new records of previously described species. Since 2015, published stygobiont richness at the SMAW has increased from 26 to 39 (Table 1).

Longley [2] hypothesized that the Edwards Aquifer foodweb is not supported by allochthonous organic matter from the surface, but rather by 'fossil' organic matter originating at depth in the saline portion of the aquifer. Longley's hypothesis foreshadowed later studies that revealed the role of chemolithoautotrophy within the aquifer. Birdwell and Engel [25] characterized microbially derived dissolved organic matter along the FWSWI with a chromophoric signature distinct from terriginous surface and soil porewaters. Gray and Engel [26] identified microbial communities along the FWSWI with taxonomic composition similar to other chemolithoautotrophic systems. Finally, Hutchins et al. [27] reported isotopic signatures of carbon that indicated that chemolithoautotrophic production, in addition to photosynthetic organic matter, supports the metazoan community at the SMAW. They also suggested that chemolithoautotrophy might facilitate reduced extinction rates during climatically unfavorable periods. Hutchins et al. [28] suggested that, as a spatially and temporally stable food source, chemolithoautotrophy might support high biological diversity by increasing resource exploitation and reducing competition. By combining stable isotope and mouthpart morphologic data, the authors illustrated trophic niche partitioning among amphipod species, making the SMAW one of a small but growing number of sites with evidence of niche partitioning among stygobionts [29–31].

Given the hydraulic and geologic complexity of the Edwards Aquifer, the SMAW likely integrates water and stygobionts from multiple flowpaths, discrete locations, and microhabitats within the aquifer. Because hydraulic conditions along any discrete flowpath vary in response to precipitation and antecedent conditions, groundwater assemblage composition at the well may vary temporally as well. Temporal dynamics in groundwater community structure have been investigated in alluvial aquifers [32], epikarst [33], and karst aquifers [34–36], although to our knowledge, studies in the latter have been limited to vadose and shallow saturated systems rather than deep, phreatic sites. In high gradient, hydraulically 'flashy' karst aquifers, studies have emphasized the role of flood pulses in the 'spatial redistribution' of species [34,35] and how species-specific responses depend on hydraulic differences in microhabitats (e.g., transmissive conduits versus peripheral fracture networks). However, as Gibert et al. [34] noted, flow conditions may be more stable in the phreatic zone. Therefore, it is unclear whether stygobionts in more hydraulically stable, phreatic aquifers exhibit similar temporal variability. If so, then from a species-accumulation perspective, sample events might capture different components of a dynamic groundwater meta-community and species richness more likely describes the meta-community at a local, rather than site scale.

At a regional scale, the SMAW and surrounding groundwater environments are part of the larger Edwards Aquifer Groundwater Ecosystem. We use the term 'Edwards Aquifer Groundwater Ecosystem' to highlight the aquifer as a spatially extensive, discrete ecosystem with unique hydrogeologic, biological, and socio-economic elements. In contrast, conceptualizations of the aquifer have tended to emphasize isolated components of the system (i.e., groundwater as a resource, single sites as critical habitat for subsets of species). Although other sites in the aquifer have not been investigated as thoroughly as the SMAW, previous work has illustrated that diverse assemblages of stygobionts occur throughout the aquifer system [17,37]. Multiple diverse groundwater sites within a contiguous but heterogeneous aquifer system afford opportunity to investigate patterns of beta-diversity at the aquifer scale. Previous work has demonstrated that globally, most stygobionts are small-range endemics [38], and that overall diversity from local to continental scales is explained in part by the species-turnover component of beta (β)-diversity [39]. Although distance between sites, coupled with small ranges and limited dispersal potential [40], is a parsimonious explanation for species turnover, environmental differences among sites have also been proposed to explain differences in species richness and composition in epikarst copepod communities [41].

Here, we present a species list for the SMAW, using published and unpublished data. We also investigate temporal dynamics in community composition at the site via high-frequency sampling across 3 years and 85 samples. We hypothesized that samples from the well do not suggest a temporally stable community, but rather, show temporal variability, as in other groundwater systems. Specifically, we hypothesized that assemblage structure at the SMAW would vary seasonally and in response to precipitation-driven hydraulic changes. To our knowledge, this is the first assessment of temporal community dynamics within a deep phreatic karst system. We also investigate species richness at the SMAW within the context of the greater Edwards Aquifer Groundwater Ecosystem, highlighting other diverse sites. We assess whether (1) regional-scale patterns of β-diversity are explained by species turnover (replacement) or regional/ site-based differences in species richness and (2) whether those dissimilarities are explained by distance or site-type (i.e., springs versus wells and caves). We predicted that species turnover rather than differences in species richness drive β-diversity patterns, and that species turnover would be affected by distance rather than site type.

2. Materials and Methods

2.1. SMAW Diversity and Temporal Dynamics

We conducted a literature review to compile a stygobiont species list for the SMAW. In a few instances, taxonomic experts were consulted to determine whether species should be considered stygobionts (as defined by Trajano and de Carvalho [42]). Unpublished, undescribed taxa were also included, based on communication with collaborators and personal observations in the authors' taxonomic area of expertise. For temporal analysis, 85 samples were collected via 60 µm drift nets attached to the outflow of the well for between 24 and 72 h between 13 February 2013 and 20 November 2015 (Table S1). Samples were preserved in 95% EtOH and sorted at $10\times$ magnification. Species were identified to the lowest taxonomic level by the authors or taxonomic experts (see acknowledgements) although some undescribed taxa were lumped as a single taxon (e.g., Microcerberidae, Trombidiformes). Voucher specimens for most taxa are retained in the Aquifer Biodiversity Collection of the EARDC, Texas State University, San Marcos, TX, USA.

Statistical tools in R v4.0.3 were applied to explore stygobiont time-series data from the SMAW. A Ward's minimum variance dendrogram based on species abundances was created using the vegan package following the method of Borcard et al. [43]. A graph of silhouette widths was visually examined to estimate the optimal number of clusters, which was statistically assessed via analysis of similarity using a Bray–Curtis similarity matrix and 9999 permutations. Species associated with each group at $p < 0.05$ were identified using Indicator Species Analysis run with 999 permutations.

To disentangle the potential effects of discharge, season, and unexplained temporal influences on community composition, redundancy analysis was performed using the vegan package. Independent variables included season, discharge, and distance-based Moran's eigenvector maps (dbMEM) derived from time (day) since the first sampling event (T_0) (Table S1). Season was coded as a categorical variable according to Kollaus and Bonner [44]: winter = December–February, spring = March–May, summer = June–August, fall = September–November. Discharge data were derived from mean daily discharge recorded by the U.S. Geological Survey at the San Marcos River gaging station USGS 08170500 [45], converted to liters per second. The gaging station is approximately 0.5 km downstream from the San Marcos Springs, and discharge at the station is a surrogate for local aquifer levels, which are correlated with average flow velocities in the Edwards Aquifer [46]. Distance-based Moran's eigenvector maps (dbMEMs) were created using the method of Legendre and Gauthier [47] as implemented using the dbmem function in the adespatial package. dbMEMs were derived from a distance matrix of the number of days between sampling events and represent a spectral decomposition of the temporal relationships among samples [47]. Significance of dbMEMs was assessed using the moran.randtest function in the adespatial package, and only dbMEMs significant at $p < 0.05$ were used in the RDA. Periodicity of significant dbMEMs was not calculated because of many missing values over the sampling period. RDAs on discharge and season only, and on the dbMEMs only, were performed prior to a global RDA with both sets of independent variables and variance partitioning. The species data matrix was Hellinger transformed and singletons and doubletons were removed, but data were not detrended prior to analysis. Significance of RDAs, independent variables, and canonical axes was assessed using the anova.cca function in vegan, with 1000 permutations. Variance explained by discharge and season versus dbMEMs was assessed using the varpart function in vegan.

2.2. Edwards Aquifer Groundwater Ecosystem β-Diversity

To place the SMAW community within the broader context of the Edwards Aquifer Groundwater Ecosystem, distance-based RDA (dbRDA) was performed to assess the influence of distance, aquifer pool, and site type on β-diversity in R v.4.0.3. A literature review identified additional Edwards Aquifer and Edwards-Trinity Aquifer sites with high diversity (with an arbitrary cut-off of 10 or more stygobiont species). Primary and grey literature resources and previously unreported records of described species represented by specimens in the collections of the author and colleagues were included (Table S2). Unpublished, undetermined taxa were not included. β-diversity (Jaccard index total dissimilarity) was estimated and partitioned into two components (replacement, richness) using the beta function in the package BAT [48]. Linear trends in β-diversity measures were assessed using linear regression against log-transformed Euclidean geographic distance between sites, with Bonferroni correction applied for multiple comparisons. Dissimilarity matrices were used as the response variable in dbRDA. Predictor variables included site type (spring or well/cave), aquifer pool, and a distance-based Moran's eigenvector map (dbMem) derived from site coordinates (UTM). Because of high multicolinearity (VIF > 10) between aquifer pool and dbMem variables, aquifer pool was removed prior to analysis. The single dbMem derived from the coordinates of the 13 assessed sites was calculated using the dbmem function in the adespatial package. Because sites are unevenly spaced and often widely distributed, the truncation threshold was set at 217039.8 m, limiting assessment of spatial structure to broad spatial scales. dbRDAs were conducted using the capscale function in the vegan package. Significance of the dbRDA and predictor terms was assessed via the anova.cca function in vegan, with 1000 replications.

3. Results

3.1. SMAW Diversity and Temporal Dynamics

Currently, 55 taxa have been documented from the SMAW, including 39 species recorded in literature (Table 1), two of which remain undescribed (*Parabogidiella* sp. Holsinger, 1980 and *Erpobdella* sp.). An additional 16 stygobiont taxa (2 described species and 14 undetermined taxa) are reported here for the first time (Table 1). Crustaceans dominate the SMAW fauna, comprising 75% of documented species. These include the only thermosbaenacean in the United States [49] and a globally significant amphipod fauna [9] of 12 species in five families. Additionally, 11 ostracod and nine isopod species occur at the site. Unique soft-bodied taxa include the only North American stygobitic leech [50], two vertebrate parasites (see below), and five species in the gastropod genus *Phreatodrobia*. The beetle, *Haideoporus texanus* (Young and Longley, 1976) is one of five species of stygobitic dytiscid beetles in the United States (four of which are associated with the Edward Aquifer [51]). A single vertebrate, the Texas blind salamander, *E. rathbuni*, occurs at the well. The SMAW is the type locality for 25 of the published taxa (64%), and eight of these (21%) are single-site endemics (Table 1). New SMAW site records, (including species descriptions and new records for described species) have accumulated unevenly over time (Figure 2).

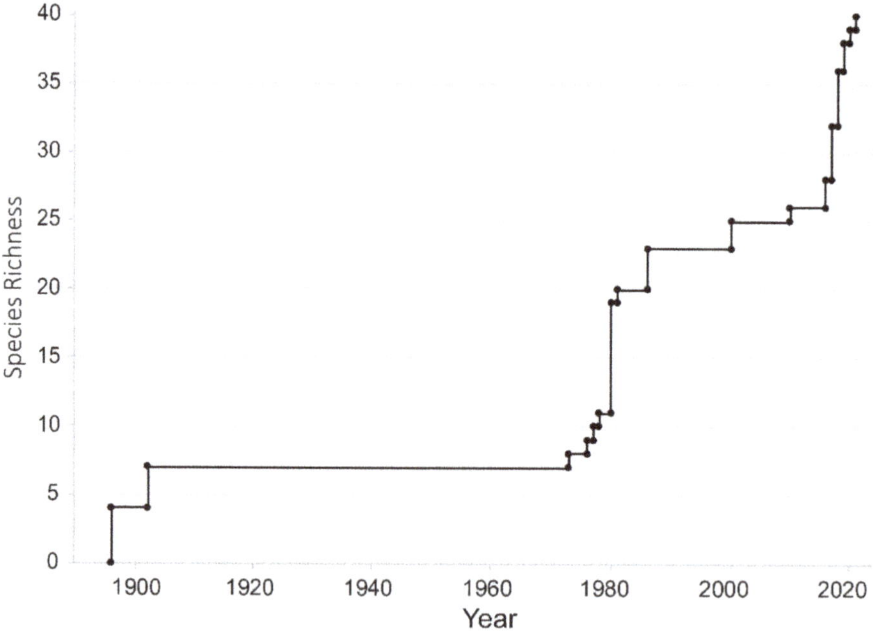

Figure 2. Published species richness at the San Marcos Artesian Well over time. Undetermined, unpublished taxa excluded.

Table 1. Stygobionts from the San Marcos Artesian Well and basis of record. § = single site endemic; † = type locality.

Phylum	Class	Order	Family	Species	Reference
Nematoda	Enoplea	Trichocephalida	Capillariidae	*Amphibiocapillaria texensis* Moravec and Huffman, 2000 †	[52]
Acanthocephala	Eoacanthocephala	Neoechinorhynchida	Dendronucleatidae	*Dendronucleata americana* Moravec and Huffman, 2000 †§	[52]
Platyhelminthes	Rhabditophora	Tricladida	Kenkiidae	*Sphalloplana mohri* Hyman, 1938	[53]
Annelida	Clitellata	Arhynchobdellida	Erpobdellidae	*Erpobdella* sp.	[2]
		Lumbriculida	Lumbriculidae	gen. sp. undet.	Steve Fend, pers. comm.
Mollusca	Gastropoda	Littorinimorpha	Cochliopidae	*Phreatodrobia micra* (Pilsbry and Ferriss, 1906)	[54]
				Phreatodrobia nugax (Pilsbry and Ferriss, 1906)	[55]
				Phreatodrobia plana Hershler and Longley, 1986	[55]
				Phreatodrobia rotunda Hershler and Longley, 1986	[55]
Arthropoda	Ostracoda	Podocopida	Candonidae	*Cabra candona mixoni* Külköylüoğlu, Yavuzatmaca, Akdemir, Schwartz and Hutchins, 2019 †§	[18]
				Candonopsis sp.	Okan Külköylüoğlu pers. comm.
				Comalcandona sp.	Okan Külköylüoğlu pers. comm.
				Cypria lacrima Külköylüoğlu, Akdemir, Yavuzatmaca, Schwartz and Hutchins, 2017 †§	[19]
				Lacrimacandona wisei Külköylüoğlu, Yavuzatmaca, Akdemir, Schwartz and Hutchins, 2017 †§	[20]
				Namiotkocypria haysensis Külköylüoğlu, 2018 †§	[56]
				Rugosuscandona scharfi Külköylüoğlu, Akdemir, Yavuzatmaca, Schwartz and Hutchins, 2017 †	[21]
				Ufocandona hannaleeae Külköylüoğlu, Yavuzatmaca, Akdemir, Schwartz and Hutchins, 2017 †§	[22]
				Schornikovcandona bellensis Külköylüoğlu, Yavuzatmaca, Akdemir, Diaz and Gibson, 2017	Okan Külköylüoğlu pers. comm.
			Darwinulidae	*Darwinula* sp.	Okan Külköylüoğlu pers. comm.
				Vestalenula sp.	Okan Külköylüoğlu pers. comm.
	Hexanauplia	Cyclopoida	Cyclopidae	*Cyclops cavernarum* Ulrich, 1902 †§	[12]
				Cyclops learii Ulrich, 1902 †§	[12]
		Harpacticoida		gen. sp. undet.	Diego Figueroa, pers. comm.
	Malacostraca	Bathynellacea	Bathynellidae	*Hobbsinella edwardensis* Camacho, Hutchins, Schwartz, Dorda, Casado and Rey, 2017 †	[57]
		Thermosbaenacea	Monodellidae	*Tethysbaena texana* (Maguire, 1965)	[17]
		Isopoda	Microcerberidae	Microcerberidae sp. 1	[58]
				Microcerberidae sp. 2	pers. obs.

Table 1. *Cont.*

Phylum	Class	Order	Family	Species	Reference
			Asellidae	*Lirceolus hardeni* Lewis and Bowman, 1996	[24]
				Lirceolus pilus (Steeves, 1968)	[24]
				Lirceolus smithii (Ulrich, 1902)[†]	[12]
				Lirceolus sp.	William Coleman pers. comm.
			Cirolanidae	*Cirolanides texensis* Benedict, 1896[†]	[13]
				Cirolanides wassenichae Schwartz, Hutchins, Schwartz, Hess and Bonett, 2019[†]	[59]
				Cirolanides sp.	pers. obs.
		Amphipoda	Artesiidae	*Artesia subterranea* Holsinger, 1980[†]	[9]
			Crangonyctidae	*Stygobromus bifurcatus* (Holsinger, 1967)	[23]
				Stygobromus flagellatus (Benedict, 1896)[†]	[13]
				Stygobromus longipes (Holsinger, 1966)	[23]
				Stygobromus russelli (Holsinger, 1967)	[9]
			Hadziidae	*Allotexiweckelia hirsuta* Holsinger, 1980[†]	[9]
				Holsingerius samacos (Holsinger, 1980)[†]	[9]
				Texiweckelia texensis (Holsinger, 1973)[†]	[15]
				Texiweckeliopsis insolita (Holsinger, 1980)[†]	[9]
			Parabogidiellidae	*Parabogidiella americana* Holsinger, 1980[†]	[9]
				Parabogidiella sp. Holsinger, 1980	[9]
			Seborgiidae	*Seborgia relicta* Holsinger, 1980[†]	[9]
		Decapoda	Palaemonidae	*Calathaemon holthuisi* (Strenth, 1976)	[60]
				Palaemon antrorum (Benedict, 1896)[†]	[13]
	Arachnida	Trombidiformes	Chappuisididae	*Chappuisides* sp.	Ian Smith, in prep.
				Uchidastygacarus sp.	Ian Smith, in prep.
			Halacaridae	gen. sp. undet.	Ian Smith, in prep.
			Mideopsidae	gen. sp. undet.	Ian Smith, in prep.
			Nudomideopsidae	*Allomideopsis wichitaensis* (Smith, 1990)	Ian Smith, in prep.
	Insecta	Coleoptera	Dytiscidae	*Haideoporus texanus* Young and Longley, 1976[†]	[16]
Chordata	Amphibia	Urodela	Plethodontidae	*Eurycea rathbuni* (Stejneger, 1896)[†]	[14]

The stygobiont community at the SMAW is characterized by some taxonomic and ecological uncertainty. In addition to the two undescribed species already mentioned, the two copepod species reported from the well are poorly described and have been designated as *nomen inquirendum* or *nomen dubium* by some [61]. Two species, the nematode *Amphibiocapillaria texensis* Moravec and Huffman, 2000 and the acanthocephalan *Dendronucleata americana* Moravec and Huffman, 2000 are parasites of the salamander *E. rathbuni*. Both parasites complete their life cycles using stygobiont invertebrates and should be considered stygobionts (David Huffman, pers. comm.). Neither parasite species have been conclusively collected from the epigean salamander *Eurycea nana* Bishop, 1941, which occurs in the San Marcos Springs and San Marcos River.

Eighty-five samples collected between 2013 and 2015 contained 42,814 individuals with an average of 275 individuals and 15 taxa per 24 h period (Table S1). Incidental epigean taxa were removed prior to temporal analysis ($n = 61$). Undetermined individuals ($n = 87$), mostly ostracods, were also removed because they could not be confidently assigned a taxonomic identity. An additional 415 juvenile and damaged individuals were assigned to taxa present in the sample.

Log-transformed total abundances show a nearly normal distribution (Figure 3), and only two species in the 85 samples (*Lirceolus pilus* (Steeves, 1968) and an unidentified nematode) were found in only one or two samples, suggesting that rare species are mostly accounted for by the sampling effort. Three species (*Phreatodrobia micra* (Pilsbry and Ferriss, 1906), *Lirceolus hardeni* Lewis and Bowman, 1996, and *Stygobromus bifurcatus* (Holsinger, 1967)) are known from the site by single specimens but were not present in any of the 85 analyzed samples. The two parasitic taxa present in the aquifer were not detected in our sampling strategy, which did not involve dissection. Additionally, copepods and mites were not identified to species, so all copepods and mites were each lumped into a single category. A histogram of the number of samples in which taxa occur shows a bimodal distribution, suggesting that most species are either common or rare (Figure 3). The shrimp, *P. antrorum*, makes up 44% of individuals, and just four taxa: *P. antrorum*, Copepoda, *Cypria lacrima* Külköylüoğlu, Akdemir, Yavuzatmaca, Schwartz and Hutchins, 2017, and *Texiweckeliopsis insolita* (Holsinger, 1980) make up over 90% of individuals. The 23 most infrequent taxa (61% of all taxa) collectively make up less than 1% of the total number of individuals.

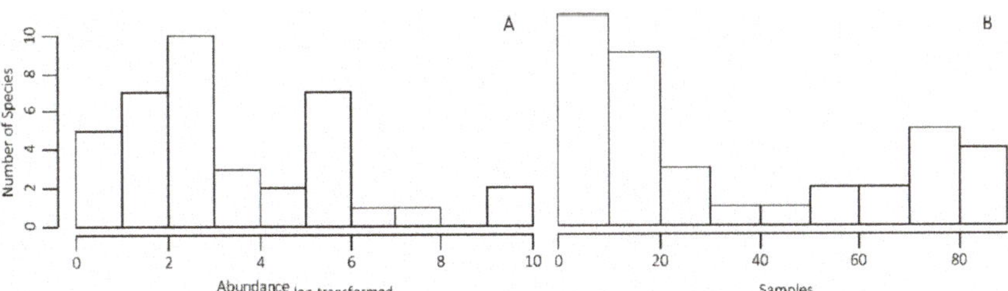

Figure 3. Histograms of (**A**) log-transformed species abundance and (**B**) frequency of species occurrences in high-frequency samples from the San Marcos Artesian Well.

Samples form significant clusters based on species abundances (Figure 4). Two clusters produced an optimal silhouette width, but between two and five clusters had similarly high silhouette widths. ANOSIM confirmed that groupings of samples into five and two clusters were both significant (R: 0.45 and $p < 0.001$; R = 0.26 and $p < 0.001$, respectively). Several taxa were significantly associated ($p < 0.05$) with one or two clusters although some species were associated with two clusters that were not nearest neighbors (i.e., species could be associated with two clusters that were otherwise compositionally dissimilar to one another).

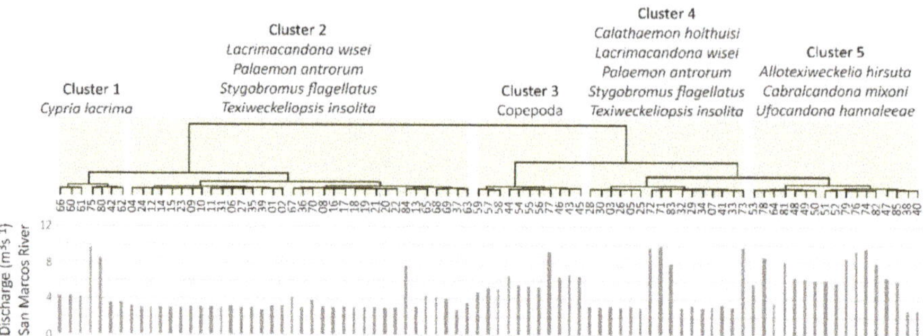

Figure 4. Ward's minimum variance dendrogram of high-frequency sample stygobiont community data from the San Marcos Artesian Well, with corresponding discharge from the nearby San Marcos River. Sample numbers (dendrogram tips) refer to sample order (e.g., 01 occurred at T_0, followed by 02) but time between consecutive samples is variable. Significant groupings are demarcated by shading, and significantly associated taxa are shown above. Discharge is not represented by a hydrograph because samples are clustered by community similarity and are not shown in chronological order.

Although there was a significant temporal trend in the community data ($p < 0.05$), only a small proportion of variance was constrained by sampling date alone. Separate redundancy analyses (RDA) using only discharge plus season and only distance-based Moran's eigenvector maps (dbMEMs) were both significant (F = 8.589, $p = 0.001$; F = 3.930, $p = 0.001$, respectively). A global RDA incorporating discharge, season, and dbMEMs was also significant (F = 9.274, $p = 0.001$). Variance partitioning between discharge plus season versus dbMEMs showed that shared explained variance was not significant. Discharge and season explained 32% of variance in community structure, and both variables were significant at $p = 0.001$ (F = 21.084 and 8.013, respectively). dbMEMs explained 18% of variance in community structure, and three of four dbMEMs were significant at $p = 0.001$ (F = 6.635–10.878). The first three axes of the global RDA were significant at $p = 0.001$ and cumulatively accounted for 45% of the total explained variance. The first axis illustrated a gradient between low-flow samples (primarily during spring and summer) with larger numbers of the shrimp *P. antrorum* and the amphipod *T. insolita* and high-flow samples (primarily during fall and winter) with higher numbers of copepods, the snail *Phreatodrobia plana* Hershler and Longley, 1986, and the amphipod *Texiweckelia texensis* (Holsinger, 1980) (Figure 5). The second axis illustrated unexplained temporal gradients (dbMEM1 and dbMEM3) between samples (primarily in the summer) with higher numbers of the ostracod *C. lacrima* and samples with higher numbers of the amphipods *S. flagellatus* and *Seborgia relicta* Holsinger, 1980 (Figure 5). Clustering of sites in RDA space (influenced by community composition and environmental variables) reflects clustering on the Ward's minimum variance dendrogram (based only on community composition, Figure 4).

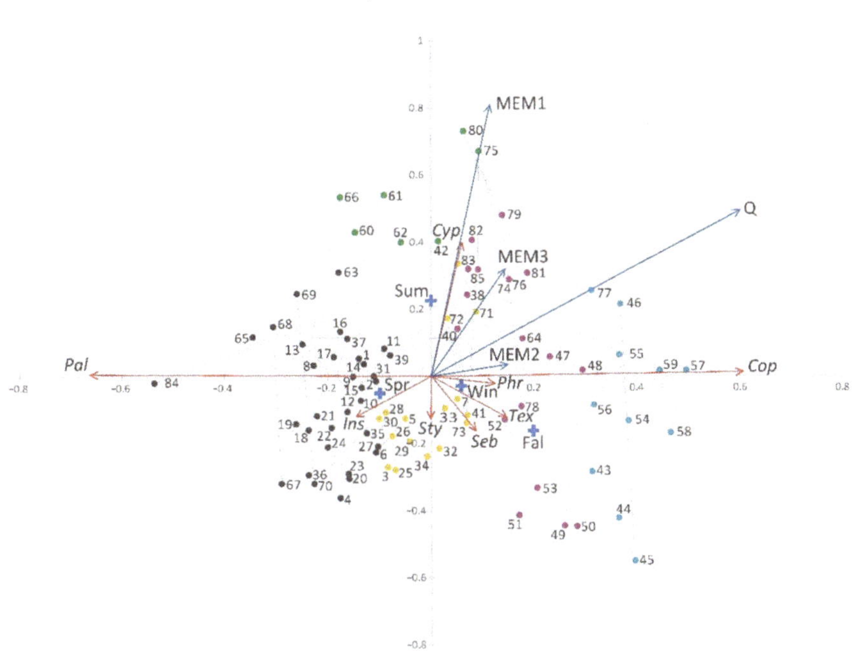

Figure 5. First two axes of global Redundancy Analysis for stygobionts from high-frequency sampling at the San Marcos Artesian Well. Sample numbers refer to sample order (e.g., 01 occurred at T_0, followed by 02) but time between consecutive samples is variable. Sample colors refer to Ward's minimum variance clustering (Figure 3): green = cluster 1, black = cluster 2, light blue = cluster 3, yellow = cluster 4, and purple = cluster 5. Faint lines connect consecutive samples. For clarity, only taxa with the highest loadings on axes 1 and 2 are shown (red arrows): *Cyp* = *Cypria lacrima*, *Phr* = *Phreatodrobia plana*, *Tex* = *Texiweckelia texensis*, *Seb* = *Seborgia relicta*, *Sty* = *Stygobromus flagellatus*, *Ins* = *Texiweckeliopsis insolita*, and *Pal* = *Palaemon antrorum*. Blue arrows show biplot scores for constraining variables: Q = discharge, MEM1–MEM3 represent significant Moran's eigenvector maps describing temporal relationships among samples at different scales. Blue pluses are centroids for categorical seasons: Sum = summer, Spr = spring, Win = winter, and Fal = fall.

3.2. Edwards Aquifer Groundwater Ecosystem β-Diversity

Thirteen sites with 10 or more stygobionts were identified across the Edwards and Edwards-Trinity Aquifers (Figure 1, Table S2). Sites included three flowing artesian wells, one cave, and nine springs varying from 1st magnitude springs (e.g., Comal Springs) to a small, intermittent spring (Sessom Creek Spring). All sites are hydraulically connected to the phreatic zone of the Edwards or Edwards-Trinity Aquifers. Total dissimilarity and replacement increased with increasing distance ($R^2 = 0.57$ and 0.32, respectively) at $p < 0.001$. Differences in species richness did not exhibit a spatial trend. dbRDA revealed significant site-type and distance-based effects on total dissimilarity (F = 2.49, $p = 0.001$, $R^2 = 0.20$) and replacement (F = 2.95, $p = 0.001$, $R^2 = 0.41$), but not on differences in species richness. Thirty-three percent of variance in total dissimilarity was constrained by the two canonical axes (axis 1: 19%, axis 2: 14%), and both site type (F = 2.14, $p = 0.008$, loadings: axis 1 = 0.15, axis 2 = -0.99) and dbMem (F = 2.84, $p = 0.001$, loadings: axis 1 = 0.89, axis 2 = 0.46) were significant terms. Thirty-seven percent of variance in replacement was constrained by the two canonical axes (axis 1: 22%, axis 2: 15%), and both site type (F = 2.47, $p = 0.003$, loadings: axis 1 = 0.09, axis 2 = -1.00) and dbMem (F = 3.43, $p = 0.001$, loadings: axis 1 = 0.91, axis 2 = 0.41) were significant terms. For each RDA, the first axis describes differences in dissimilarity explained by distance between sites and the second axis describes differences in dissimilarity explained by site type (Figure 6).

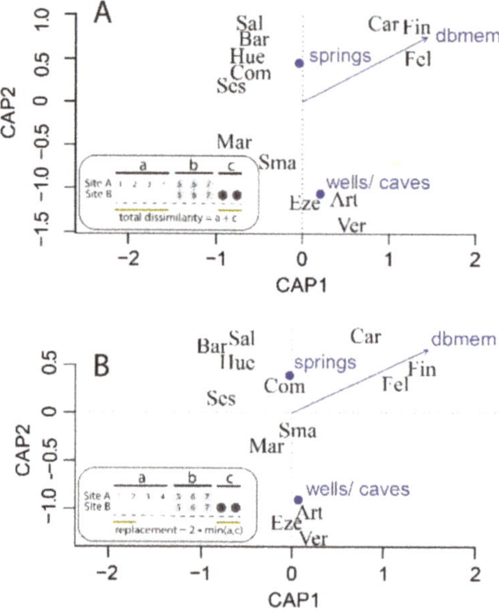

Figure 6. First two axes (canonical principal coordinates) of distance-based Redundancy Analysis of total dissimilarity (**A**) and species replacement (**B**) at diverse (> 10 stygobionts) Edwards and Edward-Trinity Aquifer sites. Blue circles are centroids for site types. dbmem is a distance-based Moran's eigenvector map derived from latitude and longitude (UTM) of sites. Sal = Salado Springs, Bar = Barton Springs, Sma = San Marcos Artesian Well, Mar = San Marcos Springs, Ses = Sessom Creek Springs, Eze = Ezell's Cave, Hue = Hueco Springs, Com = Comal Springs, Art = Artesia Pump Station Well #4, Ver = Verstraeten Well #1, Fel = San Felipe Springs, Fin = Finegan–Blue Springs, and Car = Caroline Springs. Insets are graphical representations of total dissimilarity and dissimilarity due to replacement, where numbers represent species (modified from Carvalho et al. [62]).

4. Discussion

Despite over one century of intensive sampling and study, knowledge of diversity at the SMAW remains incomplete. 'Orphan' taxa (e.g., Cyclopoida and Harpacticoida) are completely or largely unassessed. Even in better-studied groups (e.g., Isopoda), undescribed taxa have been identified. We conclude that reported species richness is underestimated. The continuing increase in species recorded at the site over time (Figure 2) and the near absence of taxa that are important elements of many groundwater communities (e.g., Copepoda) supports this assertion. The semi-normal distribution of species abundances in high-frequency samples (Figure 3A) and species accumulation curves (data not shown) suggest that sampling has been adequate for known species. Additional species discovery at the well will mostly likely result from additional taxonomic assessment of the orphan taxa discussed above, and cryptic species [23,24]. Nevertheless, with 55 groundwater-obligate taxa (Table 1), the SMAW is the most diverse groundwater site in North America and among the most diverse sites globally [63]. Proposed explanations for high-biodiversity within the Edwards Aquifer Groundwater Ecosystem include the role of marine embayments producing relic taxa (i.e., a richer colonist pool *sensu* Cardinale et al. [64]) [9] and high rates of primary productivity [27] supported, in part, by chemolithoautotrophy. Long-term productive energy, which is linked to climate and climatic-variability, has emerged as an important driver of groundwater diversity patterns in Europe and North America [65,66]. South of Pleistocene ice sheets and permafrost, Central Texas climate was cooler and wetter during glacial periods [67], potentially having a positive effect on productivity in the region. The region has become warmer and dryer since the last glacial maximum and although pronounced aridity during the mid-Holocene Altithermal Period would have reduced surface productivity, Hutchins et al. [27] hypothesized that chemolithoautotrophic production may have mitigated effects of aridity-related changes in surface productivity on stygobionts. In Europe, habitat heterogeneity is also an important predictor of stygobiont richness at mid and southern latitudes [65] and may be important in the hydrogeologically complex Edwards Aquifer.

To our knowledge, our analysis of a high-frequency dataset spanning multiple seasons and flow regimes provides the first illustration of temporal variability in a deep phreatic aquifer community. Community composition-based clustering of samples suggests that some species abundances vary in synchrony over time, which would not be expected in random samples from a temporally stable community. The bimodal distribution of taxa frequencies in repeated samples (Figure 3B) is interesting and may reflect the presence of two or more metacommunities: one comprised of species that are ubiquitous in flowpaths intersected by the well (e.g., *Lacrimacandona wisei* Külköylüoğlu, Yavuzatmaca, Akdemir, Schwartz and Hutchins, 2017, *P. antrorum*, *S. flagellatus*, *T. insolita*), and one or more communities comprised of species in more remote habitats that are only infrequently washed out, typically during high flows. Alternatively, the distribution may not reflect spatial or temporal heterogeneity in community structure, but rather, biological differences (e.g., benthic versus pelagic habitat, or swimming ability) in species' propensity to be expelled from the well, which would also produce a relationship between flow and sample composition.

Redundancy analysis illustrates that, like other groundwater systems, hydrologic regime plays an important role in structuring the SMAW stygobiont assemblage over time. Discharge was the most important predictor in the global RDA. The influence of discharge is also apparent when viewed alongside the community-based dendrogram (Figure 4), although several low-flow samples fall within otherwise high-flow clusters and vice versa. Samples 71 and 72 were high-flow samples collected immediately after an extended period of low-flow, and so may represent a 'piston-effect' in which rapid recharge by meteoric water pushes resident groundwater (and associated fauna) through the system [38]. Conversely, the low-flow samples 38, 40, and 83 all contained uncommon species. Since species richness at the SMAW is positively correlated with discharge (data not shown), uncommon species more strongly effect overall community composition in otherwise less-diverse, low-flow samples, and probably drive the clustering of these samples with more-diverse, high-flow samples. Sample 84, a high-flow sample clustering with low-flow samples, had a near absence of copepods, potentially reflecting an undetected sample processing error.

Why certain species appear to be associated with low or high flows is unclear. Gibert et al. [34] suggested that stygobionts are heterogeneously distributed across distinct microhabitats in aquifers, and flood pulses initiate transport of organisms in transmissive zones through the karst system. However, other direct and indirect mechanisms may also facilitate assemblage structure changes through demographic shifts or passive or active movement of species (Table 2). Analysis of hydrographs, geochemical dynamics, morphologic/ trophic patterns, and population dynamics within groups of concordant taxa may provide additional insight into causal relationship between hydrologic variability and stygobiont community dynamics.

Table 2. Potential mechanisms by which hydraulic changes affect stygobiont assemblage composition in deep phreatic aquifers.

	Positive Association			**Negative Association**		
	Hydraulic/ environmental effect	→	Community response	Hydraulic/ environmental effect	→	Community response
Increased flow	Activates intermittent/ alternative flowpaths and hydraulic connections	→	Animals transported from typically less-connected aquifer areas to sampling location	Increased flow within conduit	→	Animals within conduit seek refuge in adjacent fracture network
	Increased hydraulic gradient from fracture network to conduit network	→	Animals transported to sampling location from adjacent fracture network			
	Increased sediment transport and conduit boundary velocities	→	Dislodges benthic/ interstitial organisms already within sampling area	Changes in nutrient availability, community composition, and geochemistry	→	Direct and indirect demographic (e.g., reproduction) or species interaction (e.g., ecologic release) effects
	Changes in nutrient availability, community composition, and geochemistry	→	Direct and indirect demographic (e.g., reproduction) or species interaction (e.g., ecologic release) effects			
Decreased flow	Decreased flow within conduit	→	Animals within fracture network move into conduit	Changes in nutrient availability, community composition, and geochemistry	→	Direct and indirect demographic (e.g., reproduction) or species interaction (e.g., ecologic release) effects
	Changes in nutrient availability, community composition, and geochemistry	→	Direct and indirect demographic (e.g., reproduction) or species interaction (e.g., ecologic release) effects			

As Gibert et al. [34] also acknowledged, groundwater flow alone does not explain temporal dynamics of groundwater communities. The significant contribution of temporal dbMEMs in our RDA demonstrates unexplained temporal dynamics in community composition at multiple scales. We did not attempt to correlate dbMEMs with potential explanatory phenomenon because we did not have a priori predictions about potential mechanisms, and because our dataset did not span multi-year cyclical weather oscillations like the Atlantic Multi-decadal Oscillation or El Niño–Southern Oscillation.

The 55 taxa from the SMAW represent about half of the approximately 102 stygobionts recorded from the Edwards Aquifer, and most occur at other sites. Within an extensive aquifer, assessment of biodiversity at local to regional scales, rather than at single sites, makes more sense ecologically and for aquifer management. Cardoso et al. [48] discuss issues with analyses of β-diversity based on incomplete/uneven sampling, including underestimation of similarity. However, assessment of randomized accumulation curves to control for sample effort (*sensu* Cardoso et al. [68]) was not possible in this study because species lists for sites other than the SMAW were based on literature and not samples. Consequently, β-diversity data are interpreted with the acknowledgement that uneven sampling effort across sites (and across taxa) obscures patterns. Given that caveat, we did observe increasing dissimilarity and replacement with increasing distance between sites, as predicted, and total dissimilarity and replacement was greater between springs and wells/caves than within site types (although the number of diverse wells and caves was limited in number and spatial extent compared to springs). However, biplot scores showed that for both total dissimilarity and replacement, distance between sites was more important than site type. Dissimilarity among habitat types was also a relatively unimportant component of β-diversity in European stygobionts [69]. Importantly, the species richness difference component of dissimilarity did not vary by distance or site type, suggesting the lack of 'hotspot' regions within the Edwards Aquifer with respect to location or site type, despite uneven sampling effort across regions. Because of uneven site distribution across the aquifer, however, fine-scale patterns may be obscured. For example, Hutchins et al. [27] detected a positive relationship between species richness and proximity to the FWSWI, although biodiverse sites far from the FWSWI (e.g., Caroline and Finegan-Blue Springs) raise questions about the importance of that relationship. Certainly, species richness at fine scales varies in response to hydraulic and geochemical properties, as evidenced by low- and high-diversity sites in close proximity to one another (pers. obs.).

Relative to other groundwater habitats, knowledge of stygobiont diversity in deep phreatic karst aquifers is lacking. The SMAW and the Edwards Aquifer Groundwater Ecosystem illustrates that springs and wells can be particularly productive sites for sampling these habitats, and we suspect that biodiversity within the aquifer is not anomalous relative to other deep phreatic karst aquifers on a global scale. Although spatial biodiversity patterns have received a good deal of attention from groundwater ecologists, increasingly sophisticated analytical methods [70] afford more opportunity to assess spatial and temporal patterns in community structure. In the face of global climate change and increasing anthropogenic pressures on groundwater ecosystems [71], analysis of spatial and temporal trends in groundwater communities will be increasingly important.

Supplementary Materials: The following are available online at https://www.mdpi.com/article/10.3390/d13060234/s1, Table S1: Raw species data for the San Marcos Artesian Well, including variables used in redundancy analysis. MEM1-4 are distance-based Moran's Eigenvector Maps values. Table S2: Stygobiont species lists for other diverse sites (>10 stygobionts) in the Edwards and Edwards-Trinity Aquifers. Previously unpublished records of described species are included, along with specimen information, which serves as basis of the new record. Published records of undescribed/undetermined taxa are included, but unpublished records of undescribed/undetermined taxa are excluded. See below for published references.

Author Contributions: Conceptualization: B.T.H. and B.F.S.; methodology and analysis: B.T.H.; data acquisition: B.T.H., B.F.S., P.H.D., and J.R.G.; original draft preparation: B.T.H.; review and editing: B.T.H., B.F.S., P.H.D., and J.R.G. All authors have read and agreed to the published version of the manuscript.

Funding: This research had no external funding.

Institutional Review Board Statement: All research involving the collection of live animals was conducted in accordance with and approval from the Texas State University–San Marcos Institutional Animal Care and Use Committee (protocol # 0318_0428_06, approved 28 April 2014).

Informed Consent Statement: Not applicable.

Data Availability Statement: The data presented in this study are openly available as supplementary files.

Acknowledgments: We appreciate the taxonomic expertise of William Coleman, Okan Külköylüoğlu, Kathryn Perez, Anna Camacho, Steve Fend, and Ian Smith. We dedicate this paper to Glen Longley for his years uncovering the hidden species of this occult environment. The views presented herein are those of the authors and do not necessarily represent those of the U.S. Fish and Wildlife Service.

Conflicts of Interest: The authors declare no conflict of interest.

References

1. Hinojosa, J.; Green, J.; Estrada, F.; Herrera, J.; Mata, T.; Phan, D.; Tanvir Pasha, A.B.M.; Matta, A.; Johnson, D.; Kapoor, V. Determining the Primary Sources of Fecal Pollution Using Microbial Source Tracking Assays Combined with Land-use Information in the Edwards Aquifer. *Water Res.* **2020**, *184*, 116211. [CrossRef]
2. Longley, G. The Edwards Aquifer: Earth's Most Diverse Groundwater Ecosystem? *Int. J. Speleol.* **1981**, *11*, 123–128. [CrossRef]
3. Boghici, R. Hydrogeology of Trinity-Edwards Aquifer of Texas and Coahuila in the Border Region. In *Aquifers of the Edwards Plateau*; Mace, R.E., Angle, E.S., Mullican, W.F., III, Eds.; Texas Water Development Board Report 360: Austin, TX, USA, 2004; pp. 91–114.

4. Barker, R.A.; Bush, P.W.; Baker, E.T., Jr. Geologic History and Hydrogeologic Setting of the Edwards-Trinity Aquifer System, West-Central Texas. *US Geol. Surv. Water Resour. Investig. Rep.* **1994**, *94*, 1–51.

5. Maclay, R.W. Geology and Hydrology of the Edwards Aquifer in the San Antonio Area, Texas. *US Geol. Surv. Water Resour. Investig. Rep.* **1995**, *95*, 1–64.

6. Sharp, J.M.; Banner, J.L. The Edwards Aquifer: A Resource in Conflict. *GSA Today* **1997**, *7*, 1–9.

7. Schindel, G.M.; Gary, M. Hypogene Processes in the Balcones Fault Zone Segment of the Edwards Aquifer of South-Central Texas. In *Hypogene Karst Regions and Caves of the World. Cave and Karst Systems of the World*; Klimchouk, A., Palmer, A.N., De Waele, J., Auler, A.S., Audra, P., Eds.; Springer International Publishing AG: Gewerbestrasse, Switzerland, 2017; pp. 511–530.

8. Hutchins, B.T.; Schwartz, B.F.; Engel, A.S. Environmental Controls on Organic Matter Production and Transport Across Surface-Subsurface and Geochemical Boundaries in the Edwards Aquifer, Texas, USA. *Acta Carsologica* **2013**, *42*, 245–259. [CrossRef]

9. Holsinger, J.R.; Longley, G. The Subterranean Amphipod Crustacean Fauna of an Artesian Well in Texas. *Smithson. Contrib. Zool.* **1980**, *380*, 1–62. [CrossRef]

10. Ogden, A.E.; Quick, R.A.; Rothermel, S.R. Hydrochemistry of the Comal, Hueco, and San Marcos Springs, Edwards Aquifer, Texas. In *The Balcones Escarpment: Geology, Hydrology, Ecology and Social Development in Central Texas*; Abbott, P.L., Woodruff, C.M., Jr., Eds.; Geological Society of America Annual Meeting: San Antonio, TX, USA, 1986; pp. 115–130.

11. Culver, D.C.; Sket, B. Hotspots of Subterranean Biodiversity in Caves and Wells. *J. Cave Karst Stud.* **2000**, *62*, 11–17.

12. Ulrich, C.J. A Contribution to the Subterranean Fauna of Texas. *Trans. Am. Microsc. Soc.* **1902**, *23*, 83–101. [CrossRef]

13. Benedict, J.E. Preliminary Descriptions of a New Genus and Three New Species of Crustaceans from an Artesian Well at San Marcos, Texas. *Proc. U.S. Nat. Mus.* **1896**, *18*, 615–617. [CrossRef]

14. Stejneger, L. Description of a New Genus and Species of Blind Tailed Batrachians from the Subterranean Waters of Texas. *Proc. U.S. Nat. Mus.* **1896**, *18*, 619–621. [CrossRef]

15. Holsinger, J.R. Two New Species of the Subterranean Amphipod Genus *Mexiweckelia* (Gammaridae) from Mexico and Texas, with Notes on the Origin and Distribution of the Genus. *Assoc. Mex. Cave Stud. Bull.* **1973**, *5*, 1–12.

16. Young, F.L.; Longley, G. A New Subterranean Aquatic Beetle from Texas (Coleoptera: Dytiscidae-Hydroporinae). *Ann. Entomol. Soc. Am.* **1976**, *69*, 787–792. [CrossRef]

17. Karnei, H.S., Jr. A Survey of the Subterranean Aquatic Fauna of Bexar County, Texas. Master's Thesis, Southwest Texas State University, San Marcos, TX, USA, 1978.

18. Külköylüoğlu, O.; Yavuzatmaca, M.; Akdemir, D.; Schwartz, B.F.; Hutchins, B.T. Description of a New Tribe *Cabralcandonini* (Candonidae, Ostracoda) from Karst Aquifers in Central Texas, U.S.A. *J. Cave Karst Stud.* **2019**, *81*, 136–151. [CrossRef]

19. Külköylüoğlu, O.; Akdemir, D.; Yavuzatmaca, M.; Schwartz, B.F.; Hutchins, B.T. *Cypria lacrima* sp. nov. A New Ostracoda (Candonidae, Crustacea) Species from Texas, USA. *Zool. Stud.* **2017**, *56*, e15. [PubMed]

20. Külköylüoğlu, O.; Yavuzatmaca, M.; Akdemir, D.; Schwartz, B.F.; Hutchins, B.T. *Lacrimacandona* n. gen. (Crustacea: Ostracoda: Candonidae) from the Edwards Aquifer, Texas (USA). *Zootaxa* **2017**, *4277*, 261–273. [CrossRef] [PubMed]

21. Külköylüoğlu, O.; Akdemir, D.; Yavuzatmaca, M.; Schwartz, B.F.; Hutchins, B.T. *Rugosuscandona*, a New Genus of Candonidae (Crustacea: Ostracoda) from Groundwater Habitats in Texas, North America. *Species Divers.* **2017**, *22*, 175–185. [CrossRef]

22. Külköylüoğlu, O.; Yavuzatmaca, M.; Akdemir, D.; Schwartz, B.F.; Hutchins, B.T. *Ufocandona hannaleeae* gen. et sp. nov. (Crustacea, Ostracoda) from an Artesian Well in Texas, USA. *Eur. J. Taxon.* **2017**, *372*, 1–18. [CrossRef]

23. Schwartz, B.F.; Nice, C.; Jenson, A.; Gibson, J.R. *Final Report: Molecular and Morphological Analysis of Stygobromus sp. Near San Marcos, TX*; Texas Parks and Wildlife Department: Austin, TX, USA, 2016.

24. Schwartz, B.F.; Nice, C.; Coleman, W. *Final Report: Status Assessment and Ecological Characterization of the Texas Troglobitic Water Slate (Lirceolus smithii)*; Texas Parks and Wildlife Department: Austin, TX, USA, 2018.

25. Birdwell, J.E.; Engel, A.S. Variability in Terrestrial and Microbial Contributions to Dissolved Organic Matter Fluorescence in the Edwards Aquifer, Central Texas. *J. Cave Karst Stud.* **2009**, *71*, 144–156.

26. Gray, C.J.; Engel, A.S. Microbial Diversity and Impact on Carbonate Geochemistry Across a Changing Geochemical Gradient in a Karst Aquifer. *ISME J.* **2013**, *7*, 325–337. [CrossRef]

27. Hutchins, B.T.; Engel, A.S.; Nowlin, W.H.; Schwartz, B.F. Chemolithoautotrophy Supports Macroinvertebrate Food Webs and Affects Diversity and Stability in Groundwater Communities. *Ecology* **2016**, *97*, 1530–1542. [CrossRef]

28. Hutchins, B.T.; Schwartz, B.F.; Nowlin, W.H. Morphological and Trophic Specialization in a Subterranean Amphipod AssemBlage. *Freshw. Biol.* **2014**, *59*, 2447–2461. [CrossRef]

29. Fišer, C.; Blejec, A.; Trontelj, P. Niche-based Mechanisms Operating within Extreme Habitats: A Case Study of Subterranean Amphipod Communities. *Biol. Lett.* **2012**, *8*, 578–581. [CrossRef]

30. Ercoli, F.; Lefebvre, F.; Delangle, M.; Godé, N.; Caillon, M.; Raimond, R.; Souty-Grosset, C. Differing Trophic Niches of Three French Stygobionts and Their Implications for Conservation of Endemic Stygofauna. *Aquat. Conserv.* **2019**, *29*, 2193–2203. [CrossRef]

31. Chávez-Solís, E.M.; Solís, C.; Simões, N.; Mascaró, M. Distribution Patterns, Carbon Sources and Niche Partitioning in Cave Shrimps (Atyidae: *Typhlatya*). *Sci. Rep.* **2020**, *10*, 12812. [CrossRef] [PubMed]

32. Dole-Olivier, M.-J.; Marmonier, P.; Creuzé des Châtelliers, M.; Martin, D. Interstitial Fauna Associated with the Alluvial FloodPlains of the Rhône River (France). In *Groundwater Ecology*; Gibert, J., Danielopol, D.L., Stanford, J.A., Eds.; Academic Press: San Diego, CA, USA, 1994; pp. 313–346.

33. Pipan, T.; Christman, M.C.; Culver, D.C. Dynamics of Epikarst Communities: Microgeographic Pattern and Environmental Determinants of Epikarst Copepods in Organ Cave, West Virginia. *Am. Mid. Nat.* **2006**, *156*, 75–87. [CrossRef]
34. Gibert, J.; Vervier, P.; Malard, F.; Laurent, R.; Reygrobellet, J.-L. Dynamics of Communities and Ecology of Karst Ecosystems: Example of Three Karsts in Eastern and Southern France. In *Groundwater Ecology*; Gibert, J., Danielopol, D.L., Stanford, J.A., Eds.; Academic Press: San Diego, CA, USA, 1994; pp. 425–450.
35. Lorenzo, T.D.; Cipriani, D.; Fiasca, B.; Rusi, S.; Galassi, D.M.P. Groundwater Drift Monitoring as a Tool to Assess the Spatial Distribution of Groundwater Species in Karst Waters. *Hydrobiologia* **2018**, *813*, 137–156. [CrossRef]
36. Opalički Slabe, M. Patterns in Invertebrate Drift from an Alpine Karst Aquifer over a One Year Period. *Acta Carsologica* **2015**, *44*, 265–278. [CrossRef]
37. Gibson, J.R.; Harden, S.J.; Fries, J.N. Survey and Distribution of Invertebrates from Selected Springs of the Edwards Aquifer in Comal and Hays Counties, Texas. *Southwest. Nat.* **2008**, *53*, 74–84. [CrossRef]
38. Trontelj, P.; Douady, C.J.; Fišer, C.; Gibert, J.; Gorički, Š.; Lefébure, T.; Sket, B.; Zakšek, V. A Molecular Test for Cryptic Diversity in Ground Water: How Large are Ranges of Macro-stygobionts? *Freshw. Biol.* **2009**, *54*, 727–744. [CrossRef]
39. Stoch, F.; Galassi, D.M.P. Stygobiontic Crustacean Species Richness: A Question of Numbers, a Matter of Scale. *Hydrobiologia* **2010**, *653*, 217–234. [CrossRef]
40. Trontelj, P. Structure and Genetics of Cave Populations. In *Cave Ecology*; Moldovan, O.T., Kováč, L., Halse, S., Eds.; Springer Nature: Cham, Switzerland, 2018; pp. 269–296.
41. Pipan, T.; Culver, D.C.; Papi, F.; Kozel, P. Partitioning Diversity in Subterranean Invertebrates: The Epikarst Fauna of Slovenia. *PLoS ONE* **2018**, *13*, e0195991. [CrossRef]
42. Trajano, E.; de Carvalho, M.R. Towards a Biologically Meaningful Classification of Subterranean Organisms: A Critical Analysis of the Schiner-Racovitza System from a Historical Perspective, Difficulties of Its Application and Implications for Conservation. *Subterr. Biol.* **2017**, *22*, 1–26. [CrossRef]
43. Borcard, D.; Gillet, F.; Legendre, P. *Numerical Ecology with R*; Springer Science+Business Media, LLC.: New York, NY, USA, 2011.
44. Kollaus, K.A.; Bonner, T.R. Habitat Associations of a Semi-arid Fish Community in a Karst Spring-fed Stream. *J. Arid Environ.* **2012**, *76*, 72–79. [CrossRef]
45. U.S. Geological Survey. National Water Information System Data Available on the World Wide Web (USGS Water Data for the Nation). Available online: https://waterdata.usgs.gov/tx/nwis/uv/?site_no=08170500&PARAmeter_cd=00065,00060 (accessed on 19 November 2020). [CrossRef]
46. Zappitello, S.J.; Johns, D.A.; Hunt, B.B. Summary of Groundwater Tracing in the Barton Springs Edwards Aquifer from 1996 to 2017. In *Report # DR-19-04*; Barton Springs Edwards Aquifer Conservation District: Austin, TX, USA, 2019.
47. Legendre, P.; Gauthier, O. Statistical Methods for Temporal and Space-time Analysis of Community Composition Data. *Proc. R. Soc. B Biol. Sci.* **2014**, *281*, 20132728. [CrossRef] [PubMed]
48. Cardoso, P.; Rigal, F.; Carvalho, J.C. BAT–Biodiversity Assessment Tools, an R Package for the Measurement and Estimation of Alpha and Beta Taxon, Phylogenetic and Functional Diversity. *Methods Ecol. Evol.* **2015**, *6*, 232–236. [CrossRef]
49. Jaume, D. Global Diversity of Spelaeogriphaceans & Thermosbaenaceans (Crustacea; Spelaeogriphacea & Thermosbaenacea) in Freshwater. *Hydrobiologia* **2008**, *595*, 219–224.
50. Sket, B. Hirudinea. In *Stygofauna Mundi*; Botosaneanu, L., Ed.; E.J. Brill & Dr. W. Backhuys: Leiden, The Netherland, 1986; pp. 250–253.
51. Kanda, K.; Gomez, R.A.; Driesche, R.V.; Miller, K.B.; Maddison, D.R. Phylogenetic Placement of the Pacific Northwest Subterranean Endemic Diving Beetle *Stygoporus oregonensis* Larson & LaBonte (*Dytiscidae, Hydroporinae*). *Zootaxa* **2016**, *632*, 75–91.
52. Moravec, F.; Huffman, D.G. Three New Helminth Species from Two Endemic Plethodontid Salamanders, *Typhlomolge rathbuni* and *Eurycea nana*, in Central Texas. *Folia Parasitol.* **2000**, *47*, 186–194. [CrossRef]
53. Kenk, R. Freshwater Triclads (Turbellaria) of North America, IX: The Genus *Sphalloplana*. *Smithson. Contrib. Zool.* **1977**, *246*, 1–38. [CrossRef]
54. Alvear, D.; Diaz, P.H.; Gibson, J.R.; Jones, M.; Perez, K.E. An Unusually Sculptured New Species of *Phreatodrobia* Hershler & Longley (Mollusca: Caenogastropoda: Cochliopidae) from Central Texas. *Zootaxa* **2020**, *4810*, 143–152.
55. Hershler, J.R.; Longley, G. Phreatic Hydrobiids (Gastropoda: Prosobranchia) from the Edwards (Balcones Fault Zone) Aquifer Region, South-Central Texas. *Malacologia* **1986**, *27*, 127–172.
56. Külköylüoğlu, O. A New Genus and Species in the Ostracod Family Candonidae (Crustacea: Ostracoda) from Texas, USA. *J. Nat. Hist.* **2018**, *52*, 1295–1310. [CrossRef]
57. Camacho, A.I.; Hutchins, B.; Schwartz, B.F.; Dorda, B.A.; Casado, A.; Rey, I. Description of a New Genus and Species of Bathynellidae (Crustacea: Bathynellacea) from Texas Based on Morphological and Molecular Characters. *J. Nat. Hist.* **2018**, *52*, 29–51. [CrossRef]
58. Hutchins, B.T.; Schwartz, B.F.; Coleman, W.T. Three New Microcerberids (Isopoda, Microcerberidae) from Subterranean Freshwater Habitats in Texas, U.S.A. (In review)
59. Schwartz, B.F.; Hutchins, B.T.; Schwartz, Z.G.; Hess, A.J.; Bonett, R.M. *Cirolanides wassenichae* sp. nov., a Freshwater, Subterranean Cirolanidae (Isopoda, Cymothoida) with Additional Records of Other Species from Texas, United States. *Zootaxa* **2019**, *4543*, 498–514. [CrossRef]

60. Zara Environmental LLC. *Final Report: Hays County Karst Invertebrate Distribution and Cave Development*; Texas Parks and Wildlife Department: Austin, TX, USA, 2010.
61. World of Copepods. Available online: www.marinespecies.org/copepoda (accessed on 11 May 2021).
62. Carvalho, J.C.; Cardoso, P.; Borges, P.A.V.; Schmera, D.; Podani, J. Measuring Fractions of Beta Diversity and Their Relationships to Nestedness: A Theoretical and Empirical Comparison of Novel Approaches. *Oikos* **2013**, *122*, 825–834. [CrossRef]
63. Culver, D.C.; Pipan, T. *The Biology of Caves and Other Subterranean Habitats*, 2nd ed.; Oxford University Press: Oxford, UK, 2019.
64. Cardinale, B.J.; Bennett, D.M.; Nelson, C.E.; Gross, K. Does Productivity Drive Diversity or Vice Versa? A Test of the Multivariate Productivity-diversity Hypothesis in Streams. *Ecology* **2009**, *90*, 1227–1241. [CrossRef]
65. Eme, D.; Zagmajster, M.; Fišer, C.; Galassi, D.; Marmonier, P.; Stoch, F.; Cornu, J.-F.; Oberdorff, T.; Malard, F. Multi-causality and Spatial Non-stationarity in the Determinants of Groundwater Crustacean Diversity in Europe. *Ecography* **2015**, *38*, 531–540. [CrossRef]
66. Culver, D.C.; Deharveng, L.; Bedos, A.; Lewis, J.J.; Madden, M.; Reddell, J.R.; Sket, B.; Trontelj, P.; White, D. The Mid-latitude Biodiversity Ridge in Terrestrial Cave Fauna. *Ecography* **2006**, *29*, 120–128. [CrossRef]
67. Toomey, R.S., III; Blum, M.D.; Valastro, S., Jr. Late Quaternary Climates and Environments of the Edwards Plateau, Texas. *Glob. Planet. Chang.* **1993**, *7*, 299–320. [CrossRef]
68. Cardoso, P.; Borges, P.A.V.; Veech, J.A. Testing the Performance of Beta Diversity Measures Based on Incidence Data: The Robustness to Undersampling. *Divers. Distrib.* **2009**, *15*, 1081–1090. [CrossRef]
69. Malard, F.; Boutin, C.; Camacho, A.I.; Ferreira, D.; Michel, G.; Sket, B.; Stoch, F. Diversity Patterns of Stygobiotic Crustaceans Across Multiple Spatial Scales in Western Europe. *Freshw. Biol.* **2009**, *54*, 756–776. [CrossRef]
70. Cáceres, M.D.; Coll, L.; Legendre, P.; Allen, R.B.; Wiser, S.K.; Fortin, M.-J.; Condit, R.; Hubbell, S. Trajectory Analysis in Community Ecology. *Ecol. Monogr.* **2019**, *89*, e01350. [CrossRef]
71. Mammola, S.; Cardoso, P.; Culver, D.C.; Deharveng, L.; Ferreira, R.L.; Fišer, C.; Galassi, D.M.P.; Griebler, C.; Halse, S.; Humphreys, W.H.; et al. Scientists' Warning on the Conservation of Subterranean Ecosystems. *Bioscience* **2019**, *69*, 641–650. [CrossRef]

diversity

Article

Postojna-Planina Cave System in Slovenia, a Hotspot of Subterranean Biodiversity and a Cradle of Speleobiology

Maja Zagmajster [1,*], Slavko Polak [2] and Cene Fišer [1]

1   SubBioLab, Department of Biology, Biotechnical Faculty, University of Ljubljana, Jamnikarjeva 101, SI1000 Ljubljana, Slovenia; cene.fiser@bf.uni-lj.si

2   Zavod Znanje Postojna, Notranjska Museum Postojna, Kolodvorska Cesta 3, SI6230 Postojna, Slovenia; slavko.polak@notranjski-muzej.si

*   Correspondence: maja.zagmajster@bf.uni-lj.si

Abstract: The Postojna-Planina Cave System (PPCS) in central Slovenia is a globally exceptional site of subterranean biodiversity, comprised of many interconnected caves with cumulative passage length exceeding 34 km. Two rivers sink into the caves of the PPCS, called the Pivka and Rak, and join underground into Unica River, which emerges to the surface. The studies of fauna of PPCS began in the 19th century with the first scientific descriptions of specialized cave animals in the world, making it "the cradle of speleobiology". Currently, the species list of PPCS contains 116 troglobiotic animal species belonging to eight phyla, confirming its status as the richest in the world. Of these, 47 species have been scientifically described from the PPCS, and more than 10 await formal taxonomic descriptions. We expect that further sampling, detailed analyses of less studied taxa, and the use of molecular methods may reveal more species. To keep the cave animals' checklist in PPCS up to date, we have supplemented the printed checklist with an online interface. As the revised checklist is a necessary first step for further activities, we discuss the importance of PPCS in terms of future research and conservation.

Keywords: hotspot; speleobiology; subterranean biodiversity; troglobionts; Postojna-Planina Cave System; Slovenia

Citation: Zagmajster, M.; Polak, S.; Fišer, C. Postojna-Planina Cave System in Slovenia, a Hotspot of Subterranean Biodiversity and a Cradle of Speleobiology. *Diversity* **2021**, *13*, 271. https://doi.org/10.3390/d13060271

Academic Editors: Tanja Pipan, David C. Culver, Louis Deharveng and Michael Wink

Received: 4 May 2021
Accepted: 7 June 2021
Published: 15 June 2021

Publisher's Note: MDPI stays neutral with regard to jurisdictional claims in published maps and institutional affiliations.

1. Introduction

Sampling subterranean fauna is a challenging task, considering that humans have limited access to the subterranean environment. Caves are access points to reach subterranean species that can inhabit very narrow spaces in fractured rock. Inventories of subterranean biodiversity are time consuming and require technically demanding fieldwork and broad taxonomic engagement [1]. The highest conservation priority is usually given to sites with high species richness, so completing species inventories of such hotspots should be highly prioritized [2].

The Dinaric Karst in the Western Balkans in Europe is one of the global hotspots of subterranean biodiversity [3–5]. Species richness within this region is not evenly distributed [6,7]. Many caves are consistently listed among the species richest on a global scale [8]. The species richest among them is Postojna-Planina Cave System (in Slovenian: Postojnsko planinski jamski sistem; hereafter referred to as PPCS) in Slovenia (Figure 1), outstanding not only because of its biological, but also historical, touristic, and conservation importance.

The PPCS is a complex system of many caves connected by dry or flooded channels. The inner parts of the Postojnska jama ("jama" means cave in Slovenian) at the southernmost end of the PPCS were discovered in 1818 and revealed the extraordinary richness and beauty of speleothems (Figure 2). The attractive ornamentation initiated guided tours of the cave and triggered a worldwide beginning of cave tourism. The construction of the first

101

underground railway (Figure 3) and electric lighting in 1872 and 1884, respectively, promoted tourist visits to Postojnska jama. Along with tourism, scientific research developed, which laid the foundation for speleology and karstology as sciences. Biological studies of the subterranean world began with the discovery of Luka Čeč, a cave guide and assistant lamplighter, who found an unusual beetle in Postojnska jama in 1831. He gave the beetle to Count Franz Joseph Hochenwart, the curator of the Provincial Museum of Carniola in Ljubljana, who recognized the value of the find. The specimen was examined by the renowned entomologist Ferdinand Josef Schmidt, who in 1832 produced the scientific description of *Leptodirus hochenwartii* [9]. In this work he declared the animal to be a cave-adapted beetle, which was the first scientific recognition of specialized subterranean animal in the world. News of the discovery spread and in the next years, many eminent naturalists came to study the fauna of Postojnska jama, describing cave species of other animal taxa. The snowball effect triggered by the description of *L. hochenwartii* is considered the starter of speleobiology and PPCS the "cradle of speleobiology" [10].

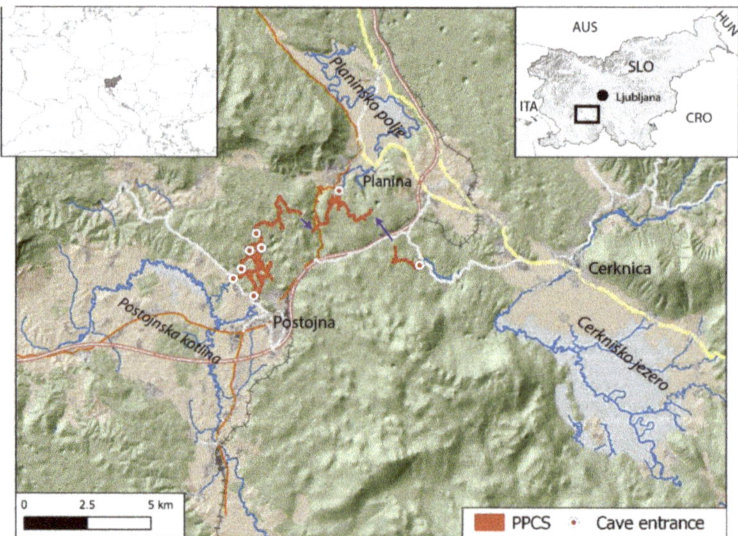

Figure 1. Location of the Postojna-Planina Cave System in central Slovenia. Top left: position of Slovenia within Europe; top right: position of PPCS within Slovenia; middle: red lines depict the planar views of all caves of the system; red dots: entrances to caves; blue arrows: parts of the system connected via subterranean rivers, but not yet passed by man.

The pace of new discoveries by numerous scientists in caves of the PPCS (see detailed review of the beginnings in [10]) produced the first checklists of the PPCS already in the 19th century. An important milestone in cataloguing was made by Benno Wolf, who collected all data on animals in caves of the world, and compiled them in Animalium Cavernarum Catalogus, issued between 1934 and 1938 [11]. However, he listed all taxa from published sources without critically evaluating their taxonomic validity, and did not distinguish the troglobiotic from non-troglobitic species. His list included 134 species for PPCS, of which about 50 could be considered troglobiotic. It was not until 30 years later, that Egon Pretner produced the next list of cave animals for PPCS, taking into account caves excluding Planinska jama in the northern part of the system [12]. Pretner listed 131 species, including about 50 troglobiotic ones. The comprehensive list of aquatic taxa from the entire PPCS, which considered the ecological status of the species, was prepared by Boris Sket in 1979 [13]. Among more than 190 species, he listed 34 aquatic troglobionts. The total number of troglobionts for PPCS, 84, was given in a comparative study of the richest subterranean sites in the world [8], but the actual species list was not included. We fill this gap with this paper, in which we have carefully evaluated and updated the list of troglobionts and discussed its importance within a broad socio-scientific-conservation context.

Figure 2. Postojnska jama is a tourist cave, but there are still many beautifully ornamented passages, where only cavers can enter—as the example of "Pisani rov". Photo: Peter Gedei.

Figure 3. As a unique tourist attraction, Postojnska jama takes tourists to deep parts of the cave with the underground railway. Photo: Slavko Polak.

Description of the PPCS

The PPCS is located in central Slovenia (Figure 1). It is the main cave system in the densely forested and uninhabited karst area between the town of Postojna in Postojnska kotlina, the town of Planina on Planinsko polje, and the town of Cerknica on Cerkniško polje (Figure 1) [14]. In the area above the PPCS, there are 16 large collapsed dolinas interrupting some of these underground passages. The cave system developed in an about 800-m-thick layer of Cretaceous limestones and dolomites, between two NW-SE oriented faults, namely the northern Idrija fault and the southern Predjama fault [15]. The thickness of the bedrock above the cave ranged from 60 to 120 m. The PPCS is typically defined as a system of six caves with large separate entrances with individual cadastre numbers in Slovenian Cadastre of Caves, namely Postojnska jama (No. 747), Otoška jama (No. 779), Pivka jama (No. 472), Črna jama (No. 471), Magdalena jama (No. 820), and Planinska jama (No. 748). Two other caves that should be considered part of the PPCS, connected via impassable flooded channels, are Lekinka (No. 1867) and Tkalca jama (No. 857) [16] (Figure 4), both of which are often excluded from biological observations of the system [12,13]. The entire cave system reaches depths of up to 115 m and includes at least 34 km of channel length (24 km of caves from Postojnska jama to Pivka jama, and approximately 10 km sum of the length of Lekinka, Planinska jama, and Tkalca jama).

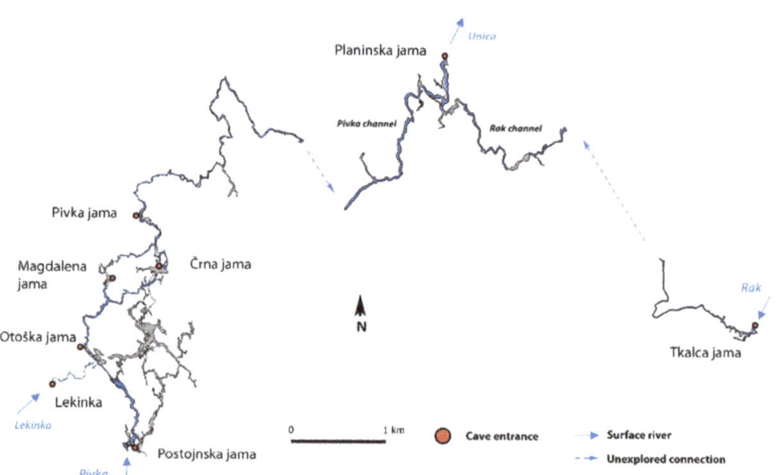

Figure 4. The plan of the Postojna-Planina Cave System, with labelled entrances to individual caves. Subterranean connections, which have not yet been explored by man, are marked with dashed blue arrow lines, while full blue arrow lines and blue names indicate surface rivers entering/leaving the system. We modified the base line outline of the caves, originally provided by the Karst Research Institute in Postojna, Slovenia.

The PPCS is located in the Black Sea drainage basin, but at least some species in the PPCS also occur in Adriatic drainage basin, indicating a complex history of drainage connectivity (Delić et al., in rev.). The PPCS contains two subterranean rivers that enter the system in the South (Pivka River) and East (Rak River). The Pivka River sinks in Postojnska jama and is directly accessible in caves up to the syphon in Pivka jama (Figure 4). It receives inflows from Lekinka and deep groundwater from the east. In 2015, parts after the siphon in Pivka jama were dived through and the length of the cave increased by about 3.5 km, leaving a section of about 800 m straight line between Pivka jama and Planinska jama unexplored [17]. The Pivka River reappears in Planinska jama, the last and northernmost cave of the PPCS. In this cave, the river flows within the Pivka channel for about 2 km, until it joins Rak River. Rak River sinks from the surface into the Tkalca jama in the east, whose channels are about 3 km long. Also in the Tkalca jama, one of the siphons of Rak River has stopped divers, leaving about 1 km of straight line distance to Planinska jama unexplored [18] (Figure 4). After the syphon, Rak River reappears in Planinska jama and flows for about 2 km through the Rak channel, and joins Pivka River. The new river, called Unica, is formed about 500 m away from the cave entrance, where it springs to Planinsko polje (Figures 1 and 4). The channels of the system along the Pivka River are vertically on two levels. The upper ones are older and dry, with many beautiful formations and speleothems (Figure 2), while the younger and lower ones called "Rov podzemne Pivke" (the channels of the subterranean Pivka River), were formed after the deepening of the riverbed.

Temperatures within the PPCS vary depending on the location of cave entrances and the distance from rivers sinks into the system. The temperature in the inner, isolated parts of Postojnska jama is about 8.5 °C, while in the parts closer to the entrances it varies mainly between 3 and 13 °C [19]. The temperature of Pivka River varies daily and seasonally, with the amplitude of fluctuations decreasing with distance from its sink [20]. Oxygen concentration in Pivka River varies similarly to surface conditions near the sink to Postojnska jama, while with distance from the sink the water becomes aerated and saturated with oxygen [20].

There are two main sources of organic material for subterranean communities. One is via the sinking rivers, which transport substantial amounts of particulate organic matter and, in the near-surface areas, also abundant plant and animal material that can become food sources. The transport capacity is strongly influenced by precipitation and is enhanced during high water levels [19]. The second source of organic matter is epikarst water, which drips from the ceiling and brings mainly diluted organic matter [21,22]. Another food source, albeit spatially limited, is bat guano and the remains of terrestrial accidentals coming into the PPCS. In general, the amount of organic matter in Pivka River decreases in the direction from Postojnska jama to Planinska jama [21].

2. Compiling the Data

We compiled the data of all troglobionts or troglobiotic populations found in PPCS (sensu Sket [23]). The latter are populations of surface species that form morphologically and ecologically distinct specialized cave populations in PPCS [24]. Non-troglobiotic species found in PPCS and dubious records were omitted from the checklist. The main source of information was published sources as well as material kept in Zoological collection of Department of Biology, Biotechnical Faculty, University of Ljubljana and Notranjska Museum Postojna. The information was collected in SubBioDB, the database on biodiversity of subterranean habitats managed in the SubBioLab at the University of Ljubljana. The data were revised and supplemented with unpublished records, either from reports or from personal communication with taxonomic authorities.

The list contains the species or the lowest taxonomic rank that could be identified. If the subspecies level was determined, we list it. We did not distinguish the level of subgenera. In some cases, taxonomic authorities specifically indicated that some individuals belonged to new species awaiting formal description. These records were added to the checklist as "*Genus* sp. n.".

We marked whether the PPCS presents a type locality for the taxon. Whenever data were available, we added information about the specific cave of the PPCS where the taxon was recorded.

We used the most current taxonomic nomenclature available for specific taxonomic groups. To list only valid species names, we consulted taxonomists or online databases maintained by specialists, namely World Register of Marine Species [25], Millibase [26], A World Catalogue of Centipedes (Chilopoda) [27], Pseudoscorpiones of the World [28], World Spider Catalogue [29], and Checklist of the Collembola [30]. To keep track of species names listed in previous checklists, we have retained the original species name in the list.

Because checklists may be outdated at the time they are published, we developed an online checklist of PPCS species (www.subbio.net/PPCS-checklist (accessed on 8 June 2021) that is fed from the SubBioDB database.

3. The Checklist of Taxa in PPCS

The checklist contains 116 species, of which 71 are aquatic and 45 terrestrial. They belong to eight phyla: 85 species of arthropods (45 crustaceans, 18 hexapods, 13 arachnids, and 9 myriapods), 12 molluscs, 11 annelids, 4 turbellarians, 1 sponge, 1 cnidarian, 1 vertebrate, and 1 protist (Table 1). Three species were found as parasitic on cave shrimp, the protist, and both thamnocephalids (Table 1). Nearly half, 47 species, have been scientifically described from PPCS (Table 1). In addition, two species remain to be identified at species level, while 12 are awaiting taxonomic description. Some of the most notable species are presented in Figure 5.

Table 1. The list of obligate subterranean (troglobiotic) species recorded in the Postojna-Planina Cave System in Slovenia. Species with type locality in PPCS are marked with [1], and troglobiotic populations of surface species are marked with *. The letter in the column A/T marks whether the species is A—aquatic or T—terrestrial. If the species was reported from a specific cave(s) of the system, the X is under the abbreviation: POJ—Postojnska jama, OJ—Otoška jama, MJ—Magdalena jama, ČJ—Črna jama, PIJ—Pivka jama, PLJ—Planinska jama, TKJ—Tkalca jama.

Taxonomic Group	Family	Species [Original Mentioning]	A/T	POJ	OJ	MJ	ČJ	PIJ	PLJ	TKJ	References
Protozoa-Ciliata											
Suctoria	Spelaeophryidae	[1] *Spelaeophrya troglocaridis* Stammer, 1935	A						X		[31]
Porifera											
Demospongia	Spongillidae	* *Ephydatia fluviatilis* (Linne, 1759)	A	X					X		[32,33]
Cnidaria											
Hydrozoa	Bougainvilliidae	[1] *Velkovrhia enigmatica* Matjašič & Sket, 1971	A						X		[13,34,35]
Turbellaria											
Temnocephalida	Scutariellidae	[1] *Bubalocerus pretneri* Matjašič, 1958	A						X		[13,36]
		[1] *Troglocaridicola capreolaria* Matjašič, 1958	A						X		[13,36]
Tricladida	Dendrocoelidae	*Dendrocoelum spelaeum* cf (Kenk, 1924)	A						X		Zoological Collection of Dept. Biology, University of Ljubljana
		[1] *Dendrocoelum tubuliferum* de Beauchamp, 1919	A	X			X				[11–13,37,38]
Mollusca											
Gastropoda	Acroloxidae	*Acroloxus tetensi* (Kuščer, 1932)	A	X					X		[12,13,39,40]
	Ancylidae	* *Ancylus fluviatilis* Mueller, 1774	A	X					X		[13,39]
	Carychiidae	*Zospeum kusceri* Wagner A.J., 1912	T				X	X			[41]
		Zospeum lautum (Frauenfeld, 1854)	T				X				[41]
		[1] *Zospeum spelaeum spelaeum* (Rossmässler, 1839) [*Carychium spleaeum; Z. alpestre rossmäsleri*]	T	X	X		X	X	X		[12,41–43]
	Hydrobiidae	* *Belgrandiella fontinalis* (Schmidt F.J., 1847)	A						X		[13,44]
		Belgrandiella schleschi Kuščer, 1932	A								[13]
		Belgrandiella kusceri (Wagner A.J., 1914)	A						X		[45]
		Hadziella ephippiostoma Kuščer, 1932	A								[13]
		Hauffenia michleri Kuščer, 1932	A						X		[13,45]
		Hauffenia subpiscinalis (Kuščer, 1932) [*Neohoratia subpiscinalis*]	A						X		[12,13,40]
		Iglica luxurians (Kuščer, 1932)	A						X		[12,13,34,45]

Table 1. *Cont.*

Taxonomic Group	Family	Species [Original Mentioning]	A/T	POJ	OJ	MJ	ČJ	PIJ	PLJ	TKJ	References
Annelida											
Hirudinea	Erpobdellidae	* *Dina krasensis* (Sket, 1968) [*Trocheta bykowskii krasense*]	A					X	X		[13,46]
	Glossiphoniidae	* *Glossiphonia complanata complanata* (Linne, 1758)	A						X		[13,46]
Oligochaeta	Lumbriculidae	*Trichodrilus ptujensis* Hrabe, 1963	A						X		[13]
		Trichodrilus strandi Hrabe, 1936	A	X				X	X		[13,47]
	Tubificidae	[1] *Epirodrilus slovenicus* Karaman Sp., 1976	A						X		[13,48,49]
		[1] *Potamothrix postojnae* Karaman Sp., 1974	A						X		[49,50]
		[1] *Psammoryctides hadzii* Karaman Sp., 1974	A						X		[49,50]
		[1] *Rhyacodrilus caudosetosus* Karaman Sp., 1983	A						X		[49,51]
		[1] *Rhyacodrilus maculatus* Karaman Sp., 1983	A						X		[49,51]
		[1] *Rhyacodrilus sketi* Karaman Sp., 1974	A						X		[49–51]
		Sketodrilus flabellisetosus (Hrabe, 1966)	A						X		[48,49]
Arthropoda-Arachnida											
Acarina	Eremaeidae	*Oppia cavatica* nomen nudum	T						X		[52]
	Eremaeidae	*Oppia malograjskiella* nomen nudum	T						X		[52]
	Labidostommidae	[1,*] *Labidostomma lyra* Willmann, 1932 [*Nicoletiella denticulata*]	T	X							[11–13,53]
Araneae	Dysderidae	*Mesostalita nocturna* (Roewer, 1931)	T	X							[12,42,54]
		[1] *Stalita taenaria* Schioedte, 1847	T	X	X	X	X		X		[11,12,42,54–60]
	Linyphiidae	[1] *Troglohyphantes excavatus* Fage, 1919 [*Troglohyphantes anellii*]	T	X							[60–62]
Palpigrada	Eukoeneniidae	*Eukoenenia austriaca austriaca* (Hansen, 1926)	T	X							[12,63–66]
Pseudoscorpiones	Chthoniidae	*Chthonius spelaeophilus* Hadži, 1930	T	X							[62]
		[1] *Chthonius cavernarum* Ellingsen, 1909	T	X							[11,12,67,68]
		Troglochthonius doratodactylus Helversen, 1968	T	X							[62]
	Neobisiidae	[1] *Neobisium pusillum* Beier, 1939	T				X				[12,13,69]
		[1] *Neobisium spelaeum spelaeum* (Schioedte, 1848)	T	X	X						[12,13,42,57,68]
		Roncus stussineri (Simon, 1881)	T								[12]
Arthropoda-Myriapoda											
Chilopoda	Lithobiidae	*Lithobius* sp.n.	T								[42]
		[1] *Lithobius stygius* Latzel, 1880	T	X	X				X		[12,42,70,71]
		[1] *Lithobius zveri* (Matić & Stentzer, 1977) [*Monotarsobius zveri*]	T						X		[72]

Table 1. *Cont.*

Taxonomic Group	Family	Species [Original Mentioning]	A/T	POJ	OJ	MJ	ČJ	PIJ	PLJ	TKJ	References
Diplopoda	Attemsiidae	[1] *Attemsia stygia* (Latzel, 1884) [*Attemsia stygium*]	T	X	X			X			[11–13,73–76]
	Haasiidae	[1] *Haasia troglodyta* (Latzel, 1884) [*Acherosoma troglodytes, Haasia troglodytes*]	T	X							[12,73,74]
	Iulidae	*Typhloiulus illirycus* (Verhoeff, 1929)	T	X	X						[11,62,76]
	Polydesmidae	[1,*] *Brachydesmus subterraneus* Heller, 1857	T	X			X	X			[11,12,77]
Symphyla	Scolopendrellidae	[1] *Scolopendreilopsis pretneri* Juberthie-Jupeau, 1963	T	X							[62,78]
	Scutigerellidae	[1] *Scutigerella hauserae* Scheller, 1990	T	X							[62,79]
Arthropoda-Crustacea Cladocera	Chydoridae	* *Chydorus sphaericus* (O.F.Mueller, 1776)	A								[80,81]
Copepoda -Calanoida	Diaptomidae	*Troglodiaptomus sketi* Petkovski, 1978 [*Troglodiaptomus sketi postojnae*]	A	X					X		[82]
Copeopoda-Cyclopoida	Cyclopidae	*Acanthocyclops kieferi* (Chappuis, 1925)	A	X					X		[82,83]
		Acanthocyclops venustus venustus (Norman & Scott, 1906) [*Acanthocyclops venustus stammeri*]	A	X					X		[82,84]
		[1] *Diacyclops charon* (Kiefer, 1931) [*Cyclops charon*]	A	X				X	X		[12,13,82,84–86]
		Diacyclops languidoides (Lilljeborg, 1901) [*Diacyclops languidoides goticus*]	A	X							[12,13,82,83]
		Diacyclops slovenicus Petkovski, 1954	A	X					X		[82,84]
		[1] *Metacyclops postojnae* Brancelj, 1987	A	X							[82,84]
		Speocyclops infernus (Kiefer, 1930)	A	X			X	X	X		[13,82–84]
Copepoda-Harpacticoida	Ameiridae	*Nitocrella* n.sp.	A	X							[83]
	Canthocamptidae	*Bryocamptus balcanicus* (Kiefer, 1933)	A	X			X	X	X		[83,87]
		Bryocamptus n.sp.	A					X			[83]
		Bryocamptus pyrenaicus (Chappuis, 1923)	A	X				X	X		[83,84,87]
		Bryocamptus typhlops (Mrazek, 1893)	A	X				X	X		[83,87]
		Bryocamptus zschokkei caucasicus Borutzky, 1960	A	X				X	X		[13,87]
		[1] *Elaphoidella cvetkae* Petkovski, 1983	A	X			X	X	X		[83,84,87,88]
		Elaphoidella elaphoides (Chappuis, 1924)	A					X	X		[87]
		[1] *Elaphoidella franci* Petkovski, 1983	A						X		[84,87,88]
		[1] *Elaphoidella jeanneli* (Chappuis, 1928)	A	X			X		X		[12,13,84,87,89]
		Elaphoidella stammeri Chappuis, 1936	A	X					X		[84,87]

Table 1. *Cont.*

Taxonomic Group	Family	Species [Original Mentioning]	A/T	POJ	OJ	MJ	ČJ	PIJ	PLJ	TKJ	References
		Maraenobiotus cf. *brucei* (Richard, 1898)	A					X			[83]
		Moraria n.sp.	A				X				[83]
		Morariopsis scotenophila (Kiefer, 1930)	A	X			X				[83,84,87]
		Pilocamptus pilosus (Douwe, 1910) [*Echinocamptus georgevitchi E. pilosus, E. unicus*]	A	X			X	X	X		[12,13,84,87]
		Stygepactophanes sp.n.	A				X				[83]
	Parastenocarididae	*Horstkurtcaris nolli alpina* (Kiefer, 1960) [*Parastenocaris nolli*]	A				X	X			[83]
		Parastenocaris sp.n. 1	A				X				[83]
		Parastenocaris sp.n. 2	A	X			X	X			[83]
Ostracoda - Podocopida	Candonidae	[1] *Typhlocypris trigonella* (Klie, 1931) [*Candona trigonella*]	A	X							[11–13,90,91]
		Typhlocypris schmeili nomen nudum	A				X				[11–13,92]
Decapoda	Atyidae	[1] *Troglocaris planinensis* Birstein, 1948 [*Troglocaris anophthalmus*]	A	X			X	X	X	X	[12,18,34,93–95]
Amphipoda	Crangonyctidae	* *Synurella ambulans* Mueller, 1846 [*Synurella jugoslavica*]	A						X		[13,34]
	Niphargidae	*Niphargus dobati* Sket 1999 [*Niphargus aquilex*]	A						X		[96,97]
		Niphargus orcinus Joseph, 1869	A								[13,97]
		[1] *Niphargus orophobata* (Sket, 1981) [*Niphargobates orophobata*]	A						X		[97,98]
		[1] *Niphargus spoeckeri* Schellenberg, 1933	A				X	X	X		[12,13,34,62,97]
		[1] *Niphargus stygius* (Schioedte, 1847)	A	X			X		X		[12,13,34,57,97]
		Niphargus sp. *stygius*-complex (Schioedte, 1847)	A	X			X		X		[12,34,57,97]
		[1] *Niphargus wolfi* Schellenberg, 1933	A	X					X		[12,13,34,97]
Isopoca	Asellidae	* *Asellus aquaticus aquaticus* (Linne, 1761)	A	X			X	X	X		[11–13,99–103]
		[1,*] *Asellus aquaticus cavernicolus* Racovitza, 1925	A	X			X	X	X		[11–13,101,102, 104,105]
		Proasellus istrianus (Stammer, 1932)	A					X			[12,13]
	Sphaeromatidae	*Monolistra racovitzai racovitzai* Strouhal, 1928	A	X					X		[106,107]
	Trichoniscidae	*Androniscus stygius* (Nemec, 1897) [*Andronicus cavernarum tschammeri*]	T	X		X	X		X		[12,108–111]
		[1] *Titanethes albus* (Koch C., 1841)	T	X	X	X	X	X	X		[12,13,42,108, 110]

109

Table 1. *Cont.*

Taxonomic Group	Family	Species [Original Mentioning]	A/T	POJ	OJ	MJ	ČJ	PIJ	PLJ	TKJ	References
Arthropoda-Hexapoda											
Collembola	Arrhopalitidae	[1] *Pygmarrhopalites postumicus* (Stach, 1945) [*Arrhopalites postumicus*]	T	X							[62,112]
	Paronellidae	*Troglopedetes pallidus* Absolon, 1907	T	X	X						[12,42,62]
	Neelidae	*Neelus* n.sp.	T	X							Lukić M., pers. comm. (2021)
	Oncopoduridae	[1] *Oncopodura cavernarum* Stach 1934	T	X							[12,13,62,113]
	Onychiuridae	*Absolonia gigantea* (Absolon, 1901) [*Onychiurus giganteus*]	T				X				[11–13,62,112–114]
		[1] *Onychiuroides postumicus* Bonet, 1931 [*Onychiurus postumicus*]	T				X				[11–13,112–114]
		[1] *Onychiurus boldorii* Denis, 1938	T	X							[12,13,114]
	Tomoceridae	*Tritomurus scutellatus* (Frauenfeld, 1854)	T	X					X		[62,114]
Diplura	Campodeidae	*Plusiocampa nivea* (Joseph, 1882) [*Campodea nivea, Plusiocampa erebophila*]	T	X							[12,13,42,115]
Coleoptera	Carabidae	*Anophthalmus amplus sedulus* (Knirsch, 1926) [*Anophthalmus pubens/pubescens*]	T						X		[116]
		Anophthalmus schmidtii schmidtii Sturm, 1844	T	X			X		X		[12,117]
	Leiodidae	[1] *Anophthalmus severi confusus* (J. Müller, 1935) [*Anophthalmus hirtus confusus*]	T			X	X				[12,116]
		Aphaobius milleri Schmidt, 1855	T				X				[12,118]
		Bathyscimorphus sagarum Bognolo, 2002	T						X		[118]
		[1] *Bathyscimorphus byssinus byssinus* (Schiödte, 1848)	T	X	X		X				[12,42,118,119]
		[1] *Bathysciotes khevenhuelleri khevenhuelleri* (L. Miller, 1852)	T	X	X	X			X		[12,42]
		[1] *Leptodirus hochenwartii hochenwartii* Schmidt, 1832	T	X	X	X	X		X		[9,12,42,56,119]
	Staphylinidae	[1] *Machaerites ravasinii* (J. Müller, 1922)	T	X					X		[12,120–123]
Vertebrata Amphibia-Urodela	Proteidae	*Proteus anguinus* Laurenti, 1768	A	X		X	X	X	X	X	[11–13,18,124]

Figure 5. The famous eight: (**a**) proteus or the olm *Proteus anguinus* Laurenti, 1768; (**b**) slenderneck beetle *Leptodirus hochenwartii hochenwartii* Schmidt, 1832; (**c**) giant cave pseudoscorpion *Neobisium spelaeum spelaeum* (Schiødte, 1848); (**d**) Postojna cave herald snail *Zospeum spelaeum spelaeum* (Rossmaessler, 1839); (**e**) symphylan *Scutigerella hauserae* Scheller, 1990; (**f**) cave hydrozoan *Velkovrhia enigmatica* Matjašič & Sket, 1971; (**g**) giant cave amphipod *Niphargus orcinus* Joseph, 1869; (**h**) Planina cave shrimp *Troglocaris planinensis* Birstein, 1948. Photo credits: (**a**–**e**)—Slavko Polak, (**f**)—Rodrigo Lopes Ferreira, (**g**)—Teo Delić, (**h**)—Rollin Verlinde.

Not all species were reported from all caves of the PPCS. The highest number was reported from Planinska jama (66 species) and Postojnska jama (64 species). From Črna jama, Pivka jama, Otoška jama Magdalena jama, and Tkalca jama, 35, 26, 11, 7, and 2 species were reported, respectively.

4. Importance of Updated Checklist

4.1. General Comments

The revised checklist revealed the presence of at least 116 troglobiotic species in PPCS, increasing the number by nearly 40 species since the last publication [8]. Not all species were reported from all caves of the PPCS. Due to the connectivity between the different parts of the system, the current differences in the number of species mainly reflect the differences in sampling, but also their positions in relation to the river sinks and surface influences. Although PPCS has the longest history of biological exploration, and the highest troglobiotic species richness in the world, new species are still expected. This is no surprise considering that PPCS is located in the heart of a region that has been consistently recognized as a global hotspot of subterranean biodiversity, where sampling is still not complete [125]. This supports the view that further studies in the system should be encouraged as they may reveal additional species for a variety of reasons: sampling of new microhabitats or more thorough sampling of less studied cave channels, use of new sampling techniques, study of currently less studied taxa [126], identification changes due to continued taxonomic activity, or introduction of molecular methods to identify cryptic species and their co-occurrence [127–129].

4.2. PPCS as a Model System for Key Biological Questions

PPCS retains its appeal for biological research, with organisms that have the potential to answer key questions in speleobiology [130]. One research direction could address cave colonization in the context of adaptation and speciation processes from surface ancestor to subterranean descendant [131]. In the PPCS, there are at least four species with extant surface ancestral populations, which provide an exceptional model system for studying these ecological and evolutionary processes during the transition from surface to subterranean environments [131–133].

The next potential of PPCS is to study the dynamics of community structure and interspecific interactions [97,134], including distances to the entrances and/or the surface. Since the main parts of PPCS are organized along the Pivka River, this gives the opportunity to observe gradual changes in surface influences downstream. An open question is how surface and subterranean species exchange and interact along this gradient, and whether the aquatic community follows this gradient. Although it has been shown that the dynamics of the relationship between surface and subterranean taxa changes from the sink of Pivka River to the deeper parts of the cave [20], their interactions are still poorly understood. Sket [20] suggested that the degree of eutrophication affects the interactions between subterranean species and accidentals and that eutrophication may favor the competitive strength of surface invaders, a hypothesis indirectly supported in other studies [135,136].

The third attractive direction is ecosystem-level oriented research on nutrient cycling, energy budget, and its top-down effects on interspecific interactions and community structure. Early studies suggest a complex pattern of organic input, likely due to the multiple windows through which PPCS communicates with the surface [21,22]. Such additional inputs in the system can affect the gradual changes of organic matter downstream. Even though more complicated, these inputs may present replicas of starting points of organic matter input and possibilities for repeated studies of changes downstream from the sources, which present an attractive venue for research.

4.3. Challenges in Cave Management and Conservation

PPCS is the system with one of the longest tourist uses in the world. Management of the PPCS must balance protection of high species richness with potential tourism pressure and other surface threats. Sustainable management is a serious challenge in show caves with such high visitor numbers as the PPCS, where Postojnska jama alone receives up to 500,000 visitors per year. Direct consequences are microclimatic changes, and also the introduction of artificial light, which promotes the growth of the so-called "lampenflora" as well as airborne bacteria of anthropogenic origin [137] with unknown effects on biotas. The negative impact of tourist use on the spatial arrangement of animals has been demonstrated, as they move to more remote dark/less disturbed passages [12,42].

A greater and less controlled threat to the PPCS comes from the surface, through agriculture and overuse of fertilizers and pesticides, and pollution from industry [138]. The Pivka River transports pollutants deep into the cave system, which is especially critical during high flow that tends to homogenize chemical and bacterial parameters throughout the river system [139]. Their impact on troglobiotic fauna is not studied and monitored, not even for the most charismatic species, the olm (*Proteus anguinus*).

In order to detect any changes in the PPCS, regular monitoring of abiotic parameters and its inhabitants should be carried out. Chemical and physical parameters of percolating water and allogenic enrichment have been monitored for decades by experts from the

Institute for Karst Research in Postojna [140]. Permanent monitoring of cave air temperature, humidity, wind flow, and CO_2 to determine human impact on the natural cave environment started in 2007. So far, only terrestrial fauna in the tourist part of Postojnska jama has been monitored since 2009. Monitoring should be extended to the entire PPCS and include aquatic fauna as well. This is especially important, considering that both *L. hochenwartii* and the olm are listed in the European Habitat's directive as species of special conservation concern, whose habitat must be protected and whose populations must be adequately monitored [141]. New methods, such as the use of DNA barcoding system for species identification [142], or protocols for metabarcoding [143] and e-DNA technologies [144,145], offer new opportunities for monitoring.

4.4. PPCS Outreach and Public Awareness

Public opinion can strengthen the long-term protection of a cave or, more generally, of subterranean habitats and their biotas. For the vast majority, tourist caves such as Postojnska jama are the only unique opportunity to personally experience the underground world. Such visits, associated with emotion, are an exceptional, albeit somewhat controversial, opportunity to engage visitors and inform them about the fragility and importance of subterranean habitats and their inhabitants, as well as about conservation issues. By restricting tourist use to a limited part of the PPCS, the rest of the system can be safeguarded from such visits, while benefitting from the personal experience gained by the visitors. This is an important prerequisite to affect their attitude toward conservation and positive view on the protection of the whole PPCS and subterranean biodiversity in general.

The fact that the PPCS is a global hotspot of subterranean biodiversity is an important opportunity to promote and present the uniqueness of subterranean environments. Steps in this direction have been made by establishing a vivarium near the entrance of Postojnska jama, an internal aquarium with olms inside the cave and two permanent exhibitions: the interactive exhibition Expo at Postojnska jama and a special speleobiological exhibition Karst Museum at Notranjska Museum Postojna.

Author Contributions: Conceptualization, M.Z., S.P. and C.F.; Data curation, M.Z., S.P. and C.F.; Formal analysis, M.Z., S.P. and C.F.; Methodology, M.Z., S.P. and C.F.; Visualization, M.Z. and S.P.; Writing–original draft, M.Z., S.P. and C.F.; Writing—review & editing, M.Z., S.P. and C.F. All authors contributed equally to the conception of the paper, data gathering and writing. All authors have read and agreed to the published version of the manuscript.

Funding: The first and last authors were funded by Slovenian Research Agency through core funding Programme P1-0184, and by the project LIFE NarcIS – Nature Conservation Information System (LIFE19 GIE/SI/000161, co-financed by EU LIFE Programme).

Institutional Review Board Statement: Not applicable.

Informed Consent Statement: Not applicable.

Data Availability Statement: All the retrieved data occur in Table 1, and on www.subbio.net/PPCS-checklist (accessed on 8 June 2021).

Acknowledgments: We thank the editors for inviting us to contribute to this Special Issue. We thank Marko Lukić (Croatian Biospeleological Society) who allowed the use of some unpublished data on cave Collembola. Teo Delić, Rodrigo Lopes Ferreira, Rollin Verlinde and Peter Gedei contributed the photographs. We are very grateful to Uroš Stepišnik (University of Ljubljana) and Matej Blatnik (Karst Research Institute at Research Centre of the Slovenian Academy of Sciences and Arts), for providing the digital maps of caves of PPCS. We thank the members of SubBioLab who contribute the data in SubBioDB via their ongoing studies. We are grateful to former SubBioLab members Simona Prevorčnik and Boris Sket, for their involvement in the early stages of list compilation.

Conflicts of Interest: The authors declare no conflict of interest.

References

1. Ficetola, G.F.; Canedoli, C.; Stoch, F. The Racovitzan impediment and the hidden biodiversity of unexplored environments. *Conserv. Biol.* **2019**, *33*, 214–216. [CrossRef]
2. Pipan, T.; Deharveng, L.; Culver, D.C. Hotspots of subterranean biodiversity. *Diversity* **2020**, *12*, 209. [CrossRef]
3. Zagmajster, M.; Eme, D.; Fišer, C.; Galassi, D.; Marmonier, P.; Stoch, F.; Cornu, J.-F.; Malard, F. Geographic variation in range size and beta diversity of groundwater crustaceans: Insights from habitats with low thermal seasonality. *Glob. Ecol. Biogeogr.* **2014**, *23*, 1135–1145. [CrossRef]
4. Eme, D.; Zagmajster, M.; Delić, T.; Fišer, C.; Flot, J.F.; Konecny-Dupre, L.; Palsson, S.; Stoch, F.; Zakšek, V.; Douady, C.J.; et al. Do cryptic species matter in macroecology? Sequencing European groundwater crustaceans yields smaller ranges but does not challenge biodiversity determinants. *Ecography* **2017**, *41*, 1–13. [CrossRef]
5. Culver, D.C.; Deharveng, L.; Bedos, A.; Lewis, J.J.; Madden, M.; Reddell, J.R.; Sket, B.; Trontelj, P.; White, D. The mid-latitude biodiversity ridge in terrestrial cave fauna. *Ecography* **2006**, *29*, 120–128. [CrossRef]
6. Bregović, P.; Fišer, C.; Zagmajster, M. Contribution of rare and common species to subterranean species richness patterns. *Ecol. Evol.* **2019**, *9*, 11606–11618. [CrossRef]
7. Zagmajster, M.; Culver, D.C.; Sket, B. Species richness patterns of obligate subterranean beetles (Insecta: Coleoptera) in a global biodiversity hotspot—Effect of scale and sampling intensity. *Divers. Distrib.* **2008**, *14*, 95–105. [CrossRef]
8. Culver, D.C.; Sket, B. Hotspots of subterranean biodiversity in caves and wells. *J. Cave Karst Stud.* **2000**, *62*, 11–17.
9. Schmidt, F. Beitrag zu Krain's fauna. *Leptodirus hochenwartii. Illyrisches Blatt* **1832**, *3*, 9–10.
10. Polak, S. Importance of discovery of the first cave beetle *Leptodirus hochenwartii* Schmidt, 1832. *Endins* **2005**, *28*, 71–80.
11. Wolf, B. *Animalium Cavernarum Catalogus*; Verlag für Naturwissenschafren: Berlin, Germany, 1934.
12. Pretner, E. Živalstvo Postojnske jame. In *150 Let Postojnske Jame 1818–1968*; Zavod Postojnske Jame: Postojna, Slovenia, 1968; pp. 59–78.
13. Sket, B. Jamska favna Notranjskega trikotnika (Cerknica—Postojna—Planina), njena ogroženost in naravovarstveni pomen. *Varst. Narave* **1979**, *13*, 45–59.
14. Gospodarič, R. The quaternary caves developement between the Pivška kotlina (Pivka basin) and Planinsko polje (polje of Planina). *Acta Carsologica* **1964**, *7*, 5–139.
15. Šebela, S. *Tectonic Structure of Postojnska Jama Cave System*; Založba ZRC: Ljubljana, Slovenia, 1998.
16. Perko, G. *Die Adelsberger Grotte in Wort und Bild*; Internationales Museum für Höhlenkunde: Postojna, Slovenia, 1910.
17. Cvjetović, S. Potapljače so Med Premagovanjem Četrtega Sifona Spremljale Kolonije Človeških Ribic. *Novice*, 14 July 2015. Available online: https://siol.net/novice/slovenija/potapljace-so-med-premagovanjem-cetrtega-sifona-spremljale-kolonije-cloveskih-ribic-199376(accessed on 31 March 2021).
18. Ilić, U. Tkalca Jama, Potop, 27.7.2013. Available online: https://www.jd-rakek.com/index.php/novice/akcije/101-tkalca-jama-potop-27-7-2013 (accessed on 31 March 2021).
19. Pipan, T.; Petrič, M.; Šebela, S.; Culver, D.C. Analyzing climate change and surface-subsurface interactions using the Postojna Planina Cave System (Slovenia) as a model system. *Reg. Environ. Chang.* **2019**, *19*, 379–389. [CrossRef]
20. Sket, B. Predhodno poročilo o ekoloških raziskavah v sistemu kraške Ljubljanice. *Biološki Vestn.* **1970**, *18*, 79–87.
21. Simon, K.S.; Pipan, T.; Culver, D.C. A conceptual model of the flow and distribution of organic carbon in caves. *J. Cave Karst Stud.* **2007**, *69*, 1–6.
22. Simon, K.S.; Pipan, T.; Ohno, T.; Culver, D.C. Spatial and temporal patterns in abundance and character of dissolved organic matter in two karst aquifers. *Fundam. Appl. Limnol.* **2010**, *177*, 81–92. [CrossRef]
23. Sket, B. Can we agree on an ecological classification of subterranean animals? *J. Nat. Hist.* **2008**, *42*, 1549–1563. [CrossRef]
24. Konec, M.; Prevorčnik, S.; Sarbu, S.M.; Verovnik, R.; Trontelj, P. Parallels between two geographically and ecologically disparate cave invasions by the same species, *Asellus aquaticus* (Isopoda, Crustacea). *J. Evol. Biol.* **2015**, *28*, 864–875. [CrossRef]
25. WoRMS Editorial Board. World Register of Marine Species. 2021. Available online: http://www.marinespecies.org (accessed on 19 March 2021).
26. Sierwald, P.; Spelda, J. MilliBase. Available online: http://www.millibase.org (accessed on 19 February 2021).
27. Bonato, L.; Chagas, A.J.; Edgecombe, G.D.; Lewis, J.G.E.; Minelli, A.; Pereira, L.A.; Shelley, R.M.; Stoev, P.; Zapparoli, M. ChiloBase 2.0—A World Catalogue of Centipedes (Chilopoda). Available online: https://chilobase.biologia.unipd.it/ (accessed on 1 March 2021).
28. Harvey, M.S. Pseudoscorpions of the World, Version 2.0. Available online: http://www.museum.wa.gov.au/catalogues/pseudoscorpions (accessed on 1 March 2021).
29. World Spider Catalog. 2021. Available online: https://wsc.nmbe.ch/ (accessed on 19 March 2021).
30. Janssen, F. Checklist of the Collembola. Available online: https://www.collembola.org/ (accessed on 19 March 2021).
31. Matjašič, J. Opazovanja na jamskem suktoriju *Spelaeophrya troglocaridis* Stammer. *Biološki Vestn.* **1956**, *5*, 71–73.
32. Velikonja, M. *Taksonomija in Sestava Favne Sladkovodnih Spužev (Porifera: Spongillidae) v Jugoslaviji s Posebnim Ozirom na Jamske in Ohridske Taksone)*; Inštitut za Biologijo in VZOZD za Biologijo: Ljubljana, Slovenia, 1990.
33. Sket, B.; Velikonja, M. Troglobitic freshwater sponges (Porifera, Spongillidae) found in Yugoslavia. *Stygologia* **1986**, *2*, 254–266.
34. Matjašič, J.; Sket, B. Jamski hidroid s slovenskega krasa. *Biološki Vestn.* **1971**, *19*, 139–145.

35. Velikonja, M. Contribution to the knowledge of the biology of the yugoslav endemic cave hydroid *Velkovrhia enigmatica* Matjašič & Sket, 1971. In Proceedings of the 9th Congreso Internacional de Espeleologia, Barcelona, Spain, 9–15 August 1986; pp. 123–125.
36. Matjašič, J. Vorläufige Mitteilungen über europäische Temnocephalen. *Biološki Vestn.* **1958**, *7*, 60–65.
37. De Beauchamp, P. Diagnoses preliminaires de Triclades obscuricoles. *Bull. Soc. Zool. Fr.* **1919**, *44*, 243–251.
38. Kenk, R. Sladkovodni trikladi iz jam severozahodnega dela Dinarskega krasa. *Prirodosl. Razpr.* **1936**, *3*, 1–29.
39. Bole, J. Rodova *Ancylus* O. F. Muell in *Acroloxus* Beck (Gastropoda, Basommatophora) v podzemeljskih vodah Jugoslavije. *Razprave-Dissertationes* **1965**, *8*, 155–175.
40. Bole, J. Recentni podzemeljski polži in razvoj nekaterih porečij na dinarskem krasu. *Razprave-Dissertationes* **1985**, *24*, 315–328.
41. Bole, J. Rod *Zospeum* Bourguignat 1856 (Gastropoda, Ellobiidae) v Jugoslaviji. *Razprave-Dissertationes* **1974**, *17*, 249–291.
42. Ramšak, L. Tourist Impact on Terrestiral Fauna Postojna and Otok Cave. Graduation Thesis, University of Ljubljana, Ljubljana, Slovenia, 2007.
43. Rossmässler, A.E. *Iconographie der Land- und Süßwasser-Mollusken, Mit Vorzügliecher Berücksichtingung der Europäischen Noch Nicht Abgebildeten Arten*; Arnoldische Buchhandlung: Dresden/Leipzig, Germany, 1839.
44. Bole, J.; Velkovrh, F. Mollusca from continental subterranean aquatic habitats. In *Stygofauna Mundi; A Faunistic, Distributional and Ecological Synthesis of the World Fauna inhabiting Subterranean Waters (including the Marine Interstitial)*; Bachuys: Leiden, Germany, 1986; pp. 177–207.
45. Kuščer, L. Höhlen- und Quellenschnecken aus dem Flussgebiet Ljubljanica. *Arch. Molluskenkd.* **1932**, *64*, 48–62.
46. Sket, B. K poznavanju favne pijavk (Hirudinea) v Jugoslaviji. *Razprave-Dissertationes* **1968**, *11*, 129–197.
47. Karaman, S. *Trichodrilus strandi* (Oligochaeta, Lumbriculidae), a new element in the fauna of Yugoslavia. *Biološki Vestn.* **1987**, *35*, 27–30.
48. Karaman, S. The second contribution to the knowledge of the freshwater Oligochaeta of Slovenia. *Biološki Vestn.* **1976**, *24*, 201–207.
49. Giani, N.; Sambugar, B.; Martínez-Ansemil, E.; Martin, P.; Schmelz, R.M. The groundwater oligochaetes (Annelida, Clitellata) of Slovenia. *Subterr. Biol.* **2011**, *9*, 85–102. [CrossRef]
50. Karaman, S. Beitrag zur Kenntnis der Süsswasseroligochaeten Sloveniens. *Biološki Vestn.* **1974**, *22*, 223–228.
51. Karaman, S. Third contribution to the knowledge of the freshwater Oligochaeta of Slovenia. *Biološki Vestn.* **1983**, *31*, 29–36.
52. Tarman, K. Prispevek k poznavanju oribatidne favne Slovenije II. *Biološki Vestn.* **1958**, *6*, 80–91.
53. Willman, C. Acari aus sudostalpinen Hohlen. Mitteilungen uber Hohlen- und Karstforsch. *Zeitschrift Hauptverbandes Dtsch. Hohlenforscher* **1932**, *4*, 158–161.
54. Caporiacco, L. Aracnidi cavernicoli e lucifugi di Postumia. *Grotte d'Italia* **1937**, *2*, 1–8.
55. Nikolić, F. Pauci iz nekih pećina Slovenije. In Proceedings of the Treći Jugoslavenski Speleološki Kongres, Sarajevo, Yugoslavia, 21–27 June 1962; Speleološki Savez Jugoslavije: Sarajevo, Yugoslavia, 1963; pp. 157–167.
56. Pretner, E. Zgodovinski pregled koleopteroloških raziskovanj v jamah Slovenije. *Acta Carsologica* **1974**, *6*, 307–316.
57. Schioedte, C.J. Undersögesler over den underjordiske Fauna i Hulerne i Krain og Istrien. Overs. over det K. danske Vidensk. *Selsk. Forh. og dets Medl. Arbeier* **1847**, *6*, 75–81.
58. Kratochvil, J. Liste générale des araignées cavernicoles en Yougoslavie. *Prirodosl. Razpr.* **1934**, *2*, 165–226.
59. Kratochvil, J. Cavernicole Dysderae. Prirodoved. *Pr. Ust. Československe Akad. ved v Brne* **1970**, *4*, 3–68.
60. Brignoli, P.M. Contributo alla conoscenza dei ragni cavernicoli della Jugoslavia (Araneae). *Fragm. Entomol.* **1971**, *7*, 103–119.
61. Deeleman-Reinhold, C.L. Revision of the cave-dwelling and related spiders of the genus *Troglohyphantes* Joseph (Linyphiidae), with special reference to the Yugoslav species. *Razprave-Dissertationes* **1978**, *23*, 1–221.
62. Polak, S. *Biološki Monitoring Postojnskega Jamskega Sistema in Predjamskega Jamskega Sistema—Končno Poročilo (Final Project Report)*; Notranjski muzej Postojna: Postojna, Slovenia, 2014.
63. Conde, B. Acquisitions recentes chez les palpigrades. *Mémoires Biospéologie* **1985**, *13*, 33–35.
64. Conde, B.; Neuherz, H. Palpigrades de la grotte de Raudner, pres de Stiwoll (Kat. Nr. 2783/04) dans le paleozoique de Graz, Styrie, Autriche. *Rev. Suisse Zool.* **1977**, *84*, 799–806. [CrossRef]
65. Conde, B. Palpigrades (Arachnida) d'Europe, des Antilles, du Paraguay et de Thailande. *Rev. Suisse Zool.* **1984**, *91*, 369–391. [CrossRef]
66. Zagmajster, M.; Kováč, Ľ. Distribution of palpigrades (Arachnida, Palpigradi) in Slovenia with a new record of *Eukoenenia austriaca* (Hansen, 1926). *Nat. Slov.* **2006**, *8*, 23–31.
67. Hadži, J. Contribution à la connaissance des Pseudoscorpiones cavernicoles. *Glas. Acad. Serbe.* **1930**, *140*, 1–36.
68. Ćurčić, B. *Catalogus faunae Jugoslaviae III/4. Arachnoidea, Pseudoscorpiones*; ZRC SAZU: Ljubljana, Slovenia, 1974.
69. Beier, M. Die Höhlenpseudoscorpione der Balkanhalbinsel. Eine auf dem Material der "Biospeologica balcanica" basierende Synopsis. Stud. Gebiet. allgem. Karstforsch.wiss. Höhlenk. Eiszeitforsch. *Nachbargebiet. Biol. Ser.* **1939**, *1*, 1–83.
70. Latzel, R. *Die Myriopoden der Österreichisch-Ungarischen Monarchie. Erste Hälfte: Die Chilopoden*; Alfred Hölder: Vienna, Austria, 1880.
71. Matić, Z.; Darabantu, C. Contributions a la connaissance des Chilopodes de Yougoslavie. *Razprave-Dissertationes* **1968**, *9*, 199–227.
72. Matić, Z.; Stentzer, I. Beitrag zur Kenntnis der Hundertfüssler (Chilopoda) aus Slowenien. *Biološki Vestn.* **1977**, *25*, 55–62.
73. Latzel, R. *Die Myriapoden der Österreichisch-Ungarischen Monarchie. Zweite Hälfte: Die Symphylen, Pauropoden und Diplopoden, Nebst Bemerkungen über Exotische und Fossile*; Alfred Holder: Vienna, Austria, 1884.
74. Strasser, K. *Catalogus faunae Jugoslaviae III/4. Diplopoda*; ZRC SAZU: Ljubljana, Slovenia, 1971.

75. Mršić, N. Attemsiidae (Diplopoda) of Yugoslavia. *Razprave-Dissertationes* **1987**, *27*, 101–168.
76. Strasser, K. Die Diplopoden Sloweniens. *Acta Carsologica* **1966**, *4*, 159–220.
77. Heller, K. Beiträge zur österreichischen Grotten -Fauna. Sitzungsberichte der Akad. der Wissenschaften Math. *Klasse* **1858**, *26*, 313–326.
78. Juberthie-Jupeau, L. Description d'une espece nouvelle de Symphyle recolte dans la grotte de Postojna (Yugoslavie). *Ann. Speleol.* **1963**, *18*, 299–304.
79. Scheller, U. The Pauropoda and Symphyla of the Geneva Museum IX. Symphyla from middle end south Europe, Turkey and Morocco (Myriapoda, Symphyla). *Rev. Suisse Zool.* **1990**, *97*, 411–425. [CrossRef]
80. Dumont, H.J.; Negrea, S. A conspectus of the Cladocera of the subterranean waters of the world. *Hydrobiologia* **1996**, *325*, 1–30. [CrossRef]
81. Brancelj, A. *Alona hercegovinae* n. sp. (Cladocera: Chydoridae), a blind cave-inhabiting Cladoceran from Hercegovina (Yugoslavia). *Hydrobiologia* **1990**, *199*, 7–16. [CrossRef]
82. Brancelj, A. Cyclopoida and Calanoida (Crustacea, Copepoda) from the Postojna-Planina cave system (Slovenia). *Biološki Vestn.* **1987**, *35*, 1–16.
83. Pipan, T.; Anton, B. Distribution patterns of copepods (crustacea: Copepoda) in percolation water. *Zool. Stud.* **2004**, *43*, 206–210.
84. Brancelj, A. Male of *Moraria radovnae* Brancelj, 1988 (Copepoda: Crustacea), and notes on endemic and rare copepod species from Slovenia and neighbouring countries. *Hydrobiologia* **2001**, *453*, 513–524. [CrossRef]
85. Brancelj, A. Microdistribution and high diversity of Copepoda (Crustacea) in a small cave in central Slovenia. *Hydrobiologia* **2002**, *477*, 59–72. [CrossRef]
86. Kiefer, F. Kurze Diagnosen neuer Süsswasser-Copepoden. *Zool. Anz.* **1931**, *94*, 219–224.
87. Brancelj, A. Rare and lesser known Harpacticoids (Copepoda Harpacticoida) from the Postojna-Planina cave system (Slovenia). *Biološki Vestn.* **1986**, *34*, 13–36.
88. Petkovski, T.K. Neue höhlenbewonende Harpacticoida (Crustacea, Copepoda) aus Slovenien. *Acta Mus. Maced. Sci. Nat.* **1983**, *16*, 177–205.
89. Chappuis, P.A. Nouveaux Copépodes cavernicoles. *Bul. Soc. St. Cluj.* **1928**, *4*, 20–34.
90. Klie, W. Zwei neue Arten der ostracoden-Gattung *Candona* aus unterirdischen Gewassern im sudostlichen Europa. *Zool. Anzeige* **1931**, *96*, 161–169.
91. Griffiths, H.I.; Brancelj, A. Preliminary list of freshwater Ostracoda (Crustacea) from Slovenia. *Ann. Istrian Mediterr. Stud.* **1996**, *9*, 201–210.
92. Mori, N.; Meisch, C. Contribution to the knowledge on the distribution of recent free-living freshwater ostracods (Podocopida, Ostracoda, Crustacea) in Slovenia. *Nat. Slov.* **2012**, *14*, 5–22.
93. Birstein, J.A. Nahoždenie peščernoj krevetki v gruntovyh vodah Macesty i svjazannye s etim voprosy. *Byull. Mosk. Obs. Isp. Prir. Otd. Biol.* **1948**, *53*, 3–10.
94. Zakšek, V.; Sket, B.; Gottstein, S.; Franjević, D.; Trontelj, P. The limits of cryptic diversity in groundwater: Phylogeography of the cave shrimp *Troglocaris anophthalmus* (Crustacea: Decapoda: Atyidae). *Mol. Ecol.* **2009**, *18*, 931–946. [CrossRef] [PubMed]
95. Jugovic, J.; Jalžić, B.; Prevorčnik, S.; Sket, B. Cave shrimps *Troglocaris* s. str. (Dormitzer, 1853), taxonomic revision and description of new taxa after phylogenetic and morphometric studies. *Zootaxa* **2012**, *3421*, 1–31. [CrossRef]
96. Sket, B. Niphargus aquilex dobati ssp. n. (Crustacea) from the karst of Slovenia. Mitt. Verb. dt. Hoehlen- u. *Karstforsch* **1999**, *45*, 54–56.
97. Trontelj, P.; Blejec, A.; Fišer, C. Ecomorphological convergence of cave communities. *Evolution* **2012**, *66*, 3852–3865. [CrossRef] [PubMed]
98. Sket, B. *Niphargobates orophobata* n.g., n.sp. (Amphipoda, Gammaridae s.l.) from cave waters in Slovenia. *Biološki Vestn.* **1981**, *29*, 105–118.
99. Verovnik, R.; Prevorčnik, S.; Jugovic, J. Description of a neotype for *Asellus aquaticus* Linné, 1758 (Crustacea: Isopoda: Asellidae), with description of a new subterranean *Asellus* species from Europe. *Zool. Anz.* **2009**, *248*, 101–118. [CrossRef]
100. Verovnik, R.; Sket, B.; Trontelj, P. The colonization of Europe by the freshwater crustacean *Asellus aquaticus* (Crustacea: Isopoda) proceeded from ancient refugia and was directed by habitat connectivity. *Mol. Ecol.* **2005**, *14*, 4355–4369. [CrossRef]
101. Verovnik, R.; Sket, B.; Trontelj, P. Phylogeography of subterranean and surface populations of water lice *Asellus aquaticus* (Crustacea: Isopoda). *Mol. Ecol.* **2004**, *13*, 1519–1532. [CrossRef]
102. Prevorčnik, S.; Blejec, A.; Sket, B. Racial differentiation in *Asellus aquaticus* (L.) (Crustacea: Isopoda: Asellidae). *Arch. Hydrobiol.* **2004**, *160*, 193–214. [CrossRef]
103. Prevorčnik, S.; Jugovic, J.; Sket, B. Geography of morphological differentiation in *Asellus aquaticus* (Crustacea : Isopoda : Asellidae). *J. Zool. Syst. Evol. Res.* **2009**, *47*, 124–131. [CrossRef]
104. Racovitza, E.G. Notes sur les Isopodes. 13. Morphologie et phylogenie des Antennes II. *Arch. Zool. Exp. Gén.* **1925**, *63*, 533–622.
105. Sket, B. Taksonomska problematika vrste *Asellus aquaticus* (L.) Rac. (Crust., Isopoda) s posebnim ozirom na populacijo v Sloveniji. Slov. Akad. Znan. Umet. razred za Prirodosl. Med. vede. *Razprave-Dissertationes* **1965**, *8*, 3–45.
106. Prevorčnik, S.; Verovnik, R.; Zagmajster, M.; Sket, B. Biogeography and phylogenetic relations within the Dinaric subgenus *Monolistra* (*Microlistra*) (Crustacea: Isopoda: Sphaeromatidae), with a description of two new species. *Zool. J. Linn. Soc.* **2010**, *159*, 1–21. [CrossRef]

107. Sket, B. Östliche gruppe der Monolistrini (Crust, Isopoda), I, Systematischer Teil. *Int. J. Speleol.* **1964**, *1*, 163–189. [CrossRef]
108. Strouhal, H. Asseln aus Balkanhohlen. *Zool. Anz.* **1938**, *124*, 269–281.
109. Strouhal, H. *Titanethes* Schiodte. Landasseln aus Balkanhohlen in der Kollektion "Biospeologica balcanica" von Prof. Dr. Absolon. 6. Mitteilung. Studien aus dem Geibete der Allg. Karstforschung, der wissenschaftlichen Hohlenkunde, der Eiszeitforsch. und den Nachbargebieten. *Biologische Ser.* **1939**, *5*, 5–34.
110. Strouhal, H. Landasseln aus Balkanhöhlen, gesammelt von Prof. Dr Karl Absolon. 10. Mitteilung.Zugleich 26. Beitrag zur Isopodenfauna des Balk. allg. Karstforsch., etc. *Briinn* **1939**, *7*, 1–37.
111. Potočnik, F.; Novak, T. Uber Landasseln (Isopoda terrestris) aus Hohlen Sloweniens. Anzeiger der math.-naturw. *Klasse Osterr. Akad. Wissenschaften Jahrgang 1980* **1980**, *5*, 75–81.
112. Stach, J. The species of the genus *Arrhopalites* Born. occurring in European caves. *Acta Musei Hist. Nat. Acad. Pol. Litt. Sci.* **1945**, *1*, 1–47.
113. Stach, J. Die Gattung *Oncopodura* Carl & Leb und eine neue Art derselben aus den Höhlen nord-östl. *Italiens. Bull. Acad. Pol. Sci. Sér. Sci. Biol.* **1934**, *2*, 1–16.
114. Denis, J.R. Collemboles d'Italie (principalement cavernicoles) (sixieme note sur la faune italienne des Collemboles). *Estratto Boll. Soc. Adriat. Sci. Nat.* **1938**, *26*, 95–165.
115. Silvestri, F. Illustrazione della *Plusiocampa (Stygiocampa) nivea* Jospeh (Campodeidae, Diplura) delle grotte di Postumia. *Boll. Lab. Entomol. Agrar. Portici* **1947**, *8*, 88–92.
116. Daffner, H. Revision der Anophthalmus—Arten und—Rassen miz lang und dicht behaarter Korperoberseite (Coleoptera, Carabidae, Trechinae). Mitteilungen der Münchener Entomol. *Gesellschaft* **1996**, *86*, 33–78.
117. Daffner, H. Die Arten und Rassen der *Anophthalmus schmidti* und -*mariae* Gruppe (Coleoptera: Carabidae: Trechinae). *Acta Entomol. Slov.* **1998**, *6*, 99–128.
118. Bognolo, M. Il genere Bathyscimorphus (Coleoptera: Cholevidae). *Coleoptera* **2002**, *6*, 1–33.
119. Pretner, E. *Catalogus Faunae Jugoslaviae III/6. Coleoptera, Catopidae, Bathysciinae*; ZRC SAZU: Ljubljana, Slovenia, 1968.
120. Broder, J. Bythoxenus subterraneus Motshcoulsky 1859 (Coleoptera. Pselaphidae) ponovno najden v Sloveniji leta 1975. *Naše Jame* **1978**, *19*, 59–61.
121. Müller, G. I Pselafidi cavernicoli del Carso Adriatico settentrionale (Venezia Giulia e Carniola). *Boll. Soc. Adriat. Sc. Nat. Trieste* **1947**, *43*, 133–146.
122. Poggi, R. Forme nuove o poco note di Pselaphidae cavernicoli del Friuli-Venezia Giulia e della Jugoslavia. *Mem. Soc. Ent. Ital.* **1991**, *70*, 201–224.
123. Nonveiller, G.; Pavićević, D. Description d'une sous-espèce nouvelle et de six espèces nouvelles du genre *Machaerites* Miller, 1855 de Slovénie et de Croatie (Coleoptera, Pselaphinae, Bythinini). *Nouv. Rev. Entomol.* **2001**, *18*, 317–333.
124. Sket, B. Distribution of *Proteus* (Amphibia: Urodela: Proteidae) and its possible explanation. *J. Biogeogr.* **1997**, *24*, 263–280. [CrossRef]
125. Zagmajster, M.; Culver, D.C.; Christman, M.C.; Sket, B. Evaluating the sampling bias in pattern of subterranean species richness: Combining approaches. *Biodivers. Conserv.* **2010**, *19*, 3035–3048. [CrossRef]
126. Du Preez, G.; Majdi, N.; Swart, A.; Traunspurger, W.; Fourie, H. Nematodes in caves: A historical perspective on their occurrence, distribution and ecological relevance. *Nematology* **2017**, *19*, 1–18. [CrossRef]
127. Delić, T.; Trontelj, P.; Rendoš, M.; Fišer, C. The importance of naming cryptic species and the conservation of endemic subterranean amphipods. *Sci. Rep.* **2017**, *7*, 3391. [CrossRef] [PubMed]
128. Lukić, M.; Delić, T.; Pavlek, M.; Deharveng, L.; Zagmajster, M. Distribution pattern and radiation of the European subterranean genus *Verhoeffiella* (Collembola, Entomobryidae). *Zool. Scr.* **2019**, *49*, 86–100. [CrossRef]
129. Fišer, C.; Robinson, C.T.; Malard, F. Cryptic species as a window into the paradigm shift of the species concept. *Mol. Ecol.* **2018**, *27*, 613–635. [CrossRef] [PubMed]
130. Mammola, S.; Amorim, I.R.; Bichuette, M.E.; Borges, P.A.V.; Cheeptham, N.; Cooper, S.J.B.; Culver, D.C.; Deharveng, L.; Eme, D.; Ferreira, R.L.; et al. Fundamental research questions in subterranean biology. *Biol. Rev.* **2020**, *95*, 1855–1872. [CrossRef] [PubMed]
131. Protas, M.; Jeffery, W.R. Evolution and development in cave animals: From fish to crustaceans. *Wiley Interdiscip. Rev. Dev. Biol.* **2012**, *1*, 823–845. [CrossRef] [PubMed]
132. Jemec, A.; Škufca, D.; Prevorčnik, S.; Fišer, Ž.; Zidar, P. Comparative study of acetylcholinesterase and glutathione S-transferase activities of closely related cave and surface *Asellus aquaticus* (Isopoda: Crustacea). *PLoS ONE* **2017**, *12*, e0176746. [CrossRef] [PubMed]
133. Fišer, Ž.; Prevorčnik, S.; Lozej, N.; Trontelj, P. No need to hide in caves: Shelter-seeking behavior of surface and cave ecomorphs of *Asellus aquaticus* (Isopoda: Crustacea). *Zoology* **2019**, *134*, 58–65. [CrossRef]
134. Pipan, T.; Christman, M.C.; Culver, D.C. Abiotic community constraints in extreme environments: Epikarst copepods as a model system. *Diversity* **2020**, *12*, 269. [CrossRef]
135. Venarsky, M.P.; Huntsman, B.M.; Huryn, A.D.; Benstead, J.P.; Kuhajda, B.R. Quantitative food web analysis supports the energy—Limitation hypothesis in cave stream ecosystems. *Oecologia* **2014**, *176*, 859–869. [CrossRef]
136. Venarsky, M.P.; Benstead, J.P.; Huryn, A.D.; Huntsman, B.M.; Edmonds, J.W.; Findlay, R.H.; Wallace, J.B.; Bruce Wallace, J. Experimental detritus manipulations unite surface and cave stream ecosystems along a common energy gradient. *Ecosystems* **2018**, *21*, 629–642. [CrossRef]

137. Mulec, J.; Oarga, A.; Šturm, S.; Tomazin, R.; Matos, T. Spacio-temporal distribution and tourist impact on airborne bacteria in a cave (Škocjan Caves, Slovenia). *Diversity* **2017**, *9*, 28. [CrossRef]
138. Mammola, S.; Cardoso, P.; Culver, D.C.; Deharveng, L.; Ferreira, R.L.; Fišer, C.; Galassi, D.M.P.; Griebler, C.; Halse, S.; Humphreys, W.F.; et al. Scientists' warning on the conservation of subterranean ecosystems. *Bioscience* **2019**, *69*, 641–650. [CrossRef]
139. Mulec, J.; Petrič, M.; Koželj, A.; Brun, C.; Batagelj, E.; Hladnik, A.; Holko, L. A multiparameter analysis of environmental gradients related to hydrological conditions in a binary karst system (underground course of the Pivka River, Slovenia). *Acta Carsologica* **2020**, *48*, 313–327. [CrossRef]
140. Šebela, S.; Turk, J. Local characteristics of Postojna cave climate, air temperature, and pressure monitoring. *Theor. Appl. Climatol.* **2011**, *105*, 371–386. [CrossRef]
141. *European Council Directive 92/43/EEC of 21 May 1992 on the conservation of natural habitats and of wild fauna and flora*; EU Council: Brussels, Belgium, 1992.
142. Tautz, D.; Arctander, P.; Minelli, A.; Thomas, R.H.; Vogler, A.P. A plea for DNA taxonomy. *Trends Ecol. Evol.* **2003**, *18*, 70–74. [CrossRef]
143. Leese, F.; Altermatt, F.; Bouchez, A.; Ekrem, T.; Hering, D.; Mergen, P.; Pawlowski, J.; Piggott, J.J.; Rimet, F.; Steinke, D.; et al. *DNAqua-Net:* Developing new genetic tools for bioassessment and monitoring of aquatic ecosystems in Europe. *Res. Ideas Outcomes* **2016**, *2*, e11321. [CrossRef]
144. Gorički, Š.; Stanković, D.; Snoj, A.; Kuntner, M.; Jeffery, W.R.; Trontelj, P.; Pavićević, M.; Grizelj, Z.; Năpăruş-Aljančič, M. Environmental DNA in subterranean biology: Range extension and taxonomic implications for *Proteus*. *Nat. Publ. Gr.* **2017**, 91–93. [CrossRef]
145. Niemiller, M.L.; Porter, M.L.; Keany, J.; Gilbert, H.; Fong, D.W.; Culver, D.C.; Hobson, C.S.; Kendall, K.D.; Davis, M.A.; Taylor, S.J. Evaluation of eDNA for groundwater invertebrate detection and monitoring: A case study with endangered *Stygobromus* (Amphipoda: Crangonyctidae). *Conserv. Genet. Resour.* **2018**, *10*, 247–257. [CrossRef]

Article

Undara Lava Cave Fauna in Tropical Queensland with an Annotated List of Australian Subterranean Biodiversity Hotspots

Stefan M. Eberhard [1],* and Francis G. Howarth [2],*

[1] Scientific Environmental Services, Subterranean Ecology Pty Ltd., 227 Coningham Road, Coningham, TAS 7054, Australia

[2] Department of Natural Sciences, B. P. Bishop Museum, Honolulu, HI 96819, USA

*   Correspondence: stefan@subterraneanecology.com.au (S.M.E.); fghcavebug@gmail.com (F.G.H.)

Abstract: The lava tubes at Undara became internationally recognised in the late 1980s, when 24 species of terrestrial cave-adapted invertebrates (troglobionts) were recorded from Bayliss Cave, making it one of the 20 richest known cave communities in the world at the time. Over the last decades, several of the Undara species have been taxonomically described and a great deal of research has been undertaken in other parts of Australia, which has revealed additional subterranean hotspots. It is therefore timely to update the list of Undara cave fauna, and to evaluate the Undara cave system in relation to other subterranean hotspots in Australia. The updated species list was compiled from the published literature and museum databases. Minimally, 78 species of arthropods have been recorded from 17 lava tube caves in the Undara Basalt. Sixteen species have been taxonomically described; 30 identified to genus and/or morpho-species; and 32 remain unidentified to species or genus level. Thirty troglobionts and one stygobiont species were recorded. Seven caves harboured obligate subterranean species; Bayliss Cave harboured the most obligate subterranean species: 23 troglobionts and one stygobiont. All these caves contained deep zone environments with high humidity, of which three also contained 'bad air' (CO_2). The unique combination of geomorphic structure and environmental parameters (high humidity) and multiple energy sources (tree roots, bats and guano, organic material wash-in) are the main factors responsible for Bayliss Cave's extraordinary local richness. Further research is needed to investigate CO_2 as a factor influencing troglobiont richness and distribution in 'bad air' caves. Undara remains the richest subterranean hotspot in humid tropical Australia; however, significantly richer subterranean assemblages are found in arid and semi-arid calcrete aquifers, karst and iron-ore terrains, mostly in Western Australia.

Keywords: basaltic lava flow; cave ecology; arthropods; tropical cave fauna; Bayliss Cave; *Psilotum*

Citation: Eberhard, S.M.; Howarth, F.G. Undara Lava Cave Fauna in Tropical Queensland with an Annotated List of Australian Subterranean Biodiversity Hotspots. *Diversity* **2021**, *13*, 326. https://doi.org/10.3390/d13070326

Academic Editors: Tanja Pipan, David C. Culver, Louis Deharveng and Michael Wink

Received: 31 May 2021
Accepted: 14 July 2021
Published: 16 July 2021

Publisher's Note: MDPI stays neutral with regard to jurisdictional claims in published maps and institutional affiliations.

1. Introduction

The lava tubes at Undara became internationally recognised when, in the late 1980s, 24 species of terrestrial cave-adapted invertebrates (troglobionts) were recorded from Bayliss Cave [1,2], making it one of the 20 richest known cave communities in the world at the time [3]. Since then, the troglobionts at Undara have been the subject of landmark studies on reductive evolutionary trends and acoustic communication in cixiid planthoppers [4,5], adaptive shift and non-relictual tropical cave-adapted animals [1], and carbon dioxide and implications for the evolution of cave-adapted animals [2,6].

The research at Undara thrust these caves into the limelight of biospeleology in Australia on four counts. Firstly, it broke the existing paradigm that mainland Australia was poor in cave-adapted animals [7]. That view was further invalidated by additional discoveries in Western Australia [8,9]. Secondly, it upended the global paradigm that cave-adapted animals were rare in tropical caves [10]. The third important aspect was that lava tubes were also considered to be devoid of cave-adapted animals [11]; and fourth, some of these tropical cave-adapted animals were non-relictual, in the sense that some cave species had closely related congeners living in surface habitats nearby.

The latter finding held important implications for understanding the evolutionary processes leading to speciation and adaptation to subterranean habitats, i.e., the adaptive shift hypothesis [12]. The situation with Undara and some other tropical cave faunas contrasts with many temperate zone cave faunas, where the surface lineages that gave rise to cave forms had long since disappeared or migrated elsewhere due to changing conditions on the surface particularly during Pleistocene glacial periods, i.e., the climatic relict hypothesis [11].

These global paradigms were also challenged after troglobionts were discovered in tropical and subtropical limestone caves and lava tubes in the Galapagos [13], Hawaii [14], Canary Islands [15,16], Jamaica [17], Congo, Thailand, Indonesia, and Central and South America [18,19]. On the other hand, recent molecular and morphological studies appear to support the climate relict hypothesis even for tropical cave animals, e.g., Soulier-Perkins [20] study of Chillagoe and Undara cixiid planthoppers and Slaney and Blair [21] study of Chillagoe and Undara ectobiid cockroaches. However, the question of which came first, isolation or cave adaptation, cannot be answered with current methods. Nonetheless, clear examples of adaptive shifts are known [22] suggesting that cave adaptation comes first at least in some cases [23].

As demonstrated in the young lava tubes on Hawaii [6,24] the primary habitat for troglobionts and cave-adapted aquatic species (stygobionts) in basaltic terrains is within intermediate sized voids (=mesocaverns). Mesocaverns (5–500 mm diameter) are distinguished from large-sized caves generally enterable by humans (>500 mm) and fine-grained interstitial habitats in porous sediments (<5 mm) [25]. Troglobionts occupy accessible cave-sized passages when environmental conditions are suitable [26]. These animals can disperse entirely underground throughout basalt flows, and they can potentially disperse into adjoining flows if there are suitable connecting voids. Since these caves comprise an interconnected lava tube system without barriers to underground dispersal, the five tube systems in the Undara Basalt are treated as one integrated cave ecosystem. Since the pool of potential colonizers is similar throughout the basalt flow, comparisons between community composition and the physical environment can provide useful insights into cave ecology.

Since publication of the first hotspot list in 2000, knowledge of global cave biodiversity has grown substantially, and the number of hotspot caves has more than doubled. However tropical hotspot caves remain in the minority with only five tropical caves harbouring 20 or more specialized cave species recognised in the last update [27]. The small representation of tropical hotspot caves may be partly because many tropical cave regions have not been adequately investigated and because many species remain undescribed. Additionally, tropical caves often contain specialized guano fauna (guanobionts) that are only weakly troglomorphic but still not known outside caves [19,28].

With publication of this special issue featuring world hotspots of subterranean biodiversity, it is thus timely to update and review the species list for Bayliss Cave and include the fauna of other cave segments in the Undara lava flow. Several of the cavernicolous species have been described in the years since the last updates by Stone [29] and Clarke [30]. The current state of knowledge on the geology, ecology, physical environment, and natural history of the subterranean animals in the Undara caves is the focus of this review. We briefly mention other biodiversity hotspots in Australia to place Undara in context and discuss the challenge of classifying cavernicoles in the face of limited taxonomic and ecological knowledge. We hope this paper stimulates renewed field sampling and taxonomic interest in Undara's remarkable tropical lava tube system.

2. Materials and Methods

The taxonomic impediment is a global problem exacerbated by reduced funding for systematic research, the 'orphaning' of taxa by retirement and passing of experts, and the increasing complexity of the discipline as new species must be compared with an increasing number of related taxa. Fontaine et al. [31] determined that the average delay between collection and description of a new species was over two decades, and that the time was further increased when the new species occurred in species-rich areas, and for poorly known taxa lacking a recent revision. These universal limitations mean that many Undara cave species have not been identified to species level and numerous putative new species await formal description.

The updated species list was compiled from the published literature and registered specimen databases of the Queensland Museum (QM) and Australian Museum (AM). In some cases, undescribed taxa reported in earlier collection lists could not be located in museum databases, probably because a substantial portion of cave collections made from Undara caves remain unsorted and unregistered. Portions of the (registered) Undara cave collections remain on loan to the BP Bishop Museum (BPBM), Hawaii, or have been loaned to taxonomists pending description. Taxa collected by Clarke [30] are deposited at various institutions including QM, AM, Australian National Insect Collection (ANIC), Tasmanian Museum and Art Gallery (TMAG) and the Northern Territory Museum and Art Gallery (NTMAG). Taxonomists holding material were contacted for information about the taxonomic and ecological status of taxa. Data on other subterranean hotspots in Australia were sourced from the published literature, unpublished consultancy reports and collection records (SE). Other collectors [32–35] list additional unidentified taxa in groups represented in our lists. Without inspecting voucher specimens, it is not possible to determine whether the taxon is already listed; therefore, the conservative option is to exclude these records from the list.

Determining whether a species is a troglobiont, exclusively restricted to caves and associated mesocaverns, rather than a soil or surface inhabitant is often problematic especially in poorly known groups and in groups that possess troglomorphic characters but are not confined to caves. To help discriminate true troglobionts from troglomorphic troglophiles and edaphophiles we also relied on field observations of behavior as well as morphology [36]. That is, in addition to reduced pigment, eyes and wings, elongated appendages, and sometimes a larger size compared to their soil or surface relatives, troglobionts generally exhibit no response to light and move comparatively slowly even when disturbed. For example, all polydesmid millipedes lack eyes, and many species occur in both caves and cryptic surface habitats. Surface dwelling polydesmids that have been studied can detect light and respond by moving into dark refuges when exposed to light [37]. Although not studied experimentally, cave-restricted polydesmids, including the pale-coloured species we observed at Undara, usually do not respond to light and move slowly.

Oxygen and carbon dioxide concentrations (as volume percentage) were measured 15 cm above the cave floor using a Draeger Multi Gas Detector with oxygen (5%/B) and carbon dioxide (0.1%/a) tubes, respectively. A few additional readings using appropriate Draeger tubes were made to test for the presence of carbon monoxide, ammonia and methane; the results were negative. Generally, two readings for carbon dioxide were taken at selected locations within each zone and corrected for barometric pressure (96.4–96.5 kPa) measured with a Thommen no. 2000 (5000 m) altimeter. Temperature and relative humidity were measured 15 cm and 2 m above the cave floor as noted using a battery-powered Bendix aspirating psychrometer. The psychrometer was accurate up to 95% RH, and the presence and thickness of fog was used to indicate higher humidity levels.

3. Geology, Geomorphology, Hydrology

'*Undara*' is an Aboriginal word meaning 'long way'. The extensive lava tube system that developed in the Undara Basalt flow is one of several basalt flows making up the McBride Volcanic Province. The province is situated on the Atherton Tableland approximately 200 km southwest of Cairns in tropical northeast Queensland. It comprises a broad topographic dome roughly 100 km in diameter, which reaches an elevation of around 1020 m descending to its margins at approximately 400 m above sea level. The dome is believed to have been built over a period of 10 million years by multiple lava flows from more than 160 vents [38] (Figure 1). Much of the basalt dome is devoid of surface watercourses. The high porosity of basalt facilitates rapid infiltration of surface water to groundwater drainages that largely correspond to now buried drainages incised in the underlying granites prior to volcanism [35]. Swampy resurgences occur at the margins where the original watercourses have been filled with basalt. Some lava tubes contain temporary or permanent perched lakes which are fed by diffuse infiltration and local surface runoff into entrances during the wet season.

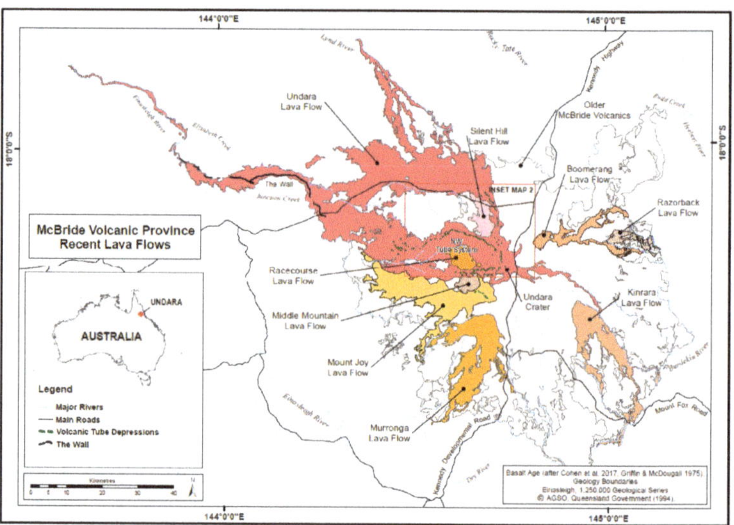

Figure 1. McBride Volcanic Province showing recent (<1 my) lava flows. Map adapted from Pearson [35] with permission from Chillagoe Caving Club. Geology boundaries from Queensland Government 1:250,000 Geological Series.

The Undara Basalt is the third youngest flow (~190,000 years, ka) [39,40], covering 1374 km^2. It is the most extensive of the recent (<1 million years, ma) basalt flows in the province. Lava flowed radially outwards from the Undara Volcano. The lava was distributed down slope by three major, and two minor, lava tube systems. The largest system is the north-western flow, which extends 176 km from the crater [35,41].

Sixty-seven arches and caves with a combined mapped length of 6.38 km are recorded, with the majority (70%) found in the north-western (NW) tube system, all of which occur within 32 km of the crater (Figure 2). The caves and arches are humanly accessible segments of the subsurface tube system, created where the roof of the tube collapsed to form an entrance. Additional caves certainly exist but are difficult to locate over the rough untracked terrain. Most of the known caves are relatively short segments of breached lava tube less than 100 m long, and the longest segment is Bayliss Cave at 1300 m. The original lava floor is rarely visible, only occurring as small patches in a few caves.

Other lava tube caves are developed in adjacent Murronga, Kinrara and Silent Hill Basalts (Figure 1), however only cave segments in Murronga and Silent Hill lava flows have been sampled for cave fauna (Table 1). These neighbouring lava flows and lava tube cave systems are important for understanding relationships between taxa and tube systems. Sampling in Murronga and Silent Hill systems has identified locally endemic troglobiont species, as well as troglobionts whose distribution range encompasses caves in the Undara Basalt tube system. As envisioned for Hawaiian cave animals [25], the cave-adapted fauna in Undara caves, may have colonised mesovoids and caves within the Undara flow from neighbouring older flows. Thus, the age of the Undara lava tube system may not indicate a maximum age of cave adaptation [42].

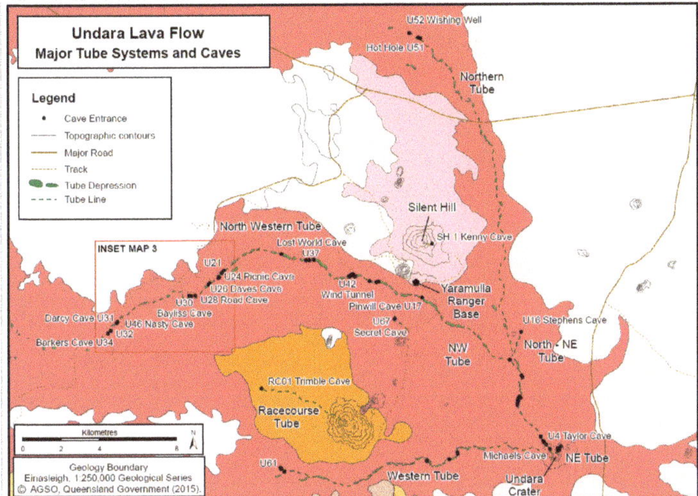

Figure 2. Lava tube systems and caves in the Undara Basalt flow. Bayliss Cave and other major cave segments are labelled. Map adapted from Pearson [35] with permission from Chillagoe Caving Club. Geology boundaries from Queensland Government 1:250,000 Geological Series.

Lava tubes at Undara and other McBride flows formed in pahoehoe basalt [43]. Pāhoehoe lava has a relatively low viscosity and flows like a river. These lava rivers crust over insulating the interior and transporting the fluid lava many kilometres downslope. These flows respond to changes in the eruption rate, composition of lava, slope, and condition of the land surface; therefore, cave formation is dynamic [44]. Overflows, from breaks in the forming roof, release lava over the cooling lava; thus, strengthening the roof downslope. These overflows build flow on flow, layer on layer [45]. These layers poorly fuse to the cooled, older surfaces so that there are numerous gaps preserved between each flow unit. Escaping gas can inflate voids within the flow and additional voids and cave-like spaces within the lava can be created by tree moulds, earthquakes, and cooling cracks. Furthermore, sections of older lava tubes are buried as the eruption continues. In this way, basaltic lava flows can cover large areas and create extensive and abundant underground habitats throughout young pāhoehoe flows [6].

Subterranean habitats in basalt are potentially as deep as the thickness of the basalt flow. Tube-fed lava flows are thicker near the vent and thin as they flow downslope. For the Undara lava flow, the minimum thickness of the habitat is indicated by the exposed lava at the base of Undara Crater, which is 50 to 60 m deep [46]. However, the actual depth of the basalt at the crater rim is likely to be much deeper. The rim of Undara Crater is 1020 m above sea level, and, at a 40-km long lava ridge ('The Wall') beginning 60 km downslope from the vent, the elevation is approximately 600 m suggesting that the basalt thickness near the rim might be 400 m. Geological drilling at "The Wall" found the thickness of basalt to be about 40 m [47]. However, the thickness at the crater may be less than 400 m since uplift of the underlying granitic inlier could affect the height of the basalt dome [46,48].

Table 1. McBride Province basalt flows (youngest to oldest) and a synopsis of their recorded lava tube systems and caves (where known) including caves which have been sampled for fauna. Geological information from [35,38,40] and $_{40}$Ar/$_{39}$Ar ages (ka/ma) from [39]; biological information compiled from [1,29,30,32–34].

Basalt Flows ($_{40}$Ar/$_{39}$Ar Age, ka/ma)	Lava Tube Systems (Length)/Number of Known Caves	Sampled Caves by Tube System from Crater	Remarks
Kinrara ($_{40}$Ar/$_{39}$Ar Age 7 ± 2 ka, 2σ)	2 tube systems, 11 caves/arches	none	Approximately 25 km SE Undara crater, potentially important for biology.
Murronga ($_{40}$Ar/$_{39}$Ar Age 153 ± 5 ka, 2σ)	2 tubes, 12 caves with combined survey length of 1.2 km	Collins No. 1, Collins No. 2, Two Ten, Long Shot	Approximately 15 km S of Undara Crater, potentially important for biology. Major caves with significant fauna.
Undara ($_{40}$Ar/$_{39}$Ar Age 189 ± 4 ka, 2σ)	5 tube systems, ~67 caves/arches with combined survey length of 6.38 km	N-NE system: Hot Hole, Wishing Well. N system: Stephens NW system: 10 caves. W system: none E system: none Other: Secret No. 1 and 2	Fauna collection records obtained from 17 caves, all except 3 sampled caves occur in NW system. 70% of known caves occur in the NW tube system.
Racecourse ~0.20 ma	1 tube, 4 caves	None sampled	Caves reported in Pearson [35], not sampled.
Boomerang ~0.23 ma	None recorded	-	-
Razorback <0.27 ma	None recorded	-	-
Silent Hill ~0.37 ma	1 tube system, 1 cave	Kenny Cave SH-1	Located within 3 km of Undara NW tube system, potentially important for biology.
Mount Joy < 0.40 ma	None recorded	-	-
Middle Mountain ~0.89 ma	None recorded	-	-
Older basalt < 3.00 ma	None recorded	-	-

4. Environment

4.1. Surface

The climate is monsoonal, characterised by hot humid summers and warm dry winters. Mean annual maximum temperature (at Mount Surprise) is 31.1 °C and mean annual minimum 16.1 °C. Mean annual rainfall is 793 mm, which falls mostly during the hotter months from November to March [49]. Rainfall sinks rapidly underground through cracks in the porous basalt. The soil layer is thin since developing soil is washed or subsides underground with rain and gravity, where it forms thick deposits on the floor of the caves. Leaflitter is rare in most Undara lava tubes, except in a few twilight zones of caves with large, exposed entrances such as Road Cave and Barkers Cave.

The surface vegetation is a dry savannah woodland, with widely spaced *Eucalytpus* trees and grassland understory (Figure 3). In stark contrast to the savannah are conspicuous patches of dark-green vine-thicket growing in larger depressions and collapsed caves (Figure 3). The vine-thicket is thought to be a remnant of a once more widespread vegetation type with strong Gondwana affinities. These botanical 'islands' generally mark the course of the major lava tubes that fed the advancing lava; some are drained lava lakes.

4.2. Underground

Globally, terrestrial cave habitats are zonal with three main zones defined by the amount of light: entrance zone, twilight zone, and dark zone. The dark zone can be further divided into two or three subzones: transition zone where diurnal and weather events on the surface affect the moisture, temperature, and airflow and deep zone where the environment is buffered from events on the surface and characterized by calm and water-saturated atmosphere. A few researchers recognize a stagnant air zone in which the environment is even more stable than in the deep zone [6,26]. The extent of each zone depends on the length and shape of the entrance(s) and passage configurations.

At Undara, CO_2 accumulates in deeper portions of most sampled caves, usually at low (<2%) but detectable levels. However, in both Bayliss Cave and Nasty Cave, CO_2 concentration approaches 6% by volume. The troglobiont species are characteristically found only in the deep and stagnant air zones. The Undara caves vary in size and shape as well as in the amount of moisture, carbon dioxide, and type and quality of nutrients present. Thus, each cave comprises a different mix of environmental variables which in turn favour differing communities of cavernicoles.

Figure 3. Oblique view looking East showing Undara Crater in the upper right corner and the entrance to Taylor Cave about one kilometre downslope (arrowed). Contrasting with the surrounding dry savanna woodland, the conspicuous ribbon of dark-green 'islands' are remnant vine-thicket vegetation growing in depressions and collapsed caves and revealing the course of the lava tube system. Picture from Google Earth. Scale not shown in this oblique view since apparent distances will vary depending on position.

5. Results

5.1. Overview of Invertebrate Sampling in the Undara Basalt Lava Tube Systems

1. Records were found for biological inventories in 17 cave segments: 14 in the northwest tube system and three in the northern and north-northeast tube systems (Table 2). Biological surveys have been conducted in a few additional caves [29,30], but the results are unpublished.
2. Minimally 78 species of arthropods have been recorded from Undara Basalt lava tubes, along with a cavernicolous fern (*Psilotum* sp.) and four bat genera (*Hipposideros, Miniopterus, Rhinolophus, Vespadelus*).
3. Of the arthropods: 16 (21%) have been taxonomically described; 30 (39%) identified to genus and/or morpho-species; and 32 (40%) remain unidentified to species or genus level.
4. Seven caves harboured obligate subterranean species; all these caves contained deep zone environments with high humidity, of which three also contained bad air (CO_2).
5. Overall, 30 troglobionts and one stygobiont species were recorded in the Undara Basalt Flow (Table 3). Bayliss Cave harboured the most obligate subterranean species: 23 troglobionts and one stygobiont.

6. Twenty-two species of arthropods and one plant are classified as troglophiles, being native cavernicoles capable of living their entire life cycle underground but populations of the same species may also be found in surface habitats (Table S1) [2,21,30,32,50–54]. Some of these may prove to be troglobionts once more is known of their biology.

7. Twenty-five species of arthropods are classified as 'visitors' (Table S2) [2,30,35,55,56]. Some habitually use caves for shelter or to find food (=trogloxenes). Others occasionally enter caves for shelter, and some wander or fall into caves accidentally (='incidental' or 'accidental' cavernicoles). In addition, unidentified mites (Arachnida: Acari) have been reported from most surveyed caves. Many are associated with guano, and a few are parasites of other cavernicoles including bats; however, because both the identity and ecological status of the mites recorded from Undara caves are unknown, they are not enumerated further in Table S2.

8. Besides Bayliss, the other caves with a high diversity of troglobionts are Nasty Cave (eight species) and Barkers Cave (seven species), which are located, respectively, 3 km and 3.7 km downflow of Bayliss (Figure 4).

Table 2. Named lava tube segments within the Undara Lava Flow from which cavernicoles have been reported, arranged by tube system, from upslope to downslope. * Indicates cave identification number, Australian Speleological Federation. See also Figure 2.

Tube System	Cave Name (No. *)	Length	Elevation	Environment	Fauna
Northwest	Michael's (no #)	15 m	975 m	All twilight	Twilight zone fauna
Northwest	Taylor (U-4)	108 m	950 m	Deep Zone	A few troglobionts, bats
Northwest	Pinwell (U-17)	150 m	850 m	Deep Zone	A few troglobionts, bats
Northwest	Wind tunnel (U-42)	293 m	800 m	Two entrances, limited deep zone	A few possible troglobionts, bats
Northwest	Lost World (U-37)	74 m	775 m	Two entrances, well-ventilated, transition zone	Blattodea: *Macropanesthia rhinoceros*, bats
Northwest	Picnic (U-24; U-25)	~450 m	740 m	Permanent pool, sump, humid, deep zone	Unidentified moths, isopods, beetles, crickets, spiders, bats
Northwest	Daves (upflow) (U-26))	50 m	730 m	Limited dark zone	Lepidoptera: *Euploia* sp. aff. *core*; unid. spider, phalangid, moths, beetles, bugs, cockroaches, ants, wasps, bats
Northwest	Daves II (U-27)	27 m	730 m	Limited dark zone	Lepidoptera: *Euploia* sp. aff. *core*; unid. spider, phalangid, moths, beetles, bugs, cockroaches, ants, wasps, bats
Northwest	Road (U-28)	220 m	700 m	Mostly twilight, intermittent stream	Stygobiont Amphipoda: *Chillagoe* sp.; many surface arthropods in leaflitter
Northwest	Bayliss (U-30)	1300 m	700 m	Hot humid bad air	23 troglobionts, one stygobiont, bats
Northwest	Nasty (U-46)	127 m	670 m	Hot, humid, bad air	8 troglobionts, bats
Northwest	Darcy (U-31)	99 m	670 m	Limited deep, elevated CO_2	trogloxenes and accidentals, bats
Northwest	Barkers (U-34)	560+ m	625 m	Deep zone, Lake at end	7 troglobionts, bats

Table 2. *Cont.*

Tube System	Cave Name (No. *)	Length	Elevation	Environment	Fauna
Tributary of Northwest	Secret Cave No. 1 (U-67)	150 m	~825 m	Deep zone	Blind cockroaches
North-NE	Stephens (=Stevens) Cave (U-16)	70 m	880 m	Mostly transition zone	Troglophilic cockroach, bats
Northern	Hot Hole (U-51)	172 m	~760 m	Warm, bad air, deep guano	Unid. spider, troglophilic cockroach, bats
Northern	Wishing Well (U-52)	104 m	~750 m	Humid, bad air	Unid. spider, troglophilic cockroach, slater, bats

Table 3. List of obligate cave species recorded from the Undara lava tube system. Troglobionts (TB), stygobionts (SB). * Indicates cave identification number, Australian Speleological Federation (ASF). Refer to Supplementary Materials for a list of troglophiles (Table S1) and trogloxenes, accidentals (Table S2).

#	SB/TB	Species	Taxonomic Classification	Caves *	References
1	SB	*Chillagoe* n. sp.	Crustacea: Amphipoda: Chillagoeidae	U-28; U-30	[57]
2	TB	Unidentified genus and species 1	Crustacea: Isopoda: Porcellionidae	U-30; U-17	[2]
3	TB	Unidentified genus and species 2	Crustacea: Isopoda: Porcellionidae	U-30; U-17	[2]
4	TB	Unidentified genus and species 3 (eyeless)	Crustacea: Isopoda: Oniscoidea	U-34	[30]
5	TB	Unidentified genus and species 1, juvenile	Arachnida: Schizomida: Hubbardiidae	U-34	[58]
6	TB	*Protochelifer* sp. nr. *cavernarum* Beier	Arachnida: Pseudoscorpionida: Cheliferidae	U-42; U-17	[30,59]
7	TB	*Amauropelma undara* Raven et al.	Arachnida: Araneae: Ctenidae	U-30; U-46	[60]
8	TB	Unidentified genus and species 1	Arachnida: Araneae: Linyphiidae	U-30	[2]
9	TB	*Nesticella* sp. 1	Arachnida: Araneae: Nesticidae	U-30	[61]
10	TB	Unidentified genus and species	Arachnida: Araneae: Oonopidae	U-30	[1]
11	TB	*Spermophora* sp. 1	Arachnida: Araneae: Pholcidae	U-30	[62] (p. 239); [2]
12	TB	*Spermophora* sp. 2	Arachnida: Araneae: Pholcidae	U-34	[30,63]
13	TB	*Dolomedes* sp. 1	Arachnida: Araneae: Pisauridae	U-34	[30]
14	TB	*Nosterella cavicola* Baehr and Jocqué.	Arachnida: Araneae: Zodariidae	U-30; U-46	[64]
15	TB	New genus and species	Chilopoda: Scutigeromorpha: Scutigeridae	U-30; U-34; U-46	[2]
16	TB	Unidentified genus and species	Diplopoda: Spirostreptida	U-30; U-46; U-34	[2]
17	TB	Unidentified genus and species 1	Diplopoda: Polydesmida	U-30	[2]
18	TB	Unidentified genus and species 2	Diplopoda: Polydesmida	U-30	[2]
19	TB	Unidentified genus and species 1	Diplopoda: Polyxenida	U-30; U-46	[2]
20	TB	*Pseudosinella* sp. 1	Collembola: Entomobryidae	U-30; U-34	[2]

Table 3. *Cont.*

#	SB/TB	Species	Taxonomic Classification	Caves *	References
21	TB	Unidentified genus and species	Collembola:	U-30; U-46	[2]
22	TB	*Neotemnopteryx baylissensis* Slaney	Insecta: Blattodea: Ectobiidae	U-30; SH-1	[65]
23	TB	*Nocticola* sp. 1	Insecta: Blattodea: Nocticolidae	U-30; U-46; U-17	[66]
24	TB	New genus and species 1	Insecta: Coleoptera: Curculionidae: Entiminae	U-30	[67]
25	TB	New genus and species 2	Insecta: Coleoptera: Curculionidae: Entiminae	U-4	[67]
26	TB	Unidentified genus and species	Insecta: Coleoptera: Pselaphinae	U-30	[2]
27	TB	Unidentified genus and species	Insecta: Coleoptera: Staphylinidae	U-30; U-34	[2]
28	TB	Unidentified genus and species	Insecta: Diplura	U-30	[2]
29	TB	*Solonaima baylissa* Hoch and Howarth	Insecta: Auchenorrhyncha: Cixiidae	U-30; U-46	[68]
30	TB	*Micropolytoxus cavicolus* Malipatil and Howarth	Insecta: Heteroptera: Reduviidae	U-30; U-34	[42]
31	TB	*Peirates* sp. 1	Insecta: Heteroptera: Reduviidae	U-30	[2]

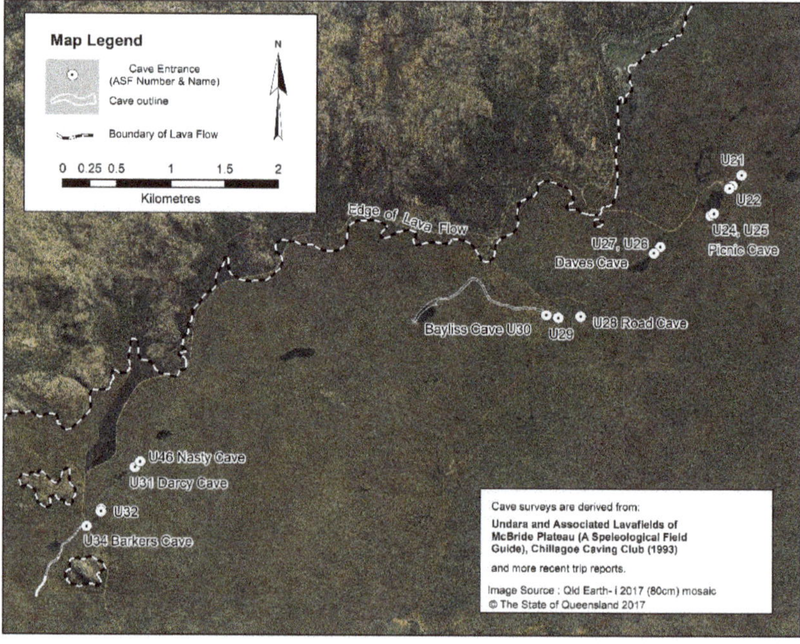

Figure 4. Aerial photograph showing cave entrances and major cave outlines in the north-western tube system at 23.5 to 31.5 km from Undara Crater. This 8 km section contains the two longest mapped tube segments in the Undara Flow, Bayliss Cave and Barkers Cave, and other biologically important segments including Nasty Cave and Road Cave (refer Table 2).

5.2. Bayliss Cave Environment

Bayliss Cave is by far the most biodiverse lava tube segment in the Undara lava flow. Therefore, we describe the cave and its environment to help understand the factors contributing to its richness. With a length of 1300 m Bayliss Cave is the longest mapped lava tube in Australia. It is entered through a narrow entrance crawlway about eight metres long and less than one metre in diameter, which opens into the main lava tube at the top of a 10 m high talus slope. From the base of the talus slope the tube descends gently downslope for 950 m to a low crawlway, only accessible when the CO_2 concentration is low, that extends the cave 350 m to a mud blockage. The average tube width and height is 8×5 m (maximum 25×11.5 m) [2,35]. The single small entrance limits air circulation and helps to maintain the high internal humidity which is sustained by condensation, infiltrating drip water, and small vadose seepages. Cave environment zones are well distinguished. The entrance zone is reduced owing to the small dimensions of the opening, and the twilight zone is limited to the eight-metre crawlway and top of the talus slope. The transition zone extends from the talus slope to the "Duck-Under", a low passage one to two metres high about 350 m into the cave; the deep and stagnant air zones are beyond the Duck-Under. The transition zone experiences the tropical winter effect, which is caused by cool, dry air entering the cave at night [69], and except for scattered ceiling drips, the floor in the transition zone is usually dry clay with a thin caliche crust. Beyond the Duck-Under, the floor is covered with moist to wet clayey soil and thin accumulations of bat guano. Thick 'curtains' of tree roots penetrate the ceiling to the floor especially beyond the Duck-Under (Figure 5).

Figure 5. Root 'curtains' of *Eucalyptus* penetrate from ceiling to floor in the deep zone about 100 m beyond the Duck-Under. Photograph by F.G. Howarth.

Savannah woodland overlies almost all the mapped cave passage, except the entrance depression and a 200 m section near the end, where the tube skirts the margin of a surface depression with vine thicket (Figure 4). The seasonally xeric environment and thin, poor soils force *Eucalyptus* woodland trees to send massive roots deep underground to obtain sufficient moisture and nutrients. The wide spacing (~9 m apart) of trees in the savannah suggests that all the root masses visible in Figure 5 connect to only one or perhaps up to three trees. Elsewhere in Australia, *Eucalyptus* spp. are renowned for their deeply penetrating root systems which may penetrate up to 70 m below the surface to tap groundwater in caves [70,71].

Between 14–15 June 1985, environmental parameters (temperature, relative humidity, carbon dioxide, oxygen), were measured at five stations located progressively deeper inside Bayliss Cave (Figure 6a,b). The atmosphere beyond the Duck-Under was foggy with water vapor condensing on surfaces; RH ranged from 98% to >100% (Figure 6a). The downward sloping tunnel accumulates carbon dioxide, which is denser than air. The concentration of CO_2 increased from about 1% at the Duck-Under to 6% by volume below The Wall > 650 m from the entrance (Figure 6b). Concomitantly oxygen concentrations decreased, indicating a biogenic origin of CO_2. Potential sources of CO_2 include respiration of tree roots, bats, invertebrates, and microorganisms. Oxygen concentration increased unexpectedly at 450 m from the entrance before decreasing again at 650 m, in parallel with increasing CO_2. The cause of the anomalous oxygen level is unknown, but this area had the highest concentration of root curtains.

The amount of moisture and CO_2 in Bayliss Cave and other Undara caves varies with the seasons and major climatic events as evidenced by the data in Table S3. Although the environmental zones in Bayliss Cave are well constrained by passage shape and length, severe floods and prolonged droughts occasionally destabilize the zones in the cave. For example, on 31 May 1986, high water marks were noted at 2 and 2.4 m above the floor in the transition zone, which indicated that at least part of the cave occasionally floods. Additionally, the caliche surface crusts characteristically form where the soil is alternately wetted then dried such as in desert regions.

In other less extensive caves with more open entrances, e.g., Barkers and Pinwill, the zones are more seasonally dynamic, with the boundary of the deep zone moving closer to the entrance during the warmer wet periods and retreating during cold spells. The dynamic nature of the cave environment was observed in Nasty Cave between 29 and 30 May 1986. The entrance to Nasty is a narrow vertical crevice about 0.75 m wide, 3 m long and 3–4 m deep. A horizontal crawlway less than one metre high at the bottom of the crevice leads into the cave. A large boulder lies across the top of the crevice and blocks access. On 29 May, one edge of the rock was raised and tied in place to gain access and conduct an environmental survey. On 29 May, the cave was mostly in the deep-stagnant air zone, and the arthropod fauna was relatively diverse and abundant. During the repeat survey conducted one day later, far fewer animals were observed, and the troglobiontic *Nocticola* and polyxenid millipedes had disappeared. The cave air had become fresher from a change in weather and increased ventilation after enlarging of the entrance. The night of 28–29 May was cloudy and relatively warm; whereas the next was clear and cold, which resulted in a significant 'winter' effect; CO_2 levels near the inner end of the cave dropped from 5.1% to 3.4% (Table S3). The boulder was replaced across the entrance at the end of the survey.

5.3. Bayliss Cave Invertebrate Fauna Distribution

A survey of arthropod diversity was undertaken at the same time and stations as the environmental parameters in June 1985. The survey documented 46 species of arthropods and inferred their ecological status [1,2] (Figure 6c above). Combined with subsequent sampling, 50 arthropod species and one plant are currently recorded from Bayliss Cave, including 24 troglobionts and one stygobiont (Table 3); 16 inferred troglophiles (Table S1), some of which may be troglobionts when more is known of their biology; and ten cave visitors including trogloxenes, accidentals (Table S2).

In the 1985 survey, all species of terrestrial troglobionts were found beyond the Duck-Under in conditions of >98% RH and elevated CO_2, and only a few individuals of eight of these species were also found in moist areas and drip holes in the transition zone (Figure 6c) [2]. The stygobiontic amphipod was found among tree root mats growing in the intermittent stream near the bottom of the entrance slope. A total of 24 arthropod species were recorded from the talus slope to the Duck-Under, of which about 33% were troglobionts; 50% were troglophiles; and 17% were trogloxenes. Thirty species were found from the Duck-Under to The Wall, of which about 67% were troglobionts; 23%

were troglophiles; and 10% were trogloxenes. From The Wall to the crawlway, there were 28 species, of which about 68% were troglobionts; 18% were troglophiles; and 14% were trogloxenes (Figure 6c). Since some species were found in more than one station, the number of observations exceeded the number of species (i.e., 82 observations of 46 species).

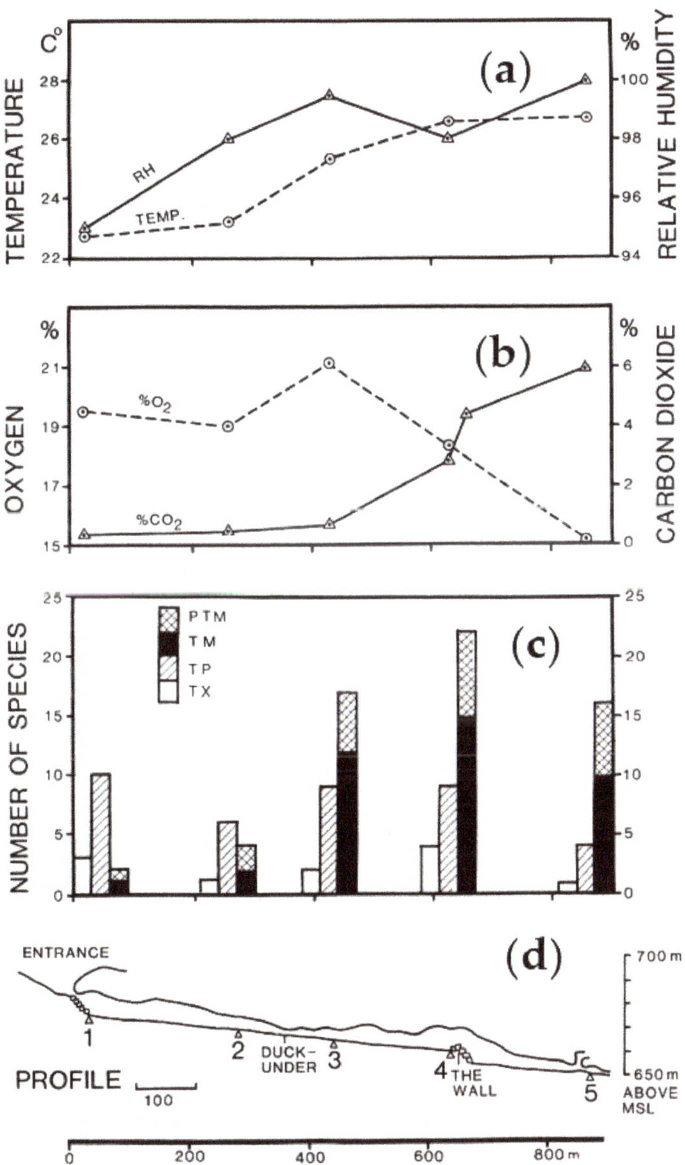

Figure 6. Environment and arthropod distribution in Bayliss Cave plots measured 14–15 June 1985 at progressively greater distances from the entrance: (**a**) temperature and relative humidity; (**b**) oxygen and carbon dioxide; (**c**) abundance of trogloxenes, troglophiles, partially troglomorphic and highly troglomorphic at the numbered positions in (**d**) map profile view with sampling stations numbered: 1—near entrance, 2—before the Duck-Under, 3—beyond the Duck-Under, 4—The Wall, 5—beyond The Wall. After [2], reproduced with author permission.

In addition to the animals (Figure 7), a primitive plant grows in the high CO_2 portion of Bayliss Cave (Figure 7). The plant is a whisk fern (Psilotales: *Psilotum* species) [72]. It is a saprophyte obtaining nutrients from the substrate. Whisk ferns are widespread in the tropics but rarely reported from caves. A different species of *Psilotum* also occurs in lava tubes on Maui and Hawaii islands.

Figure 7. *Cont.*

Figure 7. Bayliss Cave troglobionts and subterranean plant: (**a**) scutigeromorph centipede, photo F.G. Howarth; (**b**) Millipedes on bait, photo F.G. Howarth; (**c**) ectobiid cockroach *Neotemnopteryx baylissensis* Slaney, 2000, photo G. Thompson, Queensland Museum; (**d**) nocticolid cockroach *Nocticola* sp. 1, photo G. Smith; (**e**) cixiid planthopper *Solonaima baylissa* Hoch and Howarth, 1989, photo H. Reimer; (**f**) whisk fern, *Psilotum* sp. photo F.G. Howarth. (**g**) reduviid *Peirates* sp. preying on cockroach, *N. baylissensis*, photo F.G. Howarth; (**h**) atelurine silverfish, *Pseudogastrotheus undarae* Smith, 2016, photo G. Smith; (**i**) Bayliss Cave isopod, cf. Porcellionidae, photo J. Sydney; (**j**) Pinwill Cave isopod, photo J. Sydney; (**k**) pseudoscorpion, *Protochelifer* sp. nr. *cavernarum* Beier, 1967, photo A. Clarke; (**l**) Erebid moth, cf. *Schrankia* sp. adult with cocoon on tree root, photo G. Smith.

5.4. Notable Cave Species

5.4.1. Aquatic Fauna

Only one species of stygobiont is known from Undara: a blind amphipod (cf. *Chillagoe* sp.) from Road and Bayliss Caves. The nominate species, *Chillagoe thea* Barnard and Williams, 1995 was described from Chillagoe caves located 110 km north from Undara. An additional undescribed amphipod in the same genus is recorded from Camooweal caves, 600 km to the west [70].

5.4.2. Terrestrial Fauna

Spiders are well represented with at least six troglobiontic species known and several others that remain unidentified. A large eyeless ctenid, *Amauropelma undara* Raven and Gray 2001 and a tiny eyeless unidentified oonopid do not build webs but are a 'sit and wait' ambush predators. The blind zodariid spider, *Nosterella cavicola*, is also an ambush predator but females hunt from holes with raised turrets constructed in mud. An unidentified linyphiid builds intricate horizontal sheet webs across cracks and drip holes and hangs under the sheet and safely captures prey falling or landing on the sheet. A scaffold web spider, *Nesticella* species, builds tangled webs over drip holes, and prey become tangled in the loose strands of silk. Two species of long-legged spiders, *Spermophora* species 1 and 2, build loose webs on walls and in drip holes. Their webs act as tripwires and when alerted, the spider throws additional silk to envelope prey. Pisaurids are hunting spiders that prey on aquatic animals. Little has been reported on the status of the *Dolomedes* species 1 found near the lake in Barkers Cave. One potential troglobiont arachnid in Barkers Cave, a schizomid, was not detected in Bayliss.

One of the most remarkable troglobionts is the large scutigerid centipede. Troglophilic and epigean species of these long-legged centipedes escape threats and overtake prey by running extremely fast; whereas the Undara cave species (Figure 7) walks slowly (several seconds to move one body length) even when disturbed. With a body length of around seven centimetres, it is one of the largest terrestrial troglobionts known.

Two blind cockroaches live within Undara lava: the ectobiid, *Neotemnopteryx baylissensis* Slaney, 2000 and the nocticolid, *Nocticola* sp. 1 (Figure 7). Other species of *Neotemnopteryx* and *Nocticola* are recorded from other caves and tube systems in the McBride Volcanics, namely *Neotemnopteryx undarensis* Slaney, 2000 from the Hot Hole and Wishing Well in the northern tube system, and *Nocticola* sp. 2 from the Murronga flow. Two other ectobiid cockroaches are classified as troglophiles, *Paratemnopteryx stonei* Roth, 1990 in Bayliss and Barkers Caves, and *Paratemnopteryx howarthi* Roth, 1990 in Nasty Cave (Table S1).

At least four species of cave-adapted beetles are known. At time of writing the staphylinid and pselaphine have not been described; and descriptions of the two weevils, Curculionidae, are in preparation [67]. These unusual, long-legged and eyeless weevils are the first cave-adapted weevils to be described in Australia, while other undescribed subterranean Curculionidae and Pselaphinae are known from iron-ore terrains in Western Australia.

The most studied cave invertebrates at Undara (and Chillagoe) are multiple species of planthopper bugs in the family Cixiidae. Six species in the genus *Solonaima* exhibit varying degrees of cave adaptation, from epigean to troglophilic and fully troglobiontic species. At Chillagoe, two troglophilic species were generally found in the most open caves, the two moderately troglomorphic species were found in deeper caves, and the most highly modified—completely eyeless, colourless and nearly completely wingless—species, *Solonaima baylissa* Hoch and Howarth, 1989 (Figure 7) is restricted to humid cave passages with high CO_2 levels in Bayliss and Nasty caves. A second cixiid species, *Undarana rosella* Hoch and Howarth, 1989, is found closer towards the entrance in Bayliss Cave and is considered troglophilic. Nymphs of cixiids drill into tree roots and suck xylem sap, which is a dilute, nutrient-poor food, unlike phloem sap. Two other hemipteran bugs are predatory assassin bugs (Reduviidae); the smaller species is *Micropolytoxus cavicolus* Malipatil and Howarth, 1990, and the larger one is an undescribed species of *Peirates* (Figure 7). Both reduviid species have reduced eyes and do not respond to light; they also have reduced hemelytra, lack hind wings, and like other troglobionts, they move slowly.

Moths are often missed in biological surveys of caves, partly because it is difficult to collect specimens suitable for identification and because cave biologists often assume that lepidopterans use caves only for temporary refuge. However, at least two species of moths have troglomorphic populations that are restricted to caves: a tineid in the Philippines and an erebid, *Schrankia*, in Hawaii [73], while the tineid *Monopis* is a common guanophile in many Australian caves. Erebid larvae, pupae and adults found living on tree roots in Undara caves were tentatively identified as a species of *Schrankia* (Figure 7). We encourage biologists to include moths and similarly overlooked 'orphan' taxa in surveys. Additional cavernicolous moths are predicted to occur in tropical caves.

In the Murronga Basalt flow, situated about 15 km south of Undara crater, Long Shot Cave is the most diverse with at least four, possibly six, troglobionts. It is worth noting that the Murronga caves harbour troglobionts and undescribed potential troglobionts that are: (1) the same morphological species is found in both Murronga and Undara caves, namely *Solonaima baylissa*; (2) distinct congeneric species are found in the Murronga and Undara flows, namely the cockroach, *Nocticola* species; and (3) belong to taxa not detected in Undara caves, namely phalangid harvestman, *Zalmoxis lavacavernae* Hunt, 1993; the emesine bugs, *Ploiaria* spp. The range of the troglobiontic cockroach *Neotemnopteryx baylissensis* spans the Undara and Silent Hill Basalt flows.

6. Discussion

6.1. The Challenge of Classifying Cavernicoles

As Pipan et al. [28] summarised, considerable confusion exists in the literature about the terms troglobiont and stygobiont—which should be used only for species unable to survive in surface habitats, irrespective of their morphology—and troglomorph, species with reduced eyes and pigment and elongated appendages. The latter are not necessarily restricted to caves, while some species without conspicuous troglomorphic features are only found in caves and other subterranean habitats [27].

In many cases, the assignment of ecological category, especially between troglobiont versus non-troglobiont, is fraught with uncertainty because the ecology of many species is so poorly known. The dilemma is especially acute among groups displaying slight troglomorphy or belonging to groups that are primitively troglomorphic, such as silverfish. Two species of nicoletiid silverfish that are recorded from Undara caves illustrate this difficulty. *Metrinura subtropica* Smith, 2006 was initially classified as a troglobiont by Howarth and Stone [2] based on its apparent troglomorphy, behavior, and its collection deep inside Bayliss Cave. When subsequently describing this species, and another congeneric species collected on the surface at Undara, Smith [52] noted that appendage lengths and sensory appendages of *M. subtropica* were within the normal range for the genus, and therefore, he considered that this species may be a troglophile. In contrast, members of the subfamily Atelurinae are generally found living as specialised inquilines of ants or termites, and Howarth and Stone [2] considered Bayliss specimens as a trogloxene since individuals appeared to be associated with the common ant *Paratrechina* sp. However, Smith [53] later found dense populations (about 30 individuals/m^2) of *Pseudogastrotheus undarae* (Figure 7) deep inside Barkers Cave, in a zone with elevated carbon dioxide and without any obvious ant or other host present. Other species of Atelurinae have been collected without any obvious host, in soil and deep drill holes in iron-ore terrains in the Pilbara and Kimberley [74,75]. The presence of *P. undarae* deep inside the Undara caves and without hosts raises the question: Is the species troglobiontic?

While the Schiner–Racovitza system and its various derivative classification schemes are both useful, and sometimes confusing, the traditional focus of many biospeleologists on troglobionts and stygobionts, means that much cave biodiversity, biomass, and ecological function, is at risk of being under-recognised, and under-protected. In many cave ecosystems, most of the biomass and significant species richness, consists of trogloxenes, troglophiles and guanophiles. For these reasons we have chosen to list all species recorded from Undara lava caves and infer their ecological status based on expression of troglomorphy and behavior, as well as on their distribution in cave and surface collection records. However, taxonomic knowledge and field survey data in many cases is limited. With this proviso we caution that future field and taxonomic studies may determine that some taxa inferred to be troglobionts are in fact not, and vice versa.

Another dilemma facing biologists studying biodiversity is the question: What is a species? Several examples of the problem occur in this compilation of the Undara cave fauna. Isolated populations of morphologically similar individuals are often considered a single widespread species. However, more detailed study including behavior and molecular data can alter that view. For example, the pseudoscorpion previously known as *Protochelifer cavernarum* Beier, 1967 occurs in caves from southwest Australia to north Queensland, often associated with bat guano. It is recorded from the Undara and Murronga Basalts, as well as the Chillagoe karst. Subsequently, Moulds et al. [59] studied variation in DNA sequences of geographically dispersed Australian populations of *P. cavernarum* and found significant differences among the populations studied. They concluded that *P. cavernarum* was a complex of related locally endemic cave species.

Similarly, the cockroach, *Paratemnopteryx stonei*, is reported from Chillagoe and Undara. In describing the species, Roth [51] noted geographic variation calling the differences geographical races. Subsequently additional populations were collected from Broken River and Fanning River Caves. Slaney and Blair [21] analysed morphology and DNA, which confirmed the geographical differences. Later, Slaney [76] described the Fanning River population as a distinct species but left the other populations as races pending more detailed study. Other cavernicoles with geographically isolated populations may also prove to be complexes of locally endemic species.

6.2. Why Is Bayliss Cave So Rich in Troglobionts?

In terms of its species richness, Bayliss Cave stands out from all other caves in the Undara Basalt flow, and other flows in the McBride Province. There are two main reasons for the exceptional local richness in Bayliss Cave. First is size of the deep zone habitat. Its >550 m of highly suitable habitat is more than three times the size of the deep zone in the other known caves within the system. Second is the relative abundance and diversity of energy sources in the deep zone. The principal nutrients are numerous large tree root curtains, bat guano, and organic material. The latter filters in along roots, through cracks, or wanders into the cave. There is no leaflitter in Bayliss Cave as the constricted horizontal entrance limits inputs by gravity or air currents. These two factors, size and energy, allow colonization by greater numbers and diversity of cave-adapted animals from the pool of animals living in voids within the lava. The correlation between diversity and available energy agrees with the conclusions made by Brad et al. [77] in Movile Cave.

Two species of bats, *Rhinolophus megaphyllus* Gray, 1834 and *Miniopterus* species, also roost in the cave. Bats are also commonly found in most other McBride Province caves, including for example Barkers Cave, which contains many thousands of bats and a permanent lake to supply humidity. The bats and decomposing bat guano generate CO_2. However, Barkers Cave does not contain the same diversity as Bayliss Cave. This may be because of its large open entrance and shorter length (560+ m) so that most of the cave is in the transition zone, and experiences greater fluctuations in temperature and humidity that are less optimum for troglobionts. Additionally, Barkers Cave does not have such extensive tree root curtains, the food source for troglobiontic planthoppers, weevils and possibly other invertebrates.

Based on the discoveries at Undara, Howarth and Stone [2] proposed that passages with elevated levels of carbon dioxide and high humidity would be found, generally, to harbour unique communities of obligate cave species, and that these bad air zones may be the typical habitat present in mesocavernous cracks and voids. Two other lava tubes in the McBride Province corroborate this thesis. In both Nasty Cave in Undara flows and the Two Ten/Long Shot cave system in the Murronga Lava Flow, the distribution of troglobionts is correlated with CO_2 concentration [1].

The high-stress environment thesis has been sustained in the literature [19,78,79] and challenged by Humphreys [80] but never formally tested or refuted. In the years since the first observations that sparked this thesis were made in Bayliss Cave, several other karst areas in Australia with high carbon dioxide caves have been searched for troglobiont communities specifically associated with bad air zones, including: Camooweal (Queensland), Wellington and Bungonia (New South Wales), Cape Range, Roe Plains, Nullarbor Plain and Margaret River (Western Australia). Generally, bad air zones were observed to harbour troglobiont communities that were not noticeably richer or different to other deep zone habitats without bad air (S. Eberhard unpublished observations). On the Roe Plains and Nullarbor Plain, the bad air zones coincided exactly with deep zone humidity thus precluding disentanglement of these two variables by field observation. Additional research is required to resolve these conflicting observations on the role of carbon dioxide and other stressors on the biodiversity of caves.

6.3. Comparison with Other Subterranean Hotspots in Australia

Since the initial exciting discoveries made in Bayliss Cave more than 35 years ago, a great deal of research has been undertaken in other parts of Australia, especially in Western Australia where remarkably diverse stygobiont and troglobiont faunas have been revealed in arid and semi-arid zone limestone caves [81–83], calcrete and alluvial aquifers [84–86], and iron ore terrains [87,88]. Western Australia is now recognised as a globally significant hotspot for subterranean biodiversity [9]. It is therefore timely and appropriate to place the Undara cave system in context with other subterranean hotspots in Australia (Table 4).

In Australia, the terms 'troglofauna' and 'stygofauna' are commonly used, particularly in environmental impact assessments (EIA), to refer to all species collected from terrestrial and aquatic subterranean habitats, respectively. These terms do not distinguish ecological categories although they are sometimes misunderstood by non-specialists as equivalent in meaning to 'troglobiont' and 'stygobiont'. Irrespective of terminological ambiguity, accurate determination of a species ecological status is not crucial in most Australian EIA contexts, where the focus is on species conservation and extinction risk. The latter is usually assessed based on the sampled and/or interpreted distribution range of species in relation to the proposed mining footprint or other potential impact area.

Many species of troglofauna collected from calcrete and iron-ore terrains in Western Australia have typically small ranges [87]. A high proportion of these are likely to be true troglobionts, based on their troglomorphy, and/or short-range distribution and apparent absence from surface collections; a few may be soil fauna [88] and a few may be typical surface species lacking troglomorphies but occupying subterranean environments as refugia from arid surface conditions, without being present in the surrounding surface environment [8,88]. Nonetheless many of these species remain undescribed and their ecology poorly known, thus distinguishing between edaphophiles, troglophiles and troglobionts in weakly troglomorphic, or non-troglomorphic, taxa is difficult.

In compiling a preliminary list of Australia's subterranean hotspots (Table 4), we have therefore counted all recorded species, and specified the overall number of troglo/stygobionts if known. In selecting aquifer and cave/mesocavern systems to include, the geological continuity, hydrological connectivity, and overall areal extent have been taken into consideration. Some systems such as calcrete aquifers are compact and hydrogeologically well defined, whereas others such as Pilbara iron-ore formations and the Nullarbor karst (200,000 km^2) are geologically continuous over many thousands of square kilometres, yet with few if any identifiable barriers to subterranean dispersal. These large regional-scale subterranean systems are excluded as beta-diversity will be dominating alpha diversity and obscuring localised hotspots. Table 4 is not intended to be comprehensive or definitive, rather its purpose is to establish a context and framework for recognising and comparing significant hotspots and regionally significant warm spots. Patterns and highlights emerging from Table 4 include:

1. Undara remains the richest subterranean hotspot in humid tropical Australia, which far exceeds the richness of troglobionts recorded in other humid tropical Australian karsts such as Judburra-Gregory [89].
2. In the arid Yilgarn and semi-arid Pilbara regions of Western Australia significantly richer subterranean assemblages have been documented in recent decades.
3. The richest Western Australian hotspots are karstified calcrete aquifers [90–95], although mineralized iron-ore terrains [88,96,97] and 'hard-rock' limestone karsts [81–83,98,99] also harbour very-rich assemblages.
4. Most Western Australian hotspots in calcretes and iron-ore terrains do not contain enterable caves, and fauna can only be collected by sampling mesocaverns and microcaverns via constructed wells and drill holes.
5. The Yeelirrie calcrete aquifer stands out as exceptionally rich with 70 stygofauna and 45 troglofauna species, the majority being short range endemics and almost certainly obligate subterranean species based on current knowledge of the groups represented [90–92].

6. In Eastern Australia, the Jenolan karst in New South Wales recorded the highest species richness (136 taxa) however the majority of these are accidentals and troglophiles, and only 8 taxa are obligate cavernicoles [100].

7. Overall species richness in warm temperate (New South Wales) [101–103] and cool temperate (Tasmania) [104] karst areas is comparable (median 54 taxa), and these karsts harbour a much lower proportion of obligate subterranean species compared with arid and semi-arid regions, although Tasmania stands out in terms of troglobiont richness in temperate latitudes (maximum 25 species at Precipitous Bluff, Tasmania) [104–108].

8. The Peel Valley alluvial aquifer in New South Wales harbours the richest known stygofauna assemblage in eastern Australia; 54 species including 33 stygobionts [109].

9. In terms of obligate species, the richest Australian localities are in arid and semi-arid climate regions, where most of the troglobionts and stygobionts are relictual with no close surface relatives. Molecular phylogenetic studies have shown that Quaternary aridification is the likely driving mechanism for troglo/stygogenesis in these regions [110].

Table 4. Preliminary list of Australian hotspots (>30 obligate spp.) and regionally significant warm spots, grouped by geographic/climate region. Each locality is considered to represent a single subterranean ecosystem, characterised by geological and hydrological connectivity (excluding large-scale subterranean systems such as Nullarbor karst and Pilbara iron-ore ranges). Total species richness is the number of stygofauna and troglofauna (ecological categories combined) and the number in brackets (n) is the number of stygobionts (Sb) and troglobionts (Tb) where known. * For most localities in Western Australia (WA) the ecological status of many species remains unspecified however the majority are short-range endemics and almost certainly obligate subterranean species based on the groups represented.

Geographic/Climate Region	Locality Name; Geology, Hydrology Type	Total spp. Richness (No. Sb/Tb) *	Stygofauna spp. (Sb) *	Troglofauna spp. (Tb) *	Sources
Queensland, humid tropical	Undara Basalt lava tubes	77 (31)	1 (1)	76 (30)	[2], and text
Northern Territory, humid tropical	Judbarra-Gregory karst	56 (7)	3 (2)	53 (5)	[89]
Yilgarn, WA, arid	Yeelirrie calcrete aquifer	115 (*)	70 (*)	45 (*)	
Yilgarn, WA, arid	Uramurdah calcrete	45 (*)	36 (*)	9 (*)	[90–92]
Yilgarn, WA, arid	Hinkler Well calcrete	41 (*)	32 (*)	9 (*)	[95]
Yilgarn, WA, arid	Lake Violet calcrete	39 (*)	35 (*)	4 (*)	[95]
Yilgarn, WA, arid	Barwidgee calcrete	37 (*)	28 (*)	9 (*)	[95]
Pilbara, WA, arid/semi-arid	Ethel Gorge calcrete aquifer	84 (45)	84 (45)	0	[95]
Pilbara, WA, arid/semi-arid	Cape Range karst	83 (*)	42 (*)	41 (*)	
Pilbara, WA, arid/semi-arid	Barrow Island karst	74 (*)	56 (*)	18 (*)	[94]
Pilbara, WA, arid/semi-arid	Well PSS016, Robe River calcrete aquifer	54 (*)	54 (*)	0	[82,98]
Pilbara, WA, arid/semi-arid	Mesa A iron pisolite, Robe Valley	24 (*)	0	24 (*)	[83,99]
New South Wales, warm temperate	Jenolan karst	136 (8)	10 (2)	126 (6)	[85,97]
New South Wales, warm temperate	Wombeyan karst	55 (7)	5 (2)	50 (5)	[96]

Table 4. *Cont.*

Geographic/Climate Region	Locality Name; Geology, Hydrology Type	Total spp. Richness (No. Sb/Tb) *	Stygofauna spp. (Sb) *	Troglofauna spp. (Tb) *	Sources
New South Wales, warm temperate	Wee Jasper karst	53 (7)	5 (3)	48 (4)	
New South Wales, warm temperate	Peel Valley alluvial aquifer	54 (33)	54 (33)	0	[100]
Tasmania, cool temperate	Ida Bay karst	65 (18)	17 (6)	48 (12)	[101–103]
Tasmania, cool temperate	Junee Florentine karst	60 (20)	17 (8)	43 (12)	[101–103]
Tasmania, cool temperate	Precipitous Bluff karst	37 (25)	11 (11)	26 (14)	[109]

Supplementary Materials: The following are available online at https://www.mdpi.com/article/10.3390/d13070326/s1, Table S1: Troglophiles; Table S2: Trogloxenes, accidentals; Table S3: Environmental data.

Author Contributions: S.M.E. led the efforts to compile the data and completed the final formatting to comply with the journal style; S.M.E. and F.G.H. contributed equally to the conception, data analysis, and writing and editing of the drafts. All authors have read and agreed to the published version of the manuscript.

Funding: Funding for Field work in the 1980s for F.G.H. was provided by The Explorer's Club, NY, USA, through sponsorship by Bro Nicholas Sullivan and by U.S. National Science Foundation Grant BSR-85-15183 to F.G.H.

Institutional Review Board Statement: Not applicable.

Informed Consent Statement: Not applicable.

Data Availability Statement: Not applicable.

Acknowledgments: We acknowledge and greatly appreciate the terrific help from staff at the Queensland Museum and the Australian Museum who tracked down specimen records and gave advice. At the Queensland Museum: Owen Seeman (Arachnida and Myriapoda), Marissa McNamara (Crustacea), Robert Raven (Arachnida), Susan Wright (Insecta). At the Australian Museum: Helen Smith (Arachnida, Insecta, Crustacea), Mandy Reid (Malacology). Peter Bannink (Chillagoe Caving Club) generously provided much information and material including his own unpublished specimen records, and drafted the excellent maps. For providing photographs and additional information we thank Arthur Clarke, Hannelore Hoch, Greg Middleton, Graeme Smith, Joe Sydney. We appreciate comments received from three anonymous reviewers, and the academic editors, which greatly improved the manuscript. This paper is dedicated to the memory of Fred Stone, a dear colleague, mentor, caver, and a kind and gentle man.

Conflicts of Interest: The authors declare no conflict of interest.

References

1. Howarth, F.G. Environmental Ecology of North Queensland Caves: Why are there so many troglobites in Australia? In Proceedings of the Tropicon 1988, 17th Biennial Australian Speleological Conference, Tinaroo, QLD, Australia, 27–31 December 1988; pp. 76–84.
2. Howarth, F.G.; Stone, F.D. Elevated carbon dioxide levels in Bayliss Cave, Australia: Implications for the evolution of obligate cave species. *Pac. Sci.* **1990**, *44*, 207–218.
3. Culver, D.C.; Sket, B. Hotspots of subterranean biodiversity in caves and wells. *J. Cave Karst Stud.* **2000**, *62*, 11–17.
4. Hoch, H.; Howarth, F.G. The evolution of cave-adapted cixiid planthoppers in volcanic and limestone caves in north Queensland, Australia (Homoptera: Fulgoroidea). *Mem. Biospeol.* **1989**, *16*, 17–24.
5. Hoch, H.; Howarth, F.G. Reductive evolutionary trends in two new cavernicolous species of a new Australian cixiid genus (Homoptera: Fulgroidea). *Syst. Entomol.* **1989**, *14*, 179–196. [CrossRef]

6.	Howarth, F.G. High-stress subterranean habitats and evolutionary change in cave-inhabiting arthropods. *Am. Nat.* **1993**, *142*, S65–S77. [CrossRef] [PubMed]

7.	Hamilton-Smith, E. The Arthropoda of Australian caves. *J. Aust. Entomol. Soc.* **1967**, *6*, 103–118. [CrossRef]

8.	Humphreys, W.F. The significance of the subterranean fauna in biogeographical reconstruction: Examples from Cape Range Peninsula, Western Australia. *Rec. West. Aust. Mus. Suppl.* **1993**, *45*, 165–192.

9.	Guzik, M.T.; Austin, A.D.; Cooper, S.; Harvey, M.; Humpherys, W.F.; Bradford, T.; Eberhard, S.; King, R.A.; Leys, R.; Muirhead, K.A.; et al. Is the Australian subterranean fauna uniquely diverse? *Invertebr. Syst.* **2010**, *24*, 407–418. [CrossRef]

10.	Vandel, A. *Biospeleology—The Biology of Cavernicolous Animals*; Elsevier: London, UK, 1965.

11.	Barr, T.C.J. Cave ecology and the evolution of troglobites. In *Evolutionary Biology*; Hecht, M.K., Steere, W.C., Eds.; Appleton-Century-Crofts: New York, NY, USA, 1968; Volume 2, pp. 35–102.

12.	Howarth, F.G. The tropical cave environment and the evolution of troglobites. In Proceedings of the 9th Congreso Internacional de Espeleologia, Barcelona, Spain, 9–15 August 1986; pp. 153–155.

13.	Leleup, N. Premier Partie. In *Mission Zoologique Belge aux iles Galapagos et en Ecuador (N. et J. Leleup, 1964–1965) Resutats Scientifiques*; Koninklijk Museum voor Midden-Africa—Musee Royal de l'Afrique Centrale: Tervueren, Belgium, 1968; pp. 9–34.

14.	Howarth, F.G. Cavernicoles in lava tubes on the island of Hawaii. *Science* **1972**, *75*, 325–326. [CrossRef] [PubMed]

15.	Elias-Gutierrez, M.; Martinez Jeronimo, F.; Ivanova, N.V.; Valdez-Moreno, M.; Hebert, P.D.N. DNA barcodes for Cladocera and Copepoda from Mexico and Guatemala highlights and new discoveries. *Zootaxa* **2008**, *1839*, 1–42. [CrossRef]

16.	Martínez, A.; González, B.C. Volcanic Anchialine Habitats of Lanzarote. In *Cave Ecology*; Moldovan, O.T., Kovác, L., Halse, S.A., Eds.; Ecological Studies, 235; Springer: Cham, Switzerland, 2018; pp. 195–228.

17.	Peck, S.B. The Invertebrate Fauna of Tropical American Caves, Part III: Jamaica, An Introduction. *Int. J. Speleol.* **1975**, *7*, 303–326. [CrossRef]

18.	Culver, D.C.; Deharveng, L.; Bedos, A.; Lewis, J.J.; Madden, M.; Reddell, J.R.; Sket, B.; Trontelj, P.; White, D. The mid-latitude biodiversity ridge in terrestrial cave fauna. *Ecography* **2006**, *29*, 120–128. [CrossRef]

19.	Deharveng, L.; Bedos, A. The cave fauna of southeast Asia. Origin, evolution, and ecology. In *Subterranean Ecosystems*; Wilken, H., Culver, D.C., Humphreys, W.F., Eds.; Elsevier: Amsterdam, The Netherlands, 2000; pp. 603–632.

20.	Soulier-Perkins, A. Phylogenetic evidence for multiple invasions and speciation in caves: The Australian planthopper genus Solonaima (Hemiptera: Fulgoromorpha: Cixiidae). *Syst. Entomol.* **2005**, *30*, 281–288. [CrossRef]

21.	Slaney, D.P.; Blair, D. Molecules and Morphology are Concordant in Discriminating among Populations of Cave Cockroaches in the Genus Paratemnopteryx Saussure (Blattodea: Blattellidae). *Ann. Entomol. Soc. Am.* **2000**, *93*, 398–404. [CrossRef]

22.	Medeiros, M.J.; Davis, D.; Howarth, F.G.; Gillespie, R. Evolution of cave living in Hawaiian Schrankia (Lepidoptera: Noctuidae) with description of a remarkable new cave species. *Zool. J. Linn. Soc.* **2009**, *156*, 114–139. [CrossRef]

23.	Howarth, F.G.; Hoch, H.; Wessel, A. Adaptive shifts. In *Encyclopedia of Caves*, 3rd ed.; Culver, D.C., White, W., Eds.; Academic Press: Burlington, MA, USA, 2019; pp. 47–55.

24.	Howarth, F.G. A comparison of volcanic and karstic cave communities. In Proceedings of the 7th International Symposium on Vulcanospeleology, Santa Cruz de la Palma, Canary Islands, Spain, 4–11 November 1994; pp. 63–68.

25.	Howarth, F.G. Ecology of cave arthropods. *Annu. Rev. Entomol.* **1983**, *28*, 365–389. [CrossRef]

26.	Howarth, F.G.; Moldovan, O.T. Where cave animals live. In *Cave Ecology*; Moldovan, O.T., Kováč, L'., Halse, S., Eds.; Springer: Cham, Switzerland, 2018; pp. 23–37.

27.	Culver, D.; Pipan, T. *The Biology of Caves and Other Subterranean Habitats*, 2nd ed.; Oxford University Press: Oxford, UK, 2019; p. 301.

28.	Pipan, T.; Deharveng, L.; Culver, D.C. Hotspots of Subterranean Biodiversity. *Diversity* **2020**, *12*, 209. [CrossRef]

29.	Stone, F.D. Bayliss Lava Tube and the Discovery of a Rich Cave Fauna in Tropical Australia. In Proceedings of the 14th International Symposium on Vulcanospeleology, Undara National Park, Mount Surprise, QLD, Australia, 12–17 August 2010; pp. 47–58.

30.	Clarke, A. An Overview of Invertebrate Fauna Collections from the Undara Lava Tube System. In Proceedings of the 14th International Symposium on Vulcanospeleology, Undara National Park, Mount Surprise, QLD, Australia, 12–17 August 2010; pp. 59–76.

31.	Fontaine, B.; Perrad, A.; Bouchet, P.V.N.P. 21 years of shelf life between discovery and description of new species. *Curr. Biol.* **2012**, *22*, 943–944. [CrossRef]

32.	Bannink, P. List of invertebrates collected from the McBride Volcanic Province lava tubes and associated rainforest depressions. Unpublished Report. 1–12.

33.	Godwin, M.D. *Undara and Associated Lava Fields of McBride Plateau, a Speleological Field Guide*; Chillagoe Caving Club, Inc.: Cairns, Australia, 1993.

34.	Godwin, M.D.; Pearson, L.M. The Murronga lava flow. In Proceedings of the Cave Leeuwin Conference, Margaret River, WA, Australia, 30 December 1990–5 January 1991; pp. 34–54.

35.	Pearson, L.M. *Field Guide to the Lava Tubes on the McBride Volcanic Province in North Queensland*; Chillagoe Caving Club, Inc.: Cairns, Australia, 2010.

36.	Howarth, F.G.; Moldovan, O.T. The ecological classification of cave animals and their adaptations. In *Cave Ecology*; Moldovan, O.T., Kovác, L., Halse, S., Eds.; Ecological Studies; Springer: Cham, Switzerland, 2018; pp. 41–67.

37.	Cloudsley-Thompson, J.L. On the responses to environmental stimuli, and the sensory physiology of Millipedes (Diplopoda). *Proc. Zool. Soc. Lond.* **1951**, *121*, 253–277. [CrossRef]

38.	Griffin, T.J. The Geology, Mineralogy and Geochemistry of the McBride Basaltic Province, Northern Queensland. Ph.D. Thesis, James Cook University, Douglas, QLD, Australia, 1977.

39. Cohen, B.E.; Mark, D.F.; Fallon, S.J.; Stephenson, P.J. Holocene-Neogene volcanism in northeastern Australia: Chronology and eruption history. *Quat. Geochronol.* **2017**, *39*, 79–91. [CrossRef]
40. Griffin, T.G.; McDougall, I. Geochronology of the Cainozoic McBride Volcanic Province Northern Queensland. *J. Geol. Soc. Aust.* **1975**, *22*, 387–396. [CrossRef]
41. Atkinson, A. The Undara lava tube system and its caves. *Helictite* **1990**, *28*, 3–14.
42. Malipatil, M.B.; Howarth, F.G. Two new species of *Micropolytaxus* Elkins from Northern Australia (Hemiptera: Reduviidae: Saicinae). *J. Aust. Entomol. Soc.* **1990**, *29*, 37–40. [CrossRef]
43. Atkinson, A.; Atkinson, V. *Undara Volcano and Its Lava Tubes: A Geological Wonder of Australia in Undara Volcanic National Park, North Queensland*; Atkinson, Anne & Vernon: Brisbane, Australia, 1995; pp. 1–96.
44. Rein, T.; Kempe, S.; Dufresne, A. The "Cueva del Viento" on the Canaries, Spain. In Proceedings of the 17th International Vulcanspeleology Symposium Ocean View, Hawaii, HI, USA, 6–12 February 2016; pp. 1–8.
45. Peterson, D.W.; Holcomb, R.T.; Tilling, R.I.; Christiansen, R.L. Development of lava tubes in the light of observations at Mauna Ulu, Kilauea Volcano, Hawaii. *Bull. Volcanol.* **1994**, *56*, 343–360. [CrossRef]
46. Whitehead, P.W. The Regional Context of the McBride Basalt Province and the Formation of the Undara Lava Flows, Tubes, Rises and Depressions. In Proceedings of the 14th International Symposium on Vulcanospeleology, Undara National Park, Mount Surprise, QLD, Australia, 12–17 August 2010; pp. 9–18.
47. Atkinson, A.; Griffin, T.J.; Stevenson, P.J. A major lava tube system, North Queensland. *Bull. Volcanol.* **1975**, *39*, 1–28. [CrossRef]
48. Stephenson, P.J.; Griffin, T.J.; Sutherland, F.L.I.E.P. Cainozoic volcanism in north-eastern Australia. In *Geology and Geophysics of North-Eastern Australia*; Henderson, R.A., Stephenson, P.J., Eds.; Geological Association of Australia, Queensland Division: Douglas, QLD, Australia, 1980; pp. 349–374.
49. Bureau of Meteorology. Climate Statistics for Australian Locations. Available online: http://www.bom.gov.au/climate/averages/tables/cw_030036.shtml (accessed on 25 April 2021).
50. Bland, R.G.; Weinstein, P.; Slaney, D.P. Mouthpart sensilla of cave species of Australian Paratemnopteryx cockroaches (Blattaria: Blattellidae). *Int. J. Insect Morphol. Embryol.* **1998**, *16*, 291–300. [CrossRef]
51. Roth, L.M. A revision of the Australian Parcoblattini (Blattaria: Blattellidae: Blattellinae). *Mem. Qld. Mus. Nat.* **1990**, *28*, 531–596.
52. Smith, G.B. New species of *Metrinura* Mendes (Zygentoma: Nicoletiidae) from Queensland, Australia. *Aust. J. Entomol.* **2006**, *15*, 163–167. [CrossRef]
53. Smith, G.B. New Atelurinae (Zygentoma: Nicoletiidae) from Northern Australia. *Gen. Appl. Entomol.* **2016**, *44*, 21–58.
54. Stone, F.D.; (University of Hawaii). Personal Communication, 1985.
55. Davies, V.T. The huntsman spiders Heteropoda Latreille and Yiinthi gen. nov. (Araneae: Heteropodidae) in Australia. *Mem. Qld. Mus.* **1994**, *35*, 75–122.
56. Dyce, A.L.; Wellings, G. Phlebotomine sandflies (Diptera: Psychodidae) from caves in Australia. *Parasitologia* **1991**, *33*, 193–198.
57. Bradbury, J.H.; (University of Adelaide). Personal Communication, 2000.
58. Harvey, M.S. New cave-dwelling schizomids (Schizomida: Hubbardiidae) from Australia. *Rec. West. Aust. Mus. Suppl.* **2001**, *64*, 171–185. [CrossRef]
59. Moulds, T.A.; Murphy, N.; Adams, M.; Reardon, T.B.; Harvey, M.S.; Jennings, J.; Austin, A.D. Phylogeography of cave pseudoscorpions in southern Australia. *J. Biogeogr.* **2007**, *34*, 951–962. [CrossRef]
60. Raven, R.J.; Stumkat, K.; Gray, M.R. Revisions of Australian ground-hunting spiders: I. Amauropelma gen. nov. (Araneomorphae: Ctenidae). *Rec. West. Aust. Mus. Suppl.* **2001**, *64*, 187–227. [CrossRef]
61. Gray, M.R. Cavernicolous spiders (Araneae) from Undara, Queensland and Cape Range, Western Australia. *Helictite* **1989**, *27*, 87–89.
62. Main, B.Y. *Spiders*; Collins: Sydney, Australia, 1984.
63. Gray, M.R. Survey of the spider fauna of Australian caves. *Helictite* **1973**, *11*, 46–75.
64. Baehr, B.C.; Jocqué, R. The new endemic Australian genus Nosterella and a review of Nostera (Araneae: Zodariidae), including eight new species. *Mem. Qld. Mus. Nat.* **2017**, *60*, 53–76.
65. Slaney, D.P. New species of cave dwelling cockroaches in the genus Neotemnopteryx Princis (Blattaria: Blattellidae: Blattellinae). *Mem. Qld. Mus. Nat.* **2000**, *46*, 331–336.
66. Stone, F.D. The cockroaches of North Queensland caves and the evolution of tropical troglobites. In Proceedings of the Tropicon 1988, 17th Biennial Australian Speleological Conference, Tinaroo, QLD, Australia, 27–31 December 1988; pp. 88–93.
67. Escalona, H.; Oberprieler, R. *Australian National Insect Collection*; CSIRO: Canberra, Australia, 2021.
68. Hoch, H.; Howarth, F.G. Six new cavernicolous cixiid planthoppers in the genus Solonaima from Australia (Homoptera: Fulgoroidea). *Syst. Entomol.* **1989**, *14*, 377–402. [CrossRef]
69. Howarth, F.G. Bioclimatic and geologic factors governing the evolution and distribution of Hawaiian cave insects. *Entomol. Gen.* **1983**, *8*, 17–26. [CrossRef]
70. Eberhard, S.M. Nowranie Caves and the Camooweal karst area, Queensland: Hydrology, geomorphology and speleogenesis, with notes on aquatic biota. *Helictite* **2003**, *38*, 27–38.
71. Eberhard, S.M. Ecology and Hydrology of a Threatened Groundwater-Dependent Ecosystem: The Jewel Cave Karst System in Western Australia. Ph.D. Thesis, School of Environmental Science, Murdoch University, Perth, Australia, 2004.
72. Wagner, W.H. (University of Michigan, Ann Arbor, MI, USA). Personal Communication, 1988.

73. Howarth, F.G.; Medeiros, M.J.; Stone, F.D. Hawaiian lava tube cave associated Lepidoptera from the collections of Francis G Howarth and Fred D Stone. *Bish. Mus. Occas. Pap.* **2020**, *129*, 37–54.

74. Smith, G.B.; McRae, J.M. New species of subterranean silverfish (Zygentoma: Nicoletiidae: Atelurinae) from Western Australia's semi-arid Pilbara region. *Rec. West. Aust. Mus.* **2014**, *29*, 105–127. [CrossRef]

75. Smith, G.B.; McRae, J.M. Further short range endemic troglobitic silverfish (Zygentoma: Nicoletiidae; Subnicoletiinae and Coletiniinae) from north-western Australia. *Rec. West. Aust. Mus.* **2016**, *31*, 41–55. [CrossRef]

76. Slaney, D.P. New species of Australian cockroaches in the genus *Paratemnopteryx* Saussure (Blattaria, Blattellidae, Blattellinae), and a discussion of some behavioural observations with respect to the evolution and ecology of cave life. *J. Nat. Hist.* **2001**, *35*, 1001–1012. [CrossRef]

77. Brad, T.; Lepure, S.; Sarbu, S.M. The Chemoautotrophically Based Movile Cave Groundwater Ecosystem, a Hotspot of Subterranean Biodiversity. *Diversity* **2021**, *13*, 128. [CrossRef]

78. Deharveng, L.; Bedos, A. Gaz carbonique. In *Thai-Maros 85, Rapport Speleologique et Scientifique to Thailand and Sulawesi*; Association Pyreneenne de Speleologie: Toulouse, France, 1986; pp. 144–152.

79. Stone, F.D.; Howarth, F.G.; Hoch, H.; Asche, M. Root communities in lava tubes. In *Encyclopedia of Caves*, 2nd ed.; White, W.B., Culver, D.C., Eds.; Academic Press: Burlington, MA, USA, 2012; pp. 658–664.

80. Humphreys, W.F. Where angels fear to tread: Developments in cave ecology. In *Cave Ecology*; Moldovan, O.T., Kovác, L., Halse, S., Eds.; Ecological Studies; Springer: Cham, Switzerland, 2018; pp. 497–532.

81. Humphreys, W.F. (Ed.) *The Biogeography of Cape Range, Western Australia*; Records of the Western Australian Museum Supplement; Western Australian Museum: Perth, Australia, 1993; Volume 45, pp. 1–248.

82. Poore, G.C.B.; Humphreys, W.F. Bunderanthura bundera gen. et sp. nov. from Western Australia, first anchialine Leptanthuridae (Isopoda) from the Southern Hemisphere. *Rec. West. Aust. Mus.* **2013**, *28*, 21–29. [CrossRef]

83. Humphreys, G.; Alexander, J.; Harvey, M.; Humphreys, W.F. The subterranean fauna of Barrow Island, northwestern Australia. *Rec. West. Aust. Mus. Suppl.* **2013**, *83*, 145–158. [CrossRef]

84. Eberhard, S.M.; Halse, S.A.; Humphreys, W.F. Stygofauna in the Pilbara region, north-west Western Australia: A review. *J. R. Soc. West. Aust.* **2005**, *88*, 167–176.

85. Eberhard, S.M.; Halse, S.A.; Williams, M.; Scanlon, M.D.; Cocking, J.S.; Barron, H.J. Exploring the relationship between sampling efficiency and short range endemism for groundwater fauna in the Pilbara region, Western Australia. *Freshw. Biol.* **2009**, *54*, 885–901. [CrossRef]

86. Humphreys, W.F. Groundwater calcrete aquifers in the Australian arid zone: The context to an unfolding plethora of stygal diversity. *Rec. West. Aust. Mus. Suppl.* **2001**, *64*, 233–234. [CrossRef]

87. Halse, S.A. Research in calcretes and other deep subterranean habitats outside caves. In *Cave Ecology*; Moldovan, O.T., Kovác, L., Halse, S.A., Eds.; Ecological Studies; Springer: Cham, Switzerland, 2018; pp. 415–434.

88. Halse, S.A.; Pearson, G.B. Troglofauna in the vadose zone: Comparison of scraping an dtrapping results and sampling adequacy. *Subterr. Biol.* **2014**, *13*, 17–34. [CrossRef]

89. Moulds, T.; Bannink, P. Preliminary notes on the cavernicolous arthropod fauna of Judburra/Gregory karst area, northern Australia. *Helictite* **2012**, *41*, 75–85.

90. Bennelongia Pty Ltd. *Yeelirrie Subterranean Fauna Assessment*; Report Prepared for Cameco Australia Pty Ltd.; 2015/236b; Bennelongia Pty Ltd.: Perth, Australia, 2015.

91. Eberhard, S.M.; Watts, C.H.S.; Callan, S.K.; Leijs, R. Three new subterranean diving beetles (Coleoptera: Dytiscidae) from the Yeelirrie groundwater calcretes, Western Australia, and their distribution between several calcrete deposits including a potential mine site. *Rec. West. Aust. Mus.* **2016**, *31*, 27–40. [CrossRef]

92. Subterranean Ecology Pty Ltd. *Yeelirrie Subterranean Fauna Survey*; Report Prepared for BHP Billiton Yeelirrie Development Company Pty Ltd.; 2010/14; Subterranean Ecology Pty Ltd.: Perth, Australia, 2011; p. 269.

93. Subterranean Ecology Pty Ltd. *Ethel Gorge Aquifer Threatened Ecological Community—Consolidated Taxonomy*; Unpublished Report Prepared for BHP Billiton Iron Ore; Subterranean Ecology Pty Ltd.: Perth, Australia, 2013; p. 96.

94. Tang, D.; Eberhard, S.M. Two new species of Nitocrella (Crustacea, Copepoda, Harpacticoida) from groundwaters of northwestern Australia expand the geographic range of the genus in a global hotspot of subterranean biodiversity. *Subterr. Biol.* **2016**, *20*, 51–76. [CrossRef]

95. Toro Energy Limited. *Extension to the Wiluna Uranium Project: Environmental Management Plan: Subterranean Fauna Management Plan*; Toro Energy Limited: Perth, Australia, 2012.

96. Biota Environmental Services Pty Ltd. *Rio Tinto Regional Troglobitic Fauna Study*; Unpublished Report Prepared for Rio Tinto Iron Ore; Biota Environmental Services Pty Ltd.: Perth, Australia, 2013.

97. Halse, S.A.; Scanlon, M.D.; Cocking, J.S.; Barron, H.J.; Richardson, J.B.; Eberhard, S.M. Pilbara stygofauna: Deep groundwater of an arid landscape contains globally significant radiation of biodiversity. *Rec. West. Aust. Mus. Suppl.* **2014**, *78*, 443–483. [CrossRef]

98. Bennelongia Pty Ltd. *Stygofauna Survey—Exmouth Cape Aquifer: Scoping Document Describing Work Required to Determine Ecological Water Requirements for the Exmouth Cape Aquifer*; Prepared for Department of Water; Bennelongia Pty Ltd.: Perth, Australia, 2008; p. 39.

99. King, R.A.; Fagan-Jeffries, E.; Bradford, T.M.; Stringer, D.N.; Finston, T.; Halse, S.A.; Eberhard, S.M.; Humphreys, G.; Humphreys, W.F.; Austin, A.D.; et al. Cryptic Diversity down under: Defining species in the subterranean amphipod genus Nedsia Barnard and Williams (Hadzioidea: Eriopisidae) from the Pilbara, Western Australia. *Invertebr. Syst.* **2021**. in review.

100. Eberhard, S.; Smith, G.; Gibian, M.; Smith, H.; Gray, M. Invertebrate Cave Fauna of Jenolan. *Proc. Linn. Soc. N. S. W.* **2014**, *136*, 35–67.

101. Eberhard, S.M.; Spate, A. *Cave Invertebrate Survey: Toward an Atlas of NSW Cave Fauna*; Report Prepared under the NSW Heritage Assistance Program NEP 94 765; NSW Heritage Assistance Program: Canberra, Australia, 1995; pp. 1–112.
102. Thurgate, M.E.; Gough, J.S.; Clarke, A.; Serov, P.; Spate, A. Stygofauna diversity and distribution in Eastern Australian cave and karst areas. *Rec. West. Aust. Mus.* **2001**, *64*, 49–62. [CrossRef]
103. Thurgate, M.E.; Gough, J.S.; Spate, A.; Eberhard, S.M. Subterranean biodiversity in New South Wales: From rags to riches. *Rec. West. Aust. Mus. Suppl.* **2001**, *64*, 37–47. [CrossRef]
104. Eberhard, S.M.; Richardson, A.M.M.; Swain, R. *The Invertebrate Cave Fauna of Tasmania*; Report to the National Estate Office, Canberra; Zoology Department, University of Tasmania: Hobart, Australia, 1991; p. 174.
105. Ahyong, S.T. The Tasmanian Mountain Shrimps, Anaspides Thomson, 1894 (Crustacea, Syncarida, Anaspididae). *Rec. Aust. Mus.* **2016**, *68*, 313–364. [CrossRef]
106. Eberhard, S.M.; Giachino, P.M. Tasmanian Trechinae and psydrinae (Coleoptera, Carabidae): A taxonomic and biogeographic synthesis, with description of new species and evaluation of the impact of Quaternary climate changes on evolution of the subterranean fauna. *Subterr. Biol.* **2011**, *9*, 1–72.
107. Karanovic, I.; Eberhard, M.S.; Perina, G. Austromesocypris bluffensis sp. n. (Crustacea, Ostracoda, Cypridoidea, Scottiinae) from subterranean aquatic habitats in Tasmania, with a key to world species of the subfamily. *ZooKeys* **2012**. [CrossRef] [PubMed]
108. Ponder, W.F.; Clark, S.A.; Eberhard, S.M.; Studdert, J.B. A radiation of hydrobiid snails in the caves and streams at Precipitous Bluff, southwest Tasmania, Australia (Mollusca: Caenogastropoda: Rissooidea: Hydrobiidae *s.l.*). *Zootaxa* **2005**, *1074*, 1–66. [CrossRef]
109. Tomlinson, M. A Framework for Determining Environmental Water Requirements for Alluvial Aquifer Ecosystems. Ph.D. Thesis, University of New England, Armidale, Australia, 2008.
110. Humphreys, W.F. Australasian subterranean biogeography. In *Handbook of Australasian Biogeography*; Ebach, M.C., Ed.; CRC Press: Boca Raton, FL, USA, 2017; pp. 269–293.

Article

Bermuda's Walsingham Caves: A Global Hotspot for Anchialine Stygobionts

Thomas M. Iliffe * and Fernando Calderón-Gutiérrez

Department of Marine Biology, Texas A&M University at Galveston, Galveston, TX 77553-1675, USA; fercg12@tamu.edu
* Correspondence: iliffet@tamug.edu

Abstract: Bermuda is an Eocene age volcanic island in the western North Atlantic, entirely capped by Pleistocene eolian limestone. The oldest and most highly karstified limestone is a 2 km^2 outcrop of the Walsingham Formation containing most of the island's 150+ caves. Extensive networks of submerged cave passageways, flooded by saltwater, extend under the island. In the early 1980s, cave divers initially discovered an exceptionally rich and diverse anchialine community inhabiting deeper sections of the caves. The fauna inhabiting caves in the Walsingham Tract consists of 78 described species of cave-dwelling invertebrates, including 63 stygobionts and 15 stygophiles. Thus, it represents one of the world's top hotspots of subterranean biodiversity. Of the anchialine fauna, 65 of the 78 species are endemic to Bermuda, while 66 of the 78 are crustaceans. The majority of the cave species are limited in their distribution to just one or only a few adjacent caves. Due to Bermuda's high population density, water pollution, construction, limestone quarries, and trash dumping produce severe pressures on cave fauna and groundwater health. Consequently, the IUCN Red List includes 25 of Bermuda's stygobiont species as critically endangered.

Keywords: biodiversity; fauna; conservation; seamount; ecology

Citation: Iliffe, T.M.; Calderón-Gutiérrez, F. Bermuda's Walsingham Caves: A Global Hotspot for Anchialine Stygobionts. *Diversity* **2021**, *13*, 352. https://doi.org/10.3390/d13080352

Academic Editors: Tanja Pipan, David C. Culver and Louis Deharveng

Received: 10 June 2021
Accepted: 9 July 2021
Published: 30 July 2021

Publisher's Note: MDPI stays neutral with regard to jurisdictional claims in published maps and institutional affiliations.

1. Introduction

Bermuda is a small, mid-Atlantic island located 1050 km off the east coast of the United States in the Sargasso Sea, at 32°20′ N, 64°45′ W. It lies approximately equidistant by air from Boston, New York, and Atlanta, making Bermuda only a short flight away for all of the U.S. east coast. Due to its warm climate, clear tropical waters, coral reefs, pink sand beaches, and ease of access, Bermuda is a popular tourist destination.

Although the Spanish explorer Juan de Bermudez discovered these isolated islands in 1505, he did not try to land. Ten years later, he returned to Bermuda, leaving behind a dozen pigs and sows for any castaways who might become stranded there. In 1609, the English sailing ship *Sea Venture*, on its way to resupply the Jamestown Colony, was caught in a strong storm and wrecked on Bermuda's reefs. The survivors of the shipwreck were stranded on the previously uninhabited island for nine months until two new ships could be constructed from local timber. The settlement of Bermuda did not occur until 1612, when the town of St. George officially became Bermuda's first capital and the oldest continually inhabited English town in the New World [1].

Bermuda's caves have been long recognized and prominently mentioned in early written works on the island [2]. Shakespeare's play, *The Tempest*, was likely inspired by the Bermuda shipwreck and takes place in and around a cave. The first published reference to Bermuda caves was in 1623, when Captain John Smith (of Pocahontas fame) described, "in some places varye strange, darke, and cumbersome Caues." John Hardy's 1671 poetic description of Bermuda caves states [3]:

145

> "The water flowing to them [Harrington Sound] underground,
> Being most salt, and all along the shore
> There are dark caves, of a miles length or more
> Extending under ground, in which there be
> Deep holes with water, though no one can see
> A passage for it in … "

The 1872–1876 round-the-world voyage of H.M.S. *Challenger*, which marked the beginning of modern oceanography, included a stopover in Bermuda. The expedition's commander, Captain George Nares, took the opportunity to explore Paynter's Vale (a.k.a. Church) Cave, accompanied by the British Governor of Bermuda, Sir John Henry Lefroy. Nares rowed the governor across a lake at the bottom of the cave, remarking on "a slight change of level with the tide, sufficient to keep the water perfectly pure" [4].

2. Anchialine

Since Bermuda is a relatively small island with no place very far from, or very high above the sea, caves that descend deep enough, end at very clear, exceptionally blue sea-water pools. However, until the beginning of cave diving in the early 1980s, Bermuda's underwater caves were unexplored and unknown, as were the animals living in them. When cave divers discovered an amazing variety of new species, new genera, and even several new orders, Bermuda's saltwater caves clearly merited status as a unique ecosystem [5].

After examining an assortment of highly unusual caridean shrimps from tropical land-locked saltwater pools on the several Indo-West Pacific islands and from the Sinai Peninsula, Dutch carcinologist, L.B. Holthuis, recognized the significance of this habitat. He created the term "anchialine" to describe "pools with no surface connection to the sea, containing salt or brackish water, which fluctuates with the tides" [6]. At the 1984 International Symposium on the Biology of Marine Cave held in Bermuda, Holthuis's original definition was expanded and modified to include tidal, saltwater pools inside caves: "Anchialine habitats consist of bodies of haline water, usually with a restricted exposure to open air, always with more or less extensive subterranean connections to the sea, and showing noticeable marine as well as terrestrial influences" [7]. During the 2012 Second International Symposium on Anchialine Ecosystems in Croatia, the term anchialine was more broadly defined as "a tidally-influenced subterranean estuary located within crevicular and cavernous karst and volcanic terrains that extends inland to the limit of seawater penetration" [8]. The "anchialine habitat continuum", as described by van Hengstum et al. (2019), extends uninterrupted from as far inland as saline groundwater penetrates, to the offshore edge of the platform shelf. In the Pleistocene, sea level changes alternately exposed and flooded the caves, such that anchialine groundwater alternately regressed or flooded bedrock voids [9].

3. Bermuda Geography and Geology

The islands of Bermuda lie atop the Bermuda Seamount, a volcanic peak rising from a seafloor depth of more than 4000 m. A ring of coral reefs surround a central lagoon on the flattened summit of this long extinct volcano. The islands of Bermuda, on the southeast edge of the platform, enclose several harbors and small bays. Eolian limestone completely caps the Bermuda Seamount to form the island mass. Limestone stratigraphy shows numerous cycles of subaerial eolianite and shallow marine carbonate deposition during interglacial high sea stands alternating with red clay soil horizons, marking glacial episodes of lowered sea level. Eolianites represented as lithified sand dune ridges, constitute more than 90% of the limestone volume. The main eolianite units are separated by fossils soils and, listed in order of increasing age, are the Southampton, Rocky Bay, Belmont, Lower Town Hill, Upper Town Hill, and Walsingham Formations. The Walsingham Formation, made up of highly altered and very dense eolianite, was deposited ≥700,000 years ago [10].

4. Bermuda Caves

Bermuda caves are believed to have formed syngenetically by a phreatic solution of limestone during glacial low stands of sea level. When the Ice Age sea level was down, the top of the seamount was emergent, and the islands' total land mass was about 13 times as large as it is today. As a result, the sizable body of fresh groundwater necessary for cave formation was present. As post-glacial sea levels rose, large portions of the caves were drowned by the encroaching seawater as it displaced the freshwater. Continuing collapse of overlying rock into the large voids created the irregular chambers and fissure entrances characteristic of Bermuda's caves [11].

Most of Bermuda's 150+ known caves are located in the Walsingham Tract, a 4 km long by 0.5 km wide surface outcropping of the Walsingham Formation, situated between Harrington Sound and Castle Harbour (Figure 1A,B). The Walsingham Tract contains the island's longest dry caves, most notably Church, Wonderland, Admiral's, Sibley's, and Jane's Caves. Crystal, Wonderland, Walsingham, Palm Caves, and others, are isolated cave entrances, interconnected as segments of a large hydrologically linked cave system. The underwater portions of caves in the Walsingham Tract reach water depths of 24 m, but still closely resemble their dry upper levels, even to the variety of large, subaerial-formed speleothems found above and below water. Caves in the Shelly Bay area, on the opposite side of Harrington Sound, are exclusively underwater with practically no dry portions. These caves have long, nearly level, anastomosing underwater passages at 18 m depth, with entrances on Harrington Sound and passageway extending under the island. Bermuda's longest cave, the 2 km long Green Bay Cave System, is located here [9,12].

Harrington Sound is an almost totally enclosed body of saltwater, with its only connection to the sea being a narrow channel at Flatt's Inlet (Figure 1B). Due to its restricted access to the ocean, tides in Harrington Sound have only 25% the range of ocean tides and occur 3 h later. When the volume of water tidally exchanging through Flatt's Inlet is compared with the total tidal volume of Harrington Sound (area X tidal range), only about half of the tidal volume passes out through the inlet, while the remainder moves through submerged caves [13]. Tidal exchange primarily occurs in caves around the periphery of Harrington Sound where the land is the narrowest. On the east side of the Sound, exchange occurs through caves from the Walsingham Tract (Walsingham and Palm Cave Systems) to Castle Harbor, on the west side through Shelly Bay (Green Bay and Red Bay Caves) to the North Shore, and on the south side (Devil's Hole Caves) to Bermuda's South Shore. Cave pools in the Walsingham and Palm cave systems have tide ranges that decrease, while residence times increase with their relative distance away from Castle Harbour and approaching Harrington Sound. The Crystal and Wonderland Cave sections of the Walsingham System (Figure 1C) are off the main flow channels where circulation patterns are restricted. Their residence times are much longer, with the phreatic zone consisting of very clear, slowly moving or stagnant waters with a lower surface salinity [14]. Caves with greater water transport, e.g., Palm Cave System (Figure 1D), have surface waters only slightly diluted and reach normal marine salinity (35–36 ppt) below 1 m depths. The food input into caves is primarily plankton and organic matter derived from the sea itself, although primary production in open anchialine pools may provide an additional source of food in the submerged caves (Figure 2A) [15].

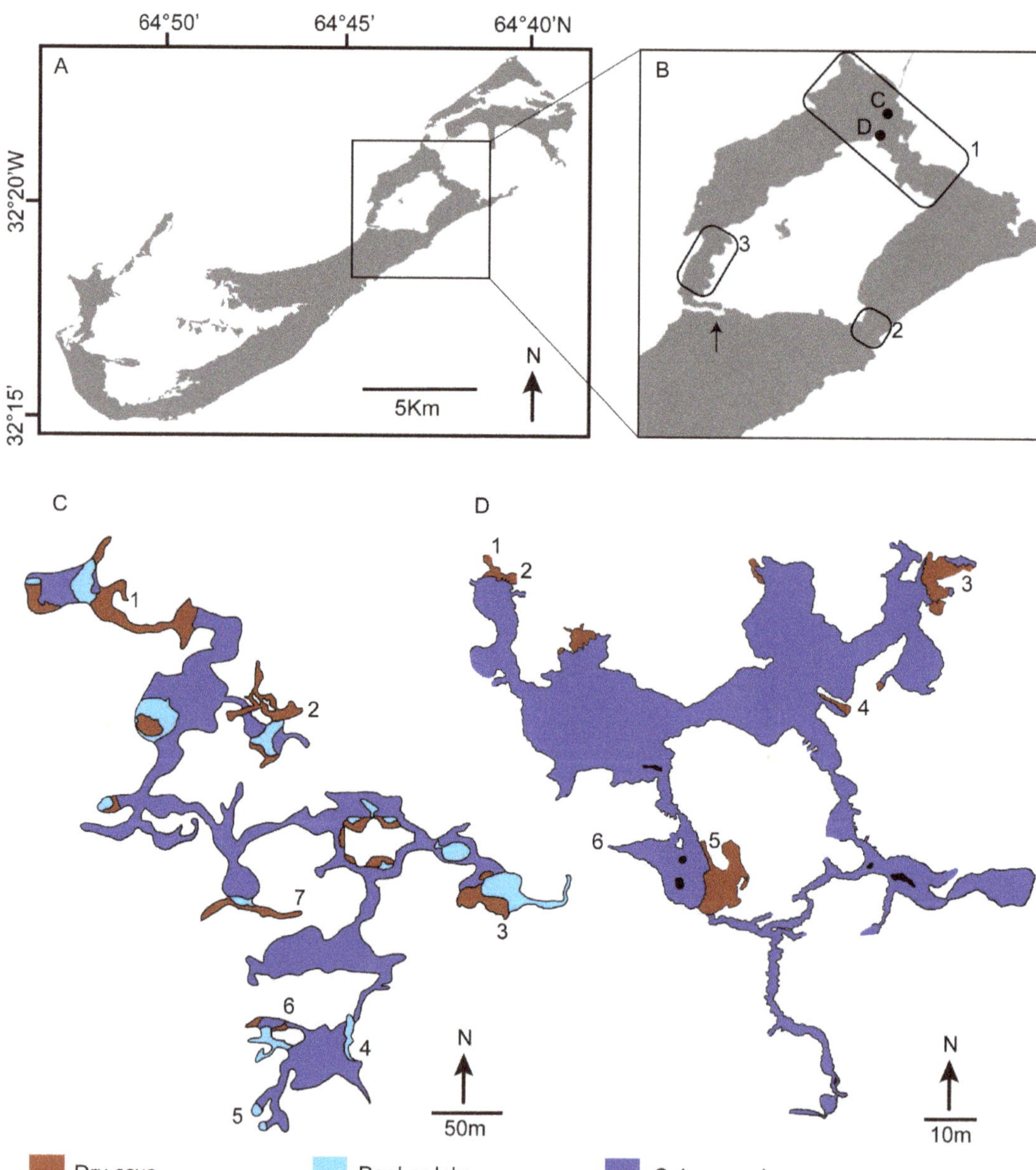

Figure 1. (**A**) Map of Bermuda in the North Atlantic Ocean, (**B**) Cave regions surrounding Harrington Sound, a large inland body of water mostly encircled by land. The arrow indicates the only opening to the sea at Flatt's Inlet, rectangles show regions of cave and karst development in (1) the Walsingham Tract (with the letters C and D indicating the location of the two main caves), (2) Devil's Hole, (3) Shelly Bay, (**C**) Walsingham Cave System with cave entrances: (1) Wonderland, (2) Crystal, (3) Walsingham, (4) Deep Blue, (5) Vine, (6) Old Horse, (7) Fern Sink (modified from map by Robert Power), (**D**) Palm Cave System with cave entrances: (1) Palm Slit, (2) Palm, (3) Strawmarket, (4) Sailor's Choice, (5) Myrtle Bank, (6) Cripple Gate (modified from map by Jason Richards).

Figure 2. Caves of the Walsingham Tract. (**A**) Deep Blue Cave entrance pool, (**B**) Cave diver between underwater stalagmites near the Crystal Cave, (**C**) Fragile saucer coral (*Agarcia fragilis* Dana, 1848) in the cavern pool of Deep Blue, (**D**) Coraline algae in the cavern ceiling of Deep Blue, (**E**) Submerged plastic garbage floating on the ceiling in the Palm Cave System, and (**F**) Rusting and disintegrating bitumen barrels in Bitumen Cave (in the Walsingham Tract). Photo credits: Tamara Thomsen (**A**); Jill Heinerth (**B,E**).

5. Cave Biology

Open water marine habitats in Bermuda have been well investigated, including numerous studies focusing on lightless environments in the deep sea and interstitial. The Bermuda Biological Station for Research (now the Bermuda Institute of Ocean Sciences) has continuously conducted marine biological research since 1903. The Bermuda Aquarium, Museum and Zoo, founded in 1926, focuses on oceanic island species and the conservation, education, and research related to them. In contrast, serious biological investigations of the underwater caves did not begin until the advent of cave diving [16].

Since Bermuda's terrestrial caves contain little organic matter, there are no known endemic terrestrial species and, in general, the fauna is sparse. Cave-dwelling bats, or other animals that bring organic material into the caves, are lacking in Bermuda. However, a rich and diverse marine biota inhabits the submarine passageways and anchialine pools of Bermuda's caves. Although many marine species are accidental or occur only at coastal cave entrances, a variety of stygobiont (i.e., aquatic cave-adapted) taxa are present, with the Walsingham Cave System—containing the greatest number of Bermuda stygobionts—included as one of the original subterranean biodiversity hotspots [17].

Anchialine cave fauna have the same adaptations to caves as freshwater and terrestrial cave organisms, i.e., reduction or loss of eyes and pigmentation, elongation of appendages, and increase in nonvisual sensory receptors [18]. Recent cave colonists tend to have pigmented body and eyes, while ancient settlers of caves have lost their eyes and pigment. The eyes of Bermuda's anchialine fauna show varying degrees of adaptations to the cave environment, ranging from the depigmented and eyeless *Mictocaris halope*, intermediate *Typhlatya iliffei* with small eyes and little pigment, to *Parhippolyte sterreri* having large eyes and bright red pigmentation (Figure 3) [19–21].

Seventy-eight cave-adapted species have been recorded in caves from the Walsingham Tract, Bermuda. The great majority of these species are crustaceans, making up 85% of the fauna: 20 copepods, 19 species of ostracods, 7 amphipods, 5 cumaceans and shrimps, 4 isopods, 2 mysids and tanaidaceans, and 1 mictacean and ingolfiellid. Non-crustacean species include five mites, three polychaetes, and two ciliates and mollusks. The poor representation of taxa other than crustaceans could be due to an effort or specialist bias [22]. Ten of the recorded species are the only representative of their genus, while *Mictocaris halope* is the only known species of its order. Twenty species belong to exclusively stygobiont taxa, while eighteen species have close relatives inhabiting the deep-sea (Table 1, Figure 3).

Figure 3. Examples of stygobiont species from the Walsingham cave system. (**A**) *Leptonerilla prospera*, (**B**) *Pseudoniphargus grandimanus*, (**C**) *Bermudamysis speluncola*, (**D**) *Mictocaris halope*, (**E**) *Arubolana aruboides*, (**F**) *Barbouria cubensis*, (**G**) *Typhlatya iliffei*, (**H**) *Procaris chacei*, and (**I**) *Parhippolyte sterreri*. Photo credits: (**A**) from Katrine Worsaae.

Table 1. Cave species inhabiting the Walsingham Tract in Bermuda. Ecology (B = Stygobiont, P = Stygophile, e = endemic to Bermuda). Species habitat (W = Water column, S = Sediment/interstitial). # Bermuda's caves (from where the species has been recorded). Monospecific taxa (belong to higher taxa to which they are the only representative, O = Order, G = Genus). Habitat and Region relationships (related species in the immediate superior taxa, "+" = numerous). Conservation status (in accordance with the IUCN Red List [23], Cr = Critically Endangered). Taxonomic status was validated in WoRMS [24].

	Ecology	Species Habitat	# Bermuda Caves	Monospecific Taxa	Species in the Genus	Habitat Relationship		Region Relationship			Conservation Status	Reference
						Cave/Interstitial	Deep Sea	Caribbean	Atlantic	Other		
Chromista												
Ciliophora, Euplotidae												
Euplotes iliffei Hill & Small in Hill, Small & Iliffe, 1986	B,e	W	1		50			X	X	X		[25]
Ciliophora, Paauronematidae												
Glauconema bermudense Small, 1986	B,e	W	4		2				X			[26]
Annelida												
Polychaeta, Nerillidae												
Leptonerilla prospera (Sterrer & Iliffe, 1982)	B,e	S	8		3	2			X	X	Cr	[27]
Speleonerilla saltatrix (Worsaae, Sterrer & Iliffe, 2004)	B,e	S	1		4	3		X	X			[28]
Oligochaeta, Tubificidae												
Phallodriloides macmasterae (Erséus, 1986)	B,e	S	1		5		2		X		Cr	[29]
Mollusca												
Gastropoda, Caecidae												
Caecum caverna Moolenbeek, Faber & Iliffe, 1988	B,e	S	2		218	1		X	X	X		[30]
Caecum troglodyta Moolenbeek, Faber & Iliffe, 1988	B,e	S	2		218	1		X	X	X		[30]

Table 1. *Cont.*

	Ecology	Species Habitat	# Bermuda Caves	Monospecific Taxa	Species in the Genus	Habitat Relationship		Region Relationship			Conservation Status	Reference
						Cave/Interstitial	Deep Sea	Caribbean	Atlantic	Other		
Arthropoda												
Arachnida												
Acari, Halacaridae												
Agauopsis bermudensis Bartsch & Iliffe, 1985	B,e	S	2		90	1		X	X	X		[31]
Agauopsis littoralis Bartsch & Iliffe, 1985	B,e	S	1		90	1		X	X	X		[31]
Copidognathus bermudensis Bartsch & Iliffe, 1985	B,e	S	2		380	2		X	X	X		[31]
Copidognathus longispinus Bartsch & Iliffe, 1985	B,e	S	1		380	2		X	X	X		[31]
Copidognathus subterraneus Bartsch & Iliffe, 1985	B,e	S	1		380	2		X	X	X		[31]
Crustacea												
Copepoda, Calanoida												
Enantiosis bermudensis Fosshagen, Boxshall & Iliffe, 2001	B,e	W	10		7	6		X		X		[32]
Epacteriscus rapax Fosshagen, 1973	P	W	9		3	2			X	X		[33]
Erebonectes nesioticus Fosshagen, in Fosshagen & Iliffe, 1985	B,e	W	4	G	1	29	1	X	X		Cr	[34]
Exumella polyarthra Fosshagen, 1970	P	W	X		4	2		X		X		[35]
Miostephos leamingtonensis Yeatman, 1980	B,e	W	1		2	1		X				[36]
Paracyclopia naessi Fosshagen in Fosshagen & Iliffe, 1985	B,e	W	5	G	1	4					Cr	[34]
Ridgewayia marki (Esterly, 1911)	P	W	2		14	4		X	X	X		[37]

153

Table 1. *Cont.*

	Ecology	Species Habitat	# Bermuda Caves	Monospecific Taxa	Species in the Genus	Habitat Relationship		Region Relationship			Conservation Status	Reference
						Cave/Interstitial	Deep Sea	Caribbean	Atlantic	Other		
Copepoda, Cyclopoida												
Halicyclops bowmani Rocha & Iliffe, 1993	B,e	W	2		89	21		X	X	X		[38]
Halicyclops herbsti Rocha & Iliffe, 1993	B,e	W	1		89	21		X	X	X		
Halicyclops ytororoma Lotufo & Rocha, 1993	P	W	2		89	21		X	X	X		[39]
Speleoithona bermudensis Rocha & Iliffe, 1993	B,e	W	2		3	2		X			Cr	[38]
Copepoda, Harpacticoida												
Intercrusia problematica Huys, 1996	B,e	W	1		2	1				X		[40]
Neoechinophora fosshageni Huys, 1996	B,e	W	2		5	4		X	X	X		[40]
Neoechinophora daltonae Huys, 1996	B,e	W	3		5	4		X	X	X		[40]
Neoechinophora jaumei Huys, 1996	B,e	W	2		5	4		X	X	X		[40]
Superornatiremis mysticus Huys, 1996	B,e	W	3		2	1				X		[40]
Copepoda, Misophrioida												
Speleophria bivexilla Boxshall & Iliffe, 1986	B,e	W	1		6	5		X	X	X	Cr	[41]
Speleophriopsis scottodicarloi (Boxshall & Iliffe, 1990)	B,e	W	2		5	4			X	X	Cr	[42]
Copepoda, Platycopioida												
Antrisocopia prehensilis Fosshagen in Fosshagen & Iliffe, 1985	B,e	W	1	G	1	1		X		X	Cr	[34]
Nanocopia minuta Fosshagen in Fosshagen & Iliffe, 1988	B,e	W	1	G	1	1		X		X	Cr	[43]

Table 1. *Cont.*

	Ecology	Species Habitat	# Bermuda Caves	Monospecific Taxa	Species in the Genus	Habitat Relationship		Region Relationship			Conservation Status	Reference
						Cave/Interstitial	Deep Sea	Caribbean	Atlantic	Other		
Ostracoda, Bairdiidae												
Aponesidea iliffei Maddocks in Maddocks & Iliffe, 1986	B,e	S	10		4					X		[44]
Havanardia keiji Maddocks in Maddocks & Iliffe, 1986	P,e	S	9		12			X	X			[44]
Neonesidea omnivaga Maddocks in Maddocks & Iliffe, 1986	P,e	S	3		99			X	X	X		[44]
Paranesidea sterreri Maddocks in Maddocks & Iliffe, 1986	P,e	S	2		41			X	X	X		[44]
Ostracoda, Candonidae												
Dolerocypria bifurca Maddocks in Maddocks & Iliffe, 1986	B,e	S	3		14	2				X		[44]
Paracypris crispa Maddocks in Maddocks & Iliffe, 1986	P,e	S	3		39			X	X	X		[44]
Ostracoda, Cytherellidae												
Cytherella bermudensis Maddocks in Maddocks & Iliffe, 1986	P,e	S	2		158		+	X	X	X		[44]
Cytherella kornickeri Maddocks in Maddocks & Iliffe, 1986	B,e	S	6		158		+	X	X	X		[44]
Ostracoda, Cylindroleberididae												
Parasterope muelleri (Skogsberg, 1920)	P		2		46				X	X		[45]
Ostracoda, Deeveyidae												
Spelaeoecia bermudensis Angel & Iliffe, 1987	B,e	W	11		11	10		X			Cr	[46]

Table 1. *Cont.*

	Ecology	Species Habitat	# Bermuda Caves	Monospecific Taxa	Species in the Genus	Habitat Relationship		Region Relationship			Conservation Status	Reference
						Cave/Interstitial	Deep Sea	Caribbean	Atlantic	Other		
Ostracoda, Halocypridae												
Metapolycope duplex Kornicker & Iliffe, 1989	B,e	S	12		10		4		X			[47]
Micropolycope eurax Kornicker & Iliffe, 1989	B,e	S	3		11	1			X	X		[47]
Micropolycope styx Kornicker & Iliffe, 1989	B,e	S	6		11	1			X	X		[47]
Ostracoda, Philomedidae												
Pseudophilomedes kylix Kornicker & Iliffe, 1989	B,e	S	5		11				X	X		[47]
Ostracoda, Polycopidae												
Polycopissa anax Kornicker & Iliffe, 1989	B,e	S	8		4				X	X		[47]
Ostracoda, Pontocyprididae												
Iliffeoecia iliffei Maddocks, 1991	B	S	2		3					X		[48]
Kareloecia minacis (Maddocks in Maddocks & Iliffe, 1986)	B,e	S	1		2	1				X		[44]
Thomontocypris lurida (Maddocks in Maddocks & Iliffe, 1986)	B,e	S	5		9		6	X	X	X		[44]
Ostracoda, Sarsiellidae												
Eusarsiella styx Kornicker & Iliffe, 1989	B,e	S	6		86	7	+	X	X	X		[47]
Caridea, Atyidae												
Typhlatya iliffei Hart & Manning, 1981	B,e	W	1		17	16		X	X	X	Cr	[19]
Caridea, Alpheidae												
Bermudacaris harti Anker & Iliffe, 2000	B,e	W	4		3					X		[20]

Table 1. *Cont.*

	Ecology	Species Habitat	# Bermuda Caves	Monospecific Taxa	Species in the Genus	Habitat Relationship		Region Relationship			Conservation Status	Reference
						Cave/Interstitial	Deep Sea	Caribbean	Atlantic	Other		
Caridea, Barbouriidae												
Barbouria cubensis (von Martens, 1872)	B	W	3	G	1	8		X	X	X	Cr	[22]
Parhippolyte sterreri (Hart & Manning, 1981)	B	W	2		6	5		X	X	X	Cr	[19]
Caridea, Procarididae												
Procaris chacei Hart & Manning, 1986	B,e	W	1		5	4		X	X	X	Cr	[49]
Isopoda, Atlantasellidae												
Atlantasellus cavernicolus Sket, 1979	B,e	S	1		2	1	1		X		Cr	[50]
Isopoda, Cirolanidae												
Arubolana aruboides (Bowman & Iliffe, 1983)	B,e	W	4		4	3		X		X	Cr	[51]
Isopoda, Leptanthuridae												
Curassanthura bermudensis Wägele & Brandt, 1985	B,e	S	1		5	4		X	X		Cr	[52]
Isopoda, Stenetriidae												
Stenobermuda iliffei Kensley, 1994	B,e	S	1		6			X	X	X		[53]
Amphipoda, Amphilochidae												
Hourstonius petulans (Karaman, 1980)	P	S	1		18		1			X		[54]
Amphipoda, Bogidiellidae												
Bermudagidiella bermudiensis (Stock, Sket & Iliffe, 1987)	B,e	S	2	G	1	115	114	X	X	X	Cr	[55]
Amphipoda, Liljeborgiidae												
Idunella sketi Karaman, 1980	P	S	1		44	1		X	X	X	Cr	[56]
Idunella verrilli Yabut & Sawicki, 2015	B,e		1		44	1		X	X	X		[56]

Table 1. *Cont.*

	Ecology	Species Habitat	# Bermuda Caves	Monospecific Taxa	Species in the Genus	Habitat Relationship		Region Relationship			Conservation Status	Reference
						Cave/Interstitial	Deep Sea	Caribbean	Atlantic	Other		
Amphipoda, Phoxocephalidae												
Cocoharpinia iliffei Karaman, 1980	B,e		1	G	1		+	X	X	X	Cr	[57]
Amphipoda, Podoceridae												
Podobothrus bermudensis Barnard & Clark, 1985	B,e		1	G	1		+	X	X	X		[58]
Amphipoda, Pseudoniphargidae												
Pseudoniphargus grandimanus Stock, Holsinger, Sket & Iliffe, 1986	B,e	W	20		72	71			X	X	Cr	[59]
Ingolfiellida, Ingolfiellidae												
Ingolfiella (Tethydiella) longipes Stock, Sket & Iliffe, 1987	B,e	S	1		44	+	+	X	X	X	Cr	[55]
Tanaidacea, Apseudidae												
Apseudes orghidani Gutu & Iliffe, 1989	B,e	W	1		44		+	X	X	X		[60]
Paradoxapseudes bermudeus (Băcescu, 1980)	P	S	1		18	3	1	X		X		[61]
Mictacea, Mictocarididae												
Mictocaris halope Bowman & Iliffe, 1985	B,e	W	4	O	1	3	3	X		X	Cr	[21]
Cumacea, Nannastacidae												
Cumella (Cumella) iliffei Băcescu, 1992	B,e	S	2		102	15		X	X	X		[62]
Cumella (Cumella) ocellata Băcescu, 1992	B,e	S	1		102	15		X	X	X		[62]
Cumella (Cumella) spinosa Băcescu & Iliffe, 1991		S	1		102	15		X	X	X		[63]

Table 1. *Cont.*

	Ecology	Species Habitat	# Bermuda Caves	Monospecific Taxa	Species in the Genus	Habitat Relationship		Region Relationship			Conservation Status	Reference
						Cave/Interstitial	Deep Sea	Caribbean	Atlantic	Other		
Cumacea, Nannastacidae												
Schizotrema agglutinanta (Băcescu, 1971)	P	S	8		20	1		X	X	X		[64]
Schizotrema wittmanni Petrescu & Sterrer, 2001	P	S	1		20	1		X	X	X		[65]
Mysida, Mysidae												
Bermudamysis speluncola Băcescu & Iliffe, 1986	B,e	W	2	G	1	2	1		X	X	Cr	[66]
Platyops sterreri Băcescu & Iliffe, 1986	B,e	W	2	G	1	2	1		X	X	Cr	[66]

Sixty-three species are stygobionts, and fifteen stygophiles (i.e., aquatic animals living in and outside caves). Most of Bermuda's endemic cave species have an extremely restricted distribution, with 26 species known only from a single cave, 14 more species limited to just two caves, and 52 species in total occurring in five or fewer caves. The ostracod *Iliffeoecia iliffei* also occurs in caves from the Galápagos Islands, while the barbouriid shrimp *Barbouria cubensis* inhabits caves from Bahamas, Caicos Islands, Cayman Islands, Cuba, Jamaica, and Mexico's Yucatan Peninsula, and *Parhippolyte sterreri* is found from caves in the Bahamas and the Yucatan Peninsula, in addition to Bermuda [67,68]. The stygophile ostracods *Havanardia keiji*, *Neonesidea omnivaga*, *Paranesidea sterreri*, *Cytherella bermudensis*, and *Paracypris crispa* were described from sample specimens collected in caves but have been also found in the island's open water environments, and are endemic to Bermuda [44]. The copepods *Epacteriscus rapax*, *Exumella polyarthra*, *Halicyclops ytororoma*, and *Ridgewayia marki*, the ostracod *Parasterope muelleri*, the cumaceans *Schizotrema agglutinanta* and *S. wittmanni*, the amphipods *Hourstonius petulans* and *Idunella sketi*, and the tanaidacean *Paradoxapseudes bermudeus* are also found from open water or interstitial environments outside Bermuda [32–34,39,65,68–71].

Considering Bermuda's isolated mid-ocean location, biogeographical affinities of the cave species may provide significant clues as to their origins. While the Gulf Stream may have transported some species to the island, others may have survived on submerged or emergent seamounts near the Mid-Atlantic Ridge, originated from Tethyan relicts, or are derived from deep-sea taxa [5]. Here are a few examples: The misophrioid copepod *Speleophria bivexilla* has congeners inhabiting caves from the Balearic Islands, Croatia, northwestern and southern Australia, and the Yucatan Peninsula, suggesting possible dispersal by plate tectonic processes [72]. The atyid shrimp *Typhlatya iliffei* belongs to an exclusively cave-adapted genus with species spread around the Caribbean and Western Mediterranean, as well as the Galapagos Islands, Ascension Island, and Madagascar. Their ancestors likely inhabited the ancient Tethys Sea with both vicariance (the opening of the Atlantic) and dispersal leading to their isolation and divergence [73]. The stygobiont amphipod *Pseudoniphargus* includes species from North Africa, the Mediterranean region and its islands, the Iberian Peninsula, the Canary Islands, Madeira, and the Azores, in addition to two species in Bermuda, on the west side of the mid-Atlantic rift, but is absent from the American continent or from the Caribbean. Based on phylogenetic analyses, the estimated age of the *Pseudoniphargus* lineage on Bermuda is relatively young, only 5 Ma [74]. The ostracod *Spelaeoecia bermudensis* belongs to a stygobiont genus with seven species from the Bahamas, two from Cuba, and one each from Jamaica, the Yucatan Peninsula, and Bermuda. The cluster of this ostracod in the Bahamas suggests possible ocean dispersal via the Gulf Stream rafting adults or larvae to Bermuda [75].

While numerous cave taxa in Bermuda are significant in their presence, the absence of several globally prominent stygobiont taxa is also remarkable. The crustacean class Remipedia inhabits anchialine caves in the Bahamas (21 spp.), Yucatan Peninsula (3 spp.), the Dominican Republic (1 sp.), Cuba (1 sp.), and Belize (1 sp.), plus the Canary Islands (2 spp.) and Australia (1 sp.) but, despite many hundreds of cave dives, has never been observed in Bermuda [76,77]. The ostracod stygobiont genus *Humphreysella* has a similar distribution and ecology to Remipedia, while being absent from Bermuda. *Humphreysella* occurs in caves in the Bahamas (3 spp.), Yucatan Peninsula (1 sp.), Cuba (1 sp.), Jamaica (1 sp.), Galápagos Islands (1 sp.), and Canary Islands (1 sp.), as well as having a closely related monotypic genus in Western Australia and Christmas Island (Indian Ocean) [24,78].

Important questions arise as to the age and origins of Bermuda's endemic cave fauna. The Bermuda seamount formed from volcanic eruptions occurring approximately 35 million years ago. Due to its attachment to the North American Plate, the seamount remained at a constant distance from the North American coast, but an ever-increasing distance from Europe as the Atlantic Ocean expanded. Limestone bedrock containing all known caves is only 1–2 million years in age. The limestone capping the summit of the Bermuda seamount extends down to about 30 m below present sea level, such that during Pleistocene regressions to 130 m, all of the island's limestone and, thus, its caves, were exposed above the ocean and were dry. The exceptionally large stalactites and stalagmites in the now underwater caves, and the level of the Pleistocene ocean, corroborate that caves must have been continuously dry for many tens thousands of years at a time (Figure 2B). It was not until about 7000 years ago that the sea level rise from the last glacial maximum (~20,000 years ago) caused Bermuda's caves to become flooded [9].

Considering the high diversity, unique ecological and taxonomic nature of the cave fauna, a relatively modern origin by cave colonization from the open ocean is highly unlikely and can be ruled out. Instead, the cave species currently found in Bermuda must have been dark-adapted organisms that moved into caves from suitable crevice or crevicular habitats in the volcanic bedrock or sides of the Bermuda seamount occurring at depths greater than where sea level was during the last glacial maximum. The presence of a highly diverse and endemic stygobiont fauna in a young (~21,000 years old) volcanic cave in the Canary Islands presents a comparable case. La Corona lava tube on Lanzarote formed during the last ice age as a dry cave that was flooded by rising sea level to a maximum water depth of 64 m [79].

6. Threats to Bermuda's Cave Fauna

Bermuda represents an extreme case of threats to marine cave fauna and a microcosm of issues facing caves around the world. The primary threats to Bermuda caves and their fauna include: (i) filling and quarrying activities, (ii) water pollution, (iii) dumping and littering, (iv) vandalism [80], and (v) climate change [81]. Since most of Bermuda's endemic stygiobionts inhabit only a single cave or cave system, pollution of these habitats can threaten entire species with extinction.

Population growth and land development have adversely affected Bermuda's caves and cave fauna. The islands have a total land area of 53.3 km^2, of which 20% (or 10 km^2) is forest and woodland [77]. As of May 2021, Bermuda's population was estimated to be 62,069, down from a high of 66,257 in 2005 [82], yielding an average population density of 1164 persons per km^2. In such a small, densely populated island, considerable human pressures have been brought to bear on the relatively unknown and poorly appreciated caves.

Intentional dumping of large quantities of refuse, raw sewage, and waste fuel oil into anchialine cave pools in Bermuda resulted in the depletion of dissolved oxygen and production of hydrogen sulfide in deep lakes from Government Quarry and Bassett's Caves [83]. The resulting anoxic conditions not only eliminated all typical cave invertebrates, but also formed black metal sulfide precipitates that substantially reduced water clarity. Such polluted waters can move for considerable distances underground and appear years later in distant cave pools. In addition, broken or missing speleothems, graffiti, names on cave walls, and litter are common sights in many dry caves in Bermuda, while sinkholes and cave entrances were long used for trash disposal [80] (Figure 2F). Tidal currents suck floating plastic bottles and bags into coastal caves (Figure 2E). Such negative impacts to caves reduce their esthetic value and, therefore, make it more difficult to justify protective measures.

Climate change is a medium-term to long-term threat, since warming climate causes an increased frequency and strength of hurricanes and sea level rise, which will directly affect the habitat availability. Furthermore, cave organisms are especially vulnerable to rapid environmental changes [81,84].

In an effort to protect caves, Bermuda's planning laws now afford caves the highest level of protection [85]. The Fourth Schedule of the Planning Act 1974 states: "The protection of caves shall take precedence over all other planning considerations and the Board shall refuse any development application or planning of subdivision if, in the opinion of the Board, the proposal will have detrimental impact on a cave entrance or underlying cave." The Protected Species Act 2003 and Protected Species Order 2012 list 22 cave-dwelling species for legal protection, while the 2014 "Management Plan for Bermuda's Critically Endangered Cave Fauna" seeks adequate protection for the entire cave habitat. The Management Plan is designed to protect caves through legislation and raised public awareness, comprehensive mapping of caves, identifying and managing point source pollution, monitoring cave air and water, examining the potential for hatchery breeding, facilitating ecological research, and undertaking active restoration of impacted caves [85].

Unfortunately, loopholes in the laws have allowed important caves with endangered and protected species to be destroyed. Wilkinson Quarry Cave at the northern end of the Walsingham Track was discovered during blasting operations in 2002. The cave contained profuse and actively growing speleothems and a large network of submerged cave passages where biological collections identified four species of stygobiont crustaceans on the IUCN Red List. While biology and conservation experts pressed for protection and preservation of the cave, consultants hired by quarry management supported its destruction arguing that the cave was (1) small, (2) structurally compromised and therefore unsafe, and (3) not ecologically or esthetically significant [86]. Despite the Planning Act law and the presence of endangered species, the Bermuda Development Applications Board approved destruction of the cave and removal of all bedrock to level off the lower quarry floor.

Another way to safeguard caves is through private and government owned nature reserves that effectively protect undeveloped woodland, karst topography, extensive cave systems, saltwater ponds, and endemic cave-adapted plants, mosses, and ferns. The 400-acre (160 ha) Walsingham Tract includes four adjacent nature reserves: the 23-acre (9.3 ha) Walsingham Trust Nature Reserve, 1.25-acre (0.5 ha) Idwal Hughes Nature Reserve, 12-acre (4.9 ha) Blue Hole Hill Park, and Crystal and Fantasy Caves operated by the Wilkinson Trust as commercial tourist attractions [87]. Since the mid-20th century, sections at the northern and southern ends of the Tract have been lost to quarrying, hotel, and residential development, increasing the importance of these nature reserves as a key conservation and restoration areas.

Thus, conservation actions to protect Bermuda's anchialine cave habitat and its unique stygobiont fauna, especially caves in the Walsingham Tract, are crucial. Further research is needed to understand the biodiversity, species biology, population sizes, and carrying capacity of the ecosystem in relation to diverse human activities.

Author Contributions: Conceptualization, T.M.I.; methodology, T.M.I.; validation, T.M.I. and F.C.-G.; investigation, T.M.I.; resources, T.M.I.; data curation, T.M.I. and F.C.-G.; writing—original draft preparation, T.M.I.; writing—review and editing, T.M.I. and F.C.-G.; visualization, T.M.I. and F.C.-G.; supervision, T.M.I.; project administration, T.M.I.; funding acquisition, T.M.I. All authors have read and agreed to the published version of the manuscript.

Funding: This research was funded by National Science Foundation grants BSR-8215672, and DEB-8001836; NOAA Ocean Exploration NA09OAR4600096; NOAA Office of Undersea Research SWS-87-23; and NOAA International Coral Reef Conservation Program NA06NOS46300-85.

Institutional Review Board Statement: Research involving the collection of live animals was conducted in accordance with and approval from the Government of Bermuda Department of Conservation Services.

Informed Consent Statement: Not applicable.

Data Availability Statement: The data presented in this study is available in the cited references.

Acknowledgments: This paper is dedicated to Wolfgang Sterrer, without whose encouragement and considerable assistance, these submarine cave biology studies in Bermuda would never have been inaugurated. We sincerely acknowledge the substantial help provided by cave divers Rob Power, Paul Hobbs, Dennis Williams, Gil Nolan, Jason and Chrissy Richards, Jill Heinerth, Paul Heinerth, and Brian Kakuk. Also, taxonomists Boris Sket, Thomas Bowman, C.W. Hart Jr., Raymond Manning, Louis Kornicker, Geoff Boxshall, John Holsinger, Mihai Băcescu, Jan Stock, Audun Fosshagen, Modest Gutu, Carlos Rocha, Iorgu Petrescu, Rosalie Maddocks, Lazare Botosaneanu, and others. We also thank the following individuals or organizations in Bermuda: Annie Glasspool, Struan Smith, the Bermuda Biological Station, the Bermuda Aquarium and Zoo, the Bermuda Zoological Society, the Bermuda Underwater Exploration Institute, the Bermuda National Trust, the Walsingham Trust, and many others.

Conflicts of Interest: The authors declare no conflict of interest. The funders had no role in the design of the study; in the collection, analyses, or interpretation of data; in the writing of the manuscript, or in the decision to publish the results.

References

1. Smithsonian.com. Bermuda—History and Heritage. Available online: https://www.smithsonianmag.com/travel/bermuda-history-and-heritage-14340790/?no-ist (accessed on 8 June 2021).
2. Iliffe, T.M. Speleological history of Bermuda. *Acta Carsologica* **1993**, *22*, 114–135.
3. Lefroy, L.H. *Memorials of the Discovery and Early Settlement of the Bermudas or Somers Islands*; Franklin Classics Trade Press: London, UK, 1877; pp. 1511–1687.
4. Thomson, C.W. *The Voyage of the "Challenger". The Atlantic; A Preliminary Account of the General Results of the Exploring Voyage of H.M.S. "Challenger" during the Year 1873 and the Early Part of the Year 1876*; Harper & Brothers: New York, NY, USA, 1878.
5. Iliffe, T.M.; Hart, C.W.; Manning, R.B. Biogeography and the caves of Bermuda. *Nature* **1983**, *302*, 141–142. [CrossRef]
6. Holthuis, L.B. Caridean shrimps found in Land-locked saltwater pools at four Indo-West Pacific localities (Sinai Peninsula, Funafuti Atoll, Maui and Hawaii Islands), with the description of one new genus and four new species. *Zool. Verh.* **1973**, *128*, 3–48.
7. Stock, J.H.; Iliffe, T.M.; Williams, D. The concept of "anchialine" reconsidered. *Stygologia* **1986**, *2*, 90–92.
8. Bishop, R.E.; Humphreys, W.F.; Cukrov, N.; Žic, V.; Boxshall, G.A.; Cukrov, M.; Iliffe, T.M.; Kršinić, F.; Moore, W.S.; Pohlman, J.W.; et al. 'Anchialine' redefined as a subterranean estuary in a crevicular or cavernous geological setting. *J. Crustac. Biol.* **2015**, *35*, 511–514. [CrossRef]
9. van Hengstum, P.J.; Cresswell, J.N.; Milne, G.A.; Iliffe, T.M. Development of anchialine cave habitats and karst subterranean estuaries since the last ice age. *Sci. Rep.* **2019**, *9*, 11907. [CrossRef] [PubMed]
10. Rowe, M.P. *The Geology of Bermuda*; bermudageology.com: Bermuda, UK, 2020.
11. Palmer, A.N.; Palmer, M.V.; Queen, J.M. Geology and origin of the caves of Bermuda. In Proceedings of the Seventh International Congress of Speleology, Sheffield, England, 11–16 September 1977; pp. 336–339.
12. Illife, T.M. The submarine caves of Bermuda. In Proceedings of the Eighth International Congress of Speleology, Bowling Green, KY, USA, 18–24 July 1981; pp. 161–163.
13. Thomas, M.L.H. *The Ecology of Harrington Sound, Bermuda*; Bermuda Zoological Society: Bermuda, UK, 2005.
14. Iliffe, T.M. Anchialine cave ecology. In *Ecosystems of the World. 30. Subterranean Ecosystems*; Wilkens, H., Culver, D.C., Humphreys, W.F., Eds.; Elsevier: Amsterdam, The Netherlands, 2000; pp. 59–76.
15. Pohlman, J.W.; Cifuentes, L.A.; Iliffe, T.M. Food Web Dynamics and Biogeochemistry of Anchialine Caves: A Stable Isotope Approach. *Ecosyst. World* **2000**, *30*, 351–363.
16. Sket, B.; Iliffe, T.M. Cave fauna of Bermuda. *Int. Rev. Gesamten Hydrobiol. Hydrogr.* **1980**, *65*, 871–882. [CrossRef]
17. Culver, D.C.; Sket, B. Hotspots of subterranean biodiversity in caves and wells. *J. Cave Karst Stud.* **2000**, *62*, 11–17.
18. Culver, D.C.; Pipan, T. *The Biology of Caves and Other Subterranean Habitats*; Oxford University Press: Oxford, UK, 2009.
19. Hart, C.W.J.; Manning, R.B. The cavernicolous caridean shrimps of Bermuda (Alpheidae, Hippolytidae, and Atyidae). *J. Crustac. Biol.* **1981**, *1*, 441–456. [CrossRef]
20. Anker, A.; Iliffe, T.M. Description of *Bermudacaris harti*, a new genus, and species (Crustacea: Decapoda: Alpheidae) from anchialine caves of Bermuda. *Proc. Biol. Soc. Wash.* **2000**, *113*, 761–775.
21. Bowman, T.E.; Iliffe, T.M. *Mictocaris halope*, a new unusual peracaridan crustacean from marine caves on Bermuda. *J. Crustac. Biol.* **1985**, *5*, 58–73. [CrossRef]
22. von Martens, E. Über Cubanische Crustaceen. *Arch. Naturgeschich.* **1872**, *38*, 77–147.
23. IUCN. The IUCN Red List of Threatened Species. Available online: http://www.iucnredlist.org/ (accessed on 30 May 2021).
24. WoRMS. Editorial Board World Register of Marine Species. Available online: http://www.marinespecies.org (accessed on 30 May 2021).
25. Hill, B.F.; Small, E.B.; Iliffe, T.M. *Euplotes iliffei* n. sp.: A new species of *Euplotes* (Ciliophora, Hypotrichida) from the marine caves of Bermuda. *J. Wash. Acad. Sci.* **1986**, *76*, 244–249.
26. Small, E.B.; Heisler, J.; Sniezek, J.; Iliffe, T.M. *Glauconema bermudense* n. sp. (Scuticociliatida, Oligohymenophorea), a troglobitic ciliophoran from Bermudian marine caves. *Stygologia* **1986**, *2*, 167–179.
27. Sterrer, W.; Iliffe, T.M. *Mesonerilla prospera*, a new archiannelid from marine caves in Bermuda. *Proc. Biol. Soc. Wash.* **1982**, *95*, 509–514.
28. Worsaae, K.; Sterrer, W.; Iliffe, T.M. *Longipalpa saltatrix*, a new genus and species of the meiofaunal family Nerillidae (Annelida: Polychaeta) from an anchihaline cave in Bermuda. *Proc. Biol. Soc. Wash.* **2004**, *117*, 346–362.
29. Erséus, C. A new species of *Phallodrilus* (Oligochaeta, Tubificidae) from a limestone cave on Bermuda. *Sarsia* **1986**, *71*, 7–9. [CrossRef]
30. Moolenbeek, R.; Faber, M.; Iliffe, T.M. Two new species of the genus *Caecum* (Gastropoda) from marine caves on Bermuda. *Stud. Honour Pieter Wagenaar Hummelinck* **1988**, *123*, 209–216.
31. Bartsch, I.; Iliffe, T.M. The halacarid fauna (Halacaridae, Acari) of Bermuda's cave. *Stygologia* **1985**, *1*, 300–321.

32. Fosshagen, A.; Boxshall, G.A.; Iliffe, T.M. The Epacteriscidae, a cave-living family of calanoid copepods. *Sarsia* **2001**, *86*, 245–318. [CrossRef]
33. Fosshagen, A. A new genus and species of bottom living calanoid (Copepoda) from Florida and Colombia. *Sarsia* **1973**, *52*, 145–154. [CrossRef]
34. Fosshagen, A.; Iliffe, T.M. Two new genera of Calanoida and a new order of Copepoda, Platycopioida, from marine caves on Bermuda. *Sarsia* **1985**, *70*, 345–358. [CrossRef]
35. Fosshagen, A. Marine Biological Investigations in the Bahamas. 15. *Ridgewayia (Copepoda, Calanoida) and two new genera of calanoids from the Bahamas. Sarsia* **1970**, *44*, 25–58. [CrossRef]
36. Yeatman, H.C. *Miostephos leamingtonensis*, a new species of copepod from Bermuda. *J. Tenn. Acad. Sci.* **1980**, *55*, 21–22.
37. Esterly, C.O. Calanoid Copepoda from the Bermuda Islands. *Proc. Am. Acad. Arts Sci.* **1911**, *47*, 219. [CrossRef]
38. Rocha, C.E.F.; Iliffe, T.M. New cyclopoids (Copepoda) from anchialine caves in Bermuda. *Sarsia* **1993**, *78*, 43–56. [CrossRef]
39. Lotufo, G.R.; da Rocha, C.E.F. Intertidal interstitial *Halicyclops* from the Brazilian coast (Copepoda: Cyclopoida). *Hydrobiologia* **1993**, *264*, 175–184. [CrossRef]
40. Huys, R. Superornatiremidae fam. nov. (Copepoda: Harpacticoida): An enigmatic family from North Atlantic anchihaline caves. *Sci. Mar.* **1996**, *60*, 497–542.
41. Boxshall, G.A.; Iliffe, T.M. New cave-dwelling misophrioids (Crustacea: Copepoda) from Bermuda. *Sarsia* **1986**, *71*, 55–64. [CrossRef]
42. Boxshall, G.A.; Iliffe, T.M. Three new species of misophrioid copepods from oceanic islands. *J. Nat. Hist.* **1990**, *24*, 595–613. [CrossRef]
43. Fosshagen, A.; Iliffe, T.M. A new genus of Platycopioida (Copepoda) from a marine cave on Bermuda. In *Biology of Copepods*; Boxshall, G.A., Schminke, H.K., Eds.; Springer: Dordrecht, The Netherlands, 1988; pp. 357–361.
44. Maddocks, R.F.; Iliffe, T.M. Podocopid Ostracoda of Bermudian caves. *Stygologia* **1986**, *2*, 26–76.
45. Skogsberg, T. Studies on marine ostracods. Part 1. (Cypridinids, Halocyprids and Polycopids). *Zool. Bidr. Fran Upps.* **1920**, 1–784. [CrossRef]
46. Angel, M.V.; Iliffe, T.M. *Spelaeoecia bermudensis*, new genus, new species, a halocyprid ostracod from marine caves in Bermuda. *J. Crustac. Biol.* **1987**, *7*, 541–553. [CrossRef]
47. Kornicker, L.S.; Iliffe, T.M. Ostracoda (Myodocopina, Cladocopina, Halocypridina) from anchialine caves in Bermuda. *Smithson. Contrib. Zool.* **1989**, *475*, 1–88. [CrossRef]
48. Maddocks, R.F. Revision of the family Pontocyprididae (Ostracoda), with new anchialine species and genera from Galapagos Islands. *Zool. J. Linn. Soc.* **1991**, *103*, 309–333. [CrossRef]
49. Hart, C.W.J.; Manning, R.B. Two new shrimps (Procarididae and Agostocarididae, new family) from marine caves of the western North Atlantic. *J. Crustac. Biol.* **1986**, *6*, 408–416. [CrossRef]
50. Sket, B. *Atlantasellus cavernicolus* n. gen., n. sp. (Isopoda Asellota, Atlantasellidae n. fam.) from Bermuda. *Biol. Vestn. Ljubl.* **1979**, *7*, 175–183.
51. Bowman, T.E.; Iliffe, T.M. *Bermudalana aruboides*, a new genus and specioes of troglobitic isopoda (Cirolanidae) from marine caves on Bermuda. *Proc. Biol. Soc. Wash.* **1983**, *96*, 291–300.
52. Wägele, J.W.; Brandt, A. New west Atlantic localities for the stygobiont paranthurid *Curassanthura* (Crustacea, Isopoda, Anthuridea) with description of *C. bermudensis* n. sp. *Bijdr. Dierkd.* **1985**, *55*, 324–330. [CrossRef]
53. Kensley, B. Records of shallow-water marine isopods from Bermuda with descriptions of four new species. *J. Crustac. Biol.* **1994**, *14*, 319–336. [CrossRef]
54. Karaman, G.S. Revision of the genus *Gitanopsis* Sars 1895 with description of new genera *Afrogitanopsis* and *Rostrogitanopsis* n. gen. (fam. Amphilochidae). *Poljopr. Sumar.* **1980**, *26*, 43–69.
55. Stock, J.H.; Sket, B.; Iliffe, T.M. Two new amphipod crustaceans from anchihaline caves in Bermuda. *Crustaceana* **1987**, *53*, 54–66. [CrossRef]
56. Karaman, G.S. Revision of genus *Idunella* Sars with description of new species, *I. sketi*, n. sp. (Fam. Liljeborgiidae). *Acta Adriat.* **1980**, *21*, 409–435.
57. Karaman, G.S. *Cocoharpinia iliffei*, new genus and species from Bermuda, with remarks to other genera and species (Fam. Phoxocephalidae). (Contribution to the Knowledge of the Amphipoda 103). *Stud. Mar.* **1980**, *9–10*, 149–175.
58. Barnard, J.L.; Clark, J. A new sea-cave amphipod from Bermuda (Dulichiidae). *Proc. Biol. Soc. Wash.* **1985**, *98*, 1048–1053.
59. Stock, J.H.; Holsinger, J.R.; Sket, B.; Iliffe, T.M. Two new species of *Pseudoniphargus* (Amphipoda), in Bermudian groundwaters. *Zool. Scr.* **1986**, *15*, 237–249. [CrossRef]
60. Gutu, M.; Iliffe, T.M. *Apseudes orghidani*, a new species of Tanaidacea (Crustacea) from an anchialine cave on Bermuda. *Trav. Mus. Natl. d'Hist. Nat. Grigore Antipa* **1989**, *30*, 161–167.
61. Băcescu, M. *Apseudes bermudeus* n.sp. from caves around Bermude [sic] Islands [*Apseudes bermudeus* n.sp. iz pecina oko Bermudskih Otoka]. *Acta Adriat.* **1980**, *21*, 401–407.
62. Băcescu, M. Deux espèces nouvelles de *Cumella* (Crustacea, Cumacea) des grottes sous-marines de Bermuda. *Trav. Mus. Natl. d'Hist. Nat. Grigore Antipa* **1992**, *32*, 257–262.
63. Băcescu, M.; Iliffe, T.M. Nouvelles espèces de *Cumella* des grottes sous-marines de Bermude. *Rev. Behav. Almyracuma Prox.* **1991**, *36*, 9–13.

64. Băcescu, M. New *Cumacea* from the littoral waters of Florida (Caribbean Sea). *Trav. Mus. Natl. d'Hist. Nat. Grigore Antipa* **1971**, *11*, 5–24.
65. Petrescu, I.; Sterrer, W. Cumacea (Crustacea) from shallow waters of Bermuda. *Ann. Nat. Naturhist. Mus. Wien Ser. B Bot. Zool.* **2001**, *103*, 89–128.
66. Băcescu, M.; Iliffe, T.M. *Bermudamysis* g.n., *Platyops* g.n. and other mysids from Bermudian caves. *Stygologia* **1986**, *2*, 93–104.
67. Ditter, R.E.; Mejía-Ortíz, L.M.; Bracken-Grissom, H.D. Anchialine adjustments: An updated phylogeny and classification for the family Barbouriidae Christoffersen, 1987 (Decapoda: Caridea). *J. Crustac. Biol.* **2020**, *40*, 401–411. [CrossRef]
68. Keyser, D.; Schöning, C. Holocene ostracoda (Crustacea) from Bermuda. *Senckenberg. Lethaea* **2000**, *80*, 567–591. [CrossRef]
69. Martín, A.; Díaz Díaz, Y.J. Biodiversidad de crustáceos peracáridos en el delta del Río Orinoco, Venezuela. *Rev. Biol. Trop.* **2006**, *55*. [CrossRef]
70. Stock, J.H.; Vonk, R. Marine interstitial Amphipoda and Isopoda (Crustacea) from Santiago, Cape Verde Islands. *Bijdr. Dierkd.* **1992**, *62*, 21–36. [CrossRef]
71. Winfield, I.; Abarca-Ávila, M.; Ortiz, M.; Cházaro-Olvera, S.; Lozano-Aburto, M.Á. Biodiversidad de los tanaidáceos (Crustacea: Peracarida: Tanaidacea) del Parque Nacional Arrecife Puerto Morelos, Quintana Roo, México. *Rev. Mex. Biodivers.* **2017**, *88*, 572–578. [CrossRef]
72. Suárez-Morales, E.; Cervantes-Martínez, A.; Gutiérrez-Aguirre, M.A.; Iliffe, T.M. A new *Speleophria* (Copepoda, Misophrioida) from an anchialine cave of the Yucatán Peninsula with comments on the biogeography of the genus. *Bull. Mar. Sci.* **2017**, *93*, 1–16. [CrossRef]
73. Jurado-Rivera, J.A.; Pons, J.; Alvarez, F.; Botello, A.; Humphreys, W.F.; Page, T.J.; Iliffe, T.M.; Willassen, E.; Meland, K.; Juan, C.; et al. Phylogenetic evidence that both ancient vicariance and dispersal have contributed to the biogeographic patterns of anchialine cave shrimps. *Sci. Rep.* **2017**, *7*, 2852. [CrossRef]
74. Stokkan, M.; Jurado-Rivera, J.A.; Oromí, P.; Juan, C.; Jaume, D.; Pons, J. Species delimitation and mitogenome phylogenetics in the subterranean genus *Pseudoniphargus* (Crustacea: Amphipoda). *Mol. Phylogenet. Evol.* **2018**, *127*, 988–999. [CrossRef]
75. Kornicker, L.S.; Iliffe, T.M.; Harrison-Nelson, E. Ostracoda (Myodocopa) from Anchialine Caves and Ocean Blue Holes. *Zootaxa* **2007**, *1565*, 1–151. [CrossRef]
76. Koenemann, S.; Iliffe, T.M. Class Remipedia Yager, 1981. In *Treatise on Zoology—Anatomy, Taxonomy, Biology: The Crustacea*; von Vaupel Klein, J.C., Charmantier-Daures, M., Schram, F.R., Eds.; Brill: Leiden, The Netherlands, 2013; pp. 125–177.
77. Ballou, L.; Iliffe, T.M.; Kakuk, B.; Gonzalez, B.C.; Osborn, K.J.; Worsaae, K.; Meland, K.; Broad, K.; Bracken-Grissom, H.; Olesen, J. Monsters in the dark: Systematics and biogeography of the stygobitic genus *Godzillius* (Crustacea: Remipedia) from the Lucayan Archipelago. *Eur. J. Taxon.* **2021**, *751*, 115–139. [CrossRef]
78. Moldovan, O.T.; Kováč, L.; Halse, S. *Ecological Studies, In Cave Ecology*; Springer: Cham, Germany, 2018.
79. Martínez, A.; Gonzalez, B.C.; Núñez, J.; Wilkens, H.; Oromí, P.; Iliffe, T.M.; Worsaae, K. *Guide to the Anchialine Ecosystems of Jameos del Agua and Túnel de la Atlántida*; Cabildo de Lanzarote: Lanzorote, Spain, 2016.
80. Iliffe, T.M. Bermuda's caves: A non-renewable resource. *Environ. Conserv.* **1979**, *6*, 181–186. [CrossRef]
81. Mammola, S.; Cardoso, P.; Culver, D.C.; Deharveng, L.; Ferreira, R.L.; Fišer, C.; Galassi, D.M.P.; Griebler, C.; Halse, S.; Humphreys, W.F.; et al. Scientists' warning on the conservation of subterranean ecosystems. *Bioscience* **2019**, *69*, 641–650. [CrossRef]
82. Worldometers.info. Bermuda Population (LIVE). Available online: https://www.worldometers.info/world-population/bermuda-population/ (accessed on 30 June 2021).
83. Iliffe, T.M.; Jickells, T.D.; Brewer, M.S. Organic pollution of an inland marine cave from Bermuda. *Mar. Environ. Res.* **1984**, *12*, 173–189. [CrossRef]
84. Parravicini, V.; Guidetti, P.; Morri, C.; Montefalcone, M.; Donato, M.; Bianchi, C.N. Consequences of sea water temperature anomalies on a Mediterranean submarine cave ecosystem. *Estuar. Coast. Shelf Sci.* **2010**, *86*, 276–282. [CrossRef]
85. Glasspool, A. *Management Plan for Bermuda's Critically Endangered Cave Fauna*; Government of Bermuda, Department of Conservation Services: Bermuda, UK, 2014.
86. Darrell, M. Letter to Bermuda Department of Planning, re: Proposed Demolition of Unstable Mound with Subterranean Void in Wilkinson Quarry, Hamilton Parish. Hamilton: Bermuda. 10 June 2020. Available online: https://www.bnt.bm/images/Education/WilkinsonQuarryCaveObjection.pdf (accessed on 28 June 2021).
87. Copeland, A. *Walsingham Trust Nature Reserve: Special Features and Potential Projects*; Museum and Zoo Report Number BAMZ#3405; Bermuda Aquarium: Bermuda, UK, 2020.

Article

Ganxiao Dong: A Hotspot of Cave Biodiversity in Northern Guangxi, China

Sunbin Huang [1,2], **Guofu Wei** [3], **Hengsong Wang** [4], **Weixin Liu** [1,*], **Anne Bedos** [5], **Louis Deharveng** [5] **and Mingyi Tian** [1]

[1] Department of Entomology, College of Plant Protection, South China Agricultural University, 483 Wushan Road, Guangzhou 510642, China; huangsunbin@163.com (S.H.); mytian@scau.edu.cn (M.T.)
[2] Mécanismes Adaptatifs et Évolution (MECADEV), UMR 7179 CNRS-MNHN, Muséum National d'Histoire Naturelle, CP50, 57 rue Cuvier, 75005 Paris, France
[3] Protection Centre of Huanjiang World Natural Heritage, Huanjiang, Hechi 547100, China; tuqinja@163.com
[4] School of Karst Science, Guizhou Normal University, 116 Baoshanbeilu, Yunyan, Guiyang 550001, China; wanghengsong796@163.com
[5] Institut de Systématique, Évolution, Biodiversité (ISYEB), UMR 7205 CNRS, Muséum National d'Histoire Naturelle, Sorbonne Université, EPHE, 45 rue Buffon, 75005 Paris, France; bedosanne@yahoo.fr (A.B.); dehar.louis@wanadoo.fr (L.D.)
*   Correspondence: da2000wei@163.com

Abstract: Located in the core zone of Mulun National Nature Reserve in northern Guangxi, the limestone cave Ganxiao Dong harbours the richest cave fauna currently known in China. In total, 26 species of cave invertebrates have been recognized so far, in spite of limited sampling efforts. Of them, 20 are troglobionts or stygobionts, including one snail, four millipedes, three spiders, one harvestman, three isopods, two springtails, two crickets, one non-glowing sticky worm, and three trechine beetles. Six other species are troglophiles. The most remarkable characteristic of this fauna is its high number of troglomorphic species, especially among millipedes, crickets and beetles.

Keywords: biodiversity; cave; hotspot; Huanjiang; Guangxi; South China Karst; troglomorphy

Citation: Huang, S.; Wei, G.; Wang, H.; Liu, W.; Bedos, A.; Deharveng, L.; Tian, M. Ganxiao Dong: A Hotspot of Cave Biodiversity in Northern Guangxi, China. *Diversity* **2021**, *13*, 355. https://doi.org/10.3390/d13080355

Academic Editor: Spyros Sfenthourakis

Received: 4 June 2021
Accepted: 28 July 2021
Published: 2 August 2021

Publisher's Note: MDPI stays neutral with regard to jurisdictional claims in published maps and institutional affiliations.

1. Introduction

The World Heritage Property of South China Karst is a cluster of seven karsts, i.e., Shilin (stone forest) in Yunnan, Shibing and Maolan in Guizhou, Jinfoshan and Wulong in Chongqing, as well as Guilin and Huanjiang in Guangxi (https://whc.unesco.org/en/list/1248/, accessed on 1 July 2021). The Huanjiang Karst component is located in the Guangxi Zhuang Autonomous Region within the boundaries of the Mulun National Nature Reserve. Actually, Maolan and Mulun are the same karst unit across the border between Guizhou and Guangxi, and the karst unit is totally included in national nature reserves. The Huanjiang karst is a large Fengcong (peak cluster) landscape covered with primary forest and hundreds of scattered caves. The cave biodiversity in this karst has been recently investigated and shown to be very rich [1–3]. The cave which will be dealt with in this paper, Ganxiao Dong, is located in the Mulun karst.

The Huanjiang Karst World Natural Heritage Site is located in Huanjiang Maonan Autonomous County, northern Guangxi, with a total area of 115.59 km^2. The area is a typical tropical-subtropical karst landscape with the main landform of so-called peak-cluster depression (Figure 1), as well as other landforms, for example, dolines, fossil valleys and cave systems. Geographically, Huanjiang Karst belongs to the slope zone from the Yunnan-Guizhou plateau to the hilly basin of Guangxi, decreasing in elevation from 1028 m a.s.l. in the northwest to 250 m a.s.l. in the southeast. The topography of the area is fragmented due to the erosion-denudation of the eastern Gubin River and the western Dagou River and the fault structure in the area, making it a typical area for the development of conical peak-cluster karst landforms in the mid-subtropical zone [4,5].

167

Figure 1. Fengcong peak cluster landscapes in Huanjiang Karst, Hechi, Guangxi, China.

Forest coverage exceeds 95% and predominantly is comprised of a mixed evergreen and deciduous broad-leaved forest. The humid subtropical climate, the diverse karst habitats and the enclosed island-like environment are conducive to the growth and reproduction of living organisms. Likewise, there are hundreds of limestone caves distributed in this area, such as Cave Ganxiao Dong, Cave Dongtu Dong, Cave Mashan Dong and so on.

Xu Xiake, the first karstologist and caver of China (1587–1641), explored or located more than 270 caves in Guangxi. After this pioneering work, nothing significant was published on the karst of China for about 300 years. Modern cave exploration and karstology developed rapidly from the 1970s onwards, first through many foreign expeditions, then through increasing efforts of Chinese cavers. Speleologically, the most impressive result of this activity is the exploration of the Shuanghe cave system in Guizhou, today the largest in China and Asia with over 300 km of passages (Jean Botazzi, pers. comm.)

Research on cave biology in Chinese caves began in 1960 with the description of five millipedes and one centipede by Loksa [6], followed in 1981 by that of the cave fish *Oreonectes anophthalmus*, described by Zheng, the first of a long series. The first cave beetle was described in 1991 (*Sinaphaenops mirabilissimus* Uéno & Wang, 1991), the first cave Collembola in 1993 (*Sinella trogla* Chen & Christiansen, 1993) and the first cave woodlice in 1995 (*Sinoniscus cavernicolus* Schultz 1995). All these animals were collected in the caves of Guangxi or Guizhou, but none of them is from the Mulun karst.

These zoological groups, fish, millipedes, woodlice, springtails and beetles, together with spiders, represent today the bulk of the cave diversity of Chinese caves, with tens of new species described since then. Three of these groups have been particularly studied in the Mulun karst: millipedes, woodlice and beetles.

Millipedes are the most common terrestrial medium- to large-size invertebrates in the caves of South China Karst [7]. However, the diversity of cave millipedes in China has been revealed and recognized only during the last few decades. Approximately 350 millipede species from China have been named at present, including at least a third only known from caves [8]. In Mulun, the first cave millipede, *Eutrichodesmus similis*, reported by Golovatch et al., 2009, was described from Gui Dong and Shenlong Dong. Since this date, 16 species of Diplopoda have been reported from this karst [8].

Oniscid isopods are the only crustaceans that are fully adapted to terrestrial habitats, mostly living in wet and often dark environments. Cave-dwelling isopods are common in southern China. Most of them are terrestrial, but a few species have returned to aquatic life [9]. Contrary to millipedes and beetles, they are represented by few genera in Chinese caves, of which *Dryadillo* is found in many provinces in China, including troglobiotic and troglophilic species, while *Trogloniscus* is endemic to southern Guizhou and northern Guangxi, with five troglobiotic or stygobiotic species.

South China (northern Guangxi and southern Guizhou) is the world hot-spot for cave Trechinae beetles, having recently overpassed the historical hotspots of the Pyrenees and Dinarides in the number of taxa and levels of troglomorphy [10]. They are diversified in many genera and species, and all narrowly endemic, including the most impressive troglomorphic beetles known so far [10]. The first troglobiotic trechine species reported in China, *Sinaphaenops mirabilissimus*, by Uéno & Wang, 1991, was discovered in a show cave in Maolan, Libo County of southern Guizhou, which is adjacent to the Mulun Reserve [11]. Subsequently, Uéno & Ran [12] reported two other species from this county: *S. gracilior* from Cave Shui Dong, and *S. wangorum* from Cave Lasuo Dong in 1998. The latter also expands into northern Guangxi, occurring in several caves of the Mulun karst of Huanjiang County [13]. Similarly, *Libotrechus nishikawai*, described by Uéno, 1998, and *Uenotrechus liboensis*, described by Deuve & Tian, 1999, were formerly found in Maolan [14,15], then collected in Mulun. In addition, *U. gejianbangi*, described by Tian & Wei, 2017, was only found from the cave Ji Dong (also called Shuiku Dong) in Mulun. The genus *Pilosaphaenops*, reported by Deuve & Tian, 2008, contains six species, of which four are distributed in the Maolan-Mulun karst: *P. hybridiformis* (Uéno, 2002) from Maolan, *P. pilosulus* (Deuve & Tian 2008), *P. whitteni* Tian, 2011 and *P. weiguofui* Huang et al., 2020 from Mulun and related karsts near Sancai and Dacai (Huanjiang). For the genus *Oodinotrechus*, reported on by Uéno, 1998, two species were reported from this area, *O. kishimotoi* Uéno, 1998 from a cave in Maolan and *O. liyoubangi* Tian, 2014 from three caves in Mulun. The Maolan-Mulun karst is therefore home to 12 cave Trechinae, and several of them are highly troglomorphic.

This paper is the first contribution dedicated to Ganxiao Dong. This cave has the richest cave fauna of the Mulun-Maolan karst and of China, according to the surveys we have conducted for 10 years, as well as Mulun being itself the richest region of China for subterranean fauna [1].

2. Materials and Methods

2.1. Research Site

The cave Ganxiao Dong is close to the provincial border between Guizhou and Guangxi (Figure 2A), in the core zone of the adjacent Maolan and Mulun National Nature Reserves, both World Natural Heritage Sites of South China Karst. The cave is located in the northwest of Xiazhai Village, Chuanshan Town, Huanjiang County (25°10′57″ N, 108°01′55″ E). It is a medium-sized cave (836 m in total length), with pools of water and various terrestrial subterranean habitats favourable for subterranean species and invertebrate communities (Figure 2B).

The main entrance (Figure 3A) of Ganxiao Dong is 16 km away from Chuanshan Town in a straight-line distance, at an altitude of 735 m. The total length of the cave is 836 m, of which the main gallery is 620 m and the branch length is 216 m. The highest point in the main gallery is 48 m and the widest section is 26 m. There are two branches after the first part of the main gallery: the western branch is a dry passage in a higher position, with a length of 73 m; the eastern branch is longer (143 m). The entrance is 20 m wide and 9 m high, orientated at 335° N with a 15–20° downward slope. Two additional cave entrances are also opened on the northern side of the hill, at the level of the western branch. Both entrances are small, accessible only by a person at the same time. Five chambers (Figure 3B) exist inside the cave.

Influenced by the fracture and the strata altitude, the cave is overall nearly north-south oriented, with many bifurcations, and most of the sub-sections facing north-west or south-east. The cave plan is "Y" shaped and extends in a right-angle bend. In the longitudinal section, the main gallery extends in a north-south upward direction, ending on a collapse of stones. Due to gravity, the roof collapsed, and formed a vaulted chamber. The main gallery of the cave is spacious, while the branches are narrower. The bottom of the cave is uneven and the ceiling obviously collapsed, which caused the development of ridges and conchoidal fractures. At the bottom of the main gallery, several depressions were found, one of them being a 1 m deep and 10 m² areal pool (Figure 3C). Cross-sections of cave galleries are mainly triangular, sub-triangular or trapezoidal, mostly of large size (one to 47 m wide and two to 30 m high).

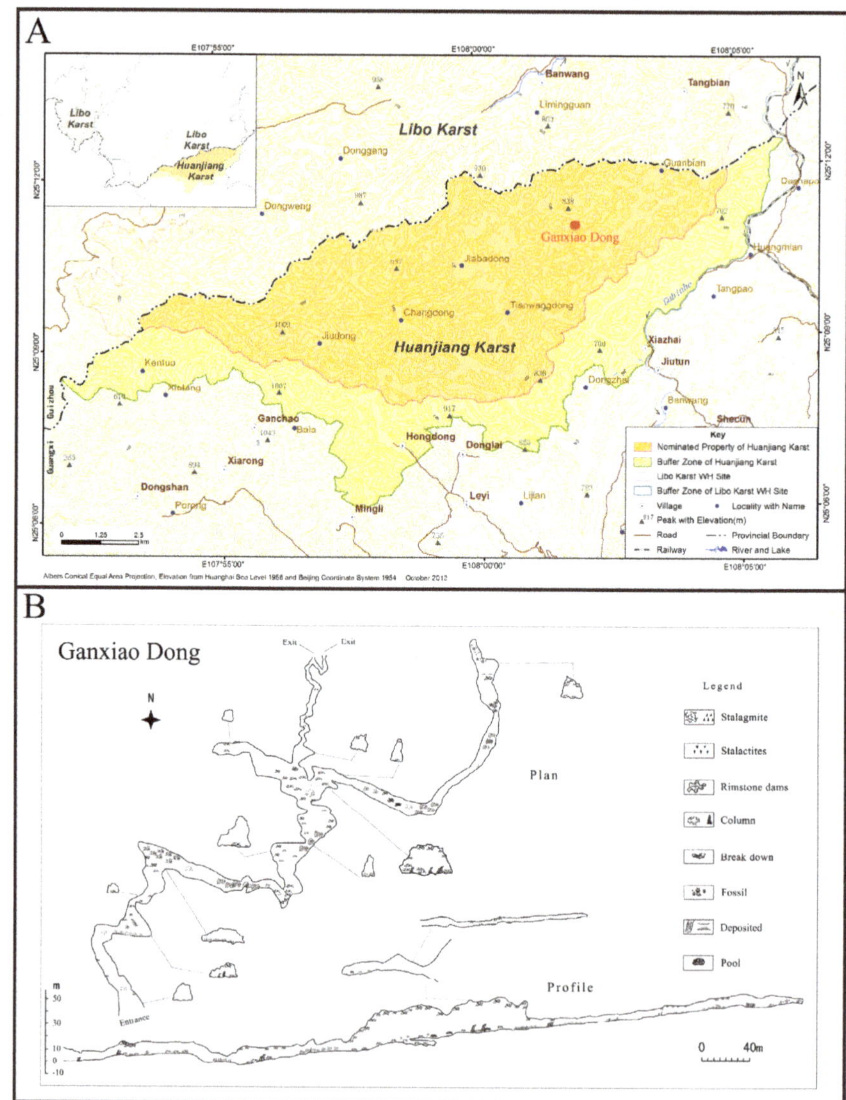

Figure 2. Location of Ganxiao Dong (**A**) and map of the cave (**B**).

The speleothems formed inside the cave are very diverse, including in particular stalagmites, stalactites, stone columns, stone mushrooms, stone curtains, stone shields, stone rimstone dams and stone pearls (Figure 3D–F). They are mainly distributed in the main gallery and the eastern branch, while the western branch has very few speleothems inside the passage because it is a high-level dry cave. Stone mushrooms, stone curtains and stone shields are only found in the main gallery, while the rimstone dams and stone pearls are only found in the eastern branch. The mechanical deposits in the cave are mainly sediments, with a few biological deposits such as bat guano.

Figure 3. Geomorphological characteristics of Ganxiao Dong. (**A**) entrance; (**B**) large chamber; (**C**) pool; (**D–F**) stalactites, rimstone dams, stalagmites and columns.

2.2. Sampling

Six biological surveys were conducted in the cave Ganxiao Dong during different seasons from 2009 to 2019. Collections were made by hand or using an aspirator and kept in 75% ethanol for morphological studies and identification or 95% ethanol for DNA sequencing and molecular analyses. Photos of the cave animals were taken by a Canon EOS 6D camera (Tokyo, Japan) with a Canon EF 100mm f/2.8L IS USM lens (Tokyo, Japan) and an adapted Meike MK-14 ext E-TTL macro flash (Hongkong, China). They were then processed using Photoshop CC 2019 (San Jose, CA, USA).

All materials studied are deposited in the animal collections of South China Agricultural University.

2.3. Terminology

Cave terrestrial animals are generally divided into three ecological categories based on their adaptation to cave life: troglobionts, troglophiles and trogloxenes [16]. Similarly, aquatic cave animals are referred to as stygobionts, stygophiles and stygoxenes. Troglobionts are fully adapted to the cave environment and are unable to survive outside the cave. Most of them are lacking in pigments, without eyes or visual organs. Troglophiles have populations in the dark cave environment but also outside (eutroglophiles) or are linked to caves only for a part of their life cycle (subtroglophiles). Trogloxenes enter the cave accidentally [17]. Troglomorphy, i.e., a set of morphological traits assumed to be linked to cave life, is the most spectacular characteristic of many cave invertebrates of South China karsts. It includes eye regression, winglessness and depigmentation, three traits observed as well in soil invertebrates, associated with body size increase and appendage elongation [18].

3. Results

3.1. Cave Fauna Composition

In total, 26 cave invertebrates have been discovered in Ganxiao Dong, including 19 troglobiotic, one stygobiotic and six troglophilic species (Table 1).

Table 1. Species list of cave-adapted animals from cave Ganxiao Dong. Column status: Tb = troglobiont; Tp = troglophile; Sb = stygobiont; * = known endemic of the Maolan-Mulun karst or of Ganxiao Dong; ? = not sure or under study.

No.	Species	Family	Order	Class	Status
1	*Sinospelaeobdella* sp.	Haemadipsidae	Gnathobdellida	Hirudinea	Tp
2	*Euplecta* sp.	Ariophatidae	Stylommatophora	Gastropoda	Tb
3	*Chalepotaxis* sp.	Helicarionidae	Stylommatophora	Gastropoda	Tp
4	*Hyleoglomeris kunnan* Golovatch, Liu & Geoffroy, 2012	Glomeridae	Glomerida	Diplopoda	Tb *
5	*Eutrichodesmus similis* Golovatch, Geoffroy, Mauries & VandenSpiegel, 2009	Haplodesmidae	Polydesmida	Diplopoda	Tb *
6	*Pacidesmus bedosae* Golovatch, Geoffroy & Mauries, 2010	Polydesmidae	Polydesmida	Diplopoda	Tb *
7	*Glyphiulus proximus* Golovatch, Geoffroy, Mauries & VandenSpiegel, 2011	Cambalopsidae	Spirostreptida	Diplopoda	Tb *
8	*Epedanidae* sp.	Epedanidae	Opiliones	Arachnida	Tb
9	*Troglocoelotes proximus* (Chen, Zhu & Kim, 2008)	Agelenidae	Araneae	Arachnida	Tb *
10	*Speleoticus libo* (Chen & Zhu, 2005)	Nesticidae	Araneae	Arachnida	Tb *
11	*Telema* sp.	Telemidae	Araneae	Arachnida	Tb
12	*Sparassidae* sp.	Sparassidae	Araneae	Arachnida	Tp
13	*Trogloniscus trilobatus* Taiti & Xue, 2012	Styloniscidae	Isopoda	Crustacea	Sb *
14	*Trogloniscus deharvengi* Taiti & Xue, 2012	Styloniscidae	Isopoda	Crustacea	Tb *
15	*Dryadillo* sp.	Armadillidae	Isopoda	Crustacea	Tb
16	*Sinella* sp.	Entomobryidae	Entomobryomorpha	Collembola	Tb
17	*Coecobrya* sp.	Entomobryidae	Entomobryomorpha	Collembola	Tb
18	*Sarasaeschna* sp.	Aeshnidae	Odonata	Insecta	Tp ?
19	*Tachycines (Gymnaeta) ferecaecus* (Gorochov, Rampini & Di Russo, 2006)	Rhaphidophoridae	Orthoptera	Insecta	Tb *
20	*Tachycines (Gymnaeta)* sp.1	Rhaphidophoridae	Orthoptera	Insecta	Tp
21	*Tachycines (Gymnaeta)* sp.2	Rhaphidophoridae	Orthoptera	Insecta	Tb
22	*Chetoneura* sp.	Keroplatidae	Diptera	Insecta	Tb
23	*Libotrechus nishikawai* Uéno, 1998	Carabidae	Coleoptera	Insecta	Tb *
24	*Sinaphaenops wangorum* Uéno & Ran, 1998	Carabidae	Coleoptera	Insecta	Tb *
25	*Pilosaphaenops hybridiformis* (Uéno, 2002)	Carabidae	Coleoptera	Insecta	Tb *
26	*Micronemadus pusillimus* (Kraatz, 1877)	Leiodidae	Coleoptera	Insecta	Tp

3.2. Notes on Cave Animals Living in Cave Ganxiao Dong

3.2.1. Leech

Sinospelaeobdella sp. occurs in Ganxiao Dong, wandering on roofs or walls (Figure 4A). Its two species, distributed in tropical continental Asia (China, Laos and Myanmar), are given in the literature as sucking the blood of different bat species, including *Rhinolophus pearsonii* Horsfield, 1851 which is present in the cave [19].

Figure 4. Cave animals in Ganxiao Dong: (**A**) leech *Sinospelaeobdella* sp.; (**B**) snail *Euplecta* sp. and (**C**) snail *Chalepotaxis* sp.; (**D**) *Pacidesmus bedosae* Golovatch et al., 2010; (**E**) *Glyphiulus proximus* Golovatch et al., 2011; (**F**) *Eutrichodesmus similis* Golovatch et al., 2009; (**G**) *Hyleoglomeris kunnan* Golovatch et al., 2012.

3.2.2. Snails

Two species of terrestrial snails are living inside Ganxiao Dong. Among them, *Euplecta* sp. (Figure 4B) is likely a troglobiont, while *Chalepotaxis* sp. (Figure 4C) is a troglophile.

3.2.3. Millipedes

Four millipede species of four different genera, families and orders, have been discovered in Ganxiao Dong [8]: *Pacidesmus bedosae* (Polydesmidae, Polydesmida) (Figure 4D), *Glyphiulus proximus* (Cambalopsidae, Spirostreptida) (Figure 4E), *Eutrichodesmus similis* (Haplodesmidae, Polydesmida)(Figure 4F) and *Hyleoglomeris kunnan* (Glomeridae, Glomerida) (Figure 4G). In China, the genus *Glyphiulus* is the richest among millipedes (43 species), followed by *Hyleoglomeris* (32 species) and *Eutrichodesmus* (24 species). All three genera, represented in China by many troglobionts that are moderately troglomorphic, are common in Ganxiao Dong, as well as in many caves of Mulun. The oligospecific genus *Pacidesmus* is less common in caves, but it encompasses impressive troglomorphic troglobionts, like *P. bedosae* of Ganxiao Dong with very long antennae and legs. The co-occurrence of several troglobiotic millipede species in this cave is not unusual. In Mulun, up to 6 unambiguously troglobiotic and often troglomorphic species may be found in a same cave [1]. Actually, Diplopoda represents the most diversified invertebrate group of the Mulun karst with 16 troglobionts recorded so far, a richness unmatched anywhere else in the world [8].

3.2.4. Harvestman

Only an unidentified harvestman species lives in Gánxiao Dong, belonging to the family Epedanidae (Figure 5A). It is omnivorous and considered to be a troglobiont due to noticeable depigmentation.

Figure 5. Harvestman and spiders in Ganxiao Dong. (**A**) Epedanidae sp.; (**B**) *Troglocoelotes proximus* (Chen, Zhu & Kim, 2008); (**C**) Sparassidae sp.

3.2.5. Spiders

Four spider species are found in Ganxiao Dong. Three of them, namely *Troglocoelotes proximus* (Figure 5B), *Speleoticus libo* and *Telema* sp. are troglobiotic. An unidentified species of Sparassidae, a troglophile, is also found inside this cave (Figure 5C).

3.2.6. Woodlice

Three species of troglobiotic woodlice (Isopoda) occur inside Ganxiao Dong. The genus *Trogloniscus* has two species, blind and depigmented, of which one, *T. trilobatus* is a stygobiont (Figure 6A) and the other one, *T. deharvengi*, is a troglobiont (Figure 6B) (Taiti & Xue, 2012). These two species are very similar in habitus, but easily distinguished on small somatic characters. It is very unusal to have, in a same well defined genus, species with such different ecological life styles, especially because both are abundant in the cave. Further investigations on the ecology of these species are clearly needed; in particular, the aquatic could be amphibious, like it happens in other Styloniscidae of SE Asia. The third species, *Dryadillo* sp. (Figure 6C) is also a troglobiont but has some remains of eyes. All three species are saprophagous and rather abundant in the cave.

Figure 6. Woodlice in Ganxiao Dong. (**A**) *Trogloniscus trilobatus* Taiti & Xue, 2012; (**B**) *Trogloniscus deharvengi* Taiti & Xue, 2012; (**C**) *Dryadillo* sp.

3.2.7. Springtails

Two springtails (Collembola), *Sinella* sp. and *Coecobrya* sp., occur in sympatry within Ganxiao Dong. They are small, whitish and good at jumping, feeding on dead wood, litter, or animal dung in the cave. Springtails play an important role in the cave food chain as they are the prey to many predators such as ground beetles and spiders. The two genera have several troglobionts and surface species, mostly undescribed, and are often abundant in caves of Mulun and China. Morphologically, they exhibit various degrees of troglomorphy, up to extreme appendage elongation.

3.2.8. Dragonfly

An unexpected discovery was a larva of the dragonfly *Sarasaeschna* sp. (Aeshnidae, Figure 7A), collected in a pool far inside the cave, suggesting some links to the cave environment (possibly troglophily) and currently under study (Haomiao Zhang, pers. comm.).

Figure 7. Dragonfly and crickets in Ganxiao Dong. (**A**) Nymph of *Sarasaeschna* sp.; (**B**) *Tachycines (Gymnaeta)* sp.1; (**C**) *Tachycines (Gymnaeta) ferecaecus* (Gorochov, Rampini & Di Russo, 2006); (**D**) *Tachycines (Gymnaeta)* sp.2.

3.2.9. Crickets

Three species of cave crickets (Rhaphidophoridae) are living in Ganxiao Dong. This remarkable diversity is not exceptional in southern China karst, where two and sometimes three species of Rhaphidophoridae may co-occur. It is the case in the Parking cave of Huoyan (Longshan in Hunan province) which hosts *Tachycines (Gymnaeta) omninocaecus* (Gorochov et al., 2006), *T. (G.) solidus* (Gorochov et al., 2006) and *Eutachycines crenatus* (Gorochov et al., 2006), each species occupying a different section of the cave with limited spatial overlap (Gorochov et al., 2006 and unpublished data). Among the Ganxiao Dong crickets, two are troglobionts with depigmented bodies: *Tachycines (Gymnaeta) ferecaecus* with very reduced eyes (Figure 7C), originally described from a cave in Maolan, and the blind species illustrated in Figure 7D. The third species, *T. (G.)* sp.1 (Figure 7B), has medium-size eyes and a coloured body, and is likely a troglophile. The three species have no wings, but well-developed jumping legs and are omnivorous, feeding on bat guano, fungi or preying on other small invertebrates.

3.2.10. Non-Glowing Sticky Worm

One species of non-glowing sticky worm, *Chetoneura* sp. (Figure 8A), occurs in Ganxiao Dong. *Chetoneura* catches their prey by means of a special curtain-like trap which is composed of threads with dew sticked by the worms (Figure 8B,C). A recent biological study [20] shows that *Chetoneura shennonggongensis* Amorim & Niu 2008, described from China, may spend its entire life within the cave, as adults have a very short life span, do not feed and have very limited flying ability. Therefore, in spite of the large size of their eyes, they can be considered as troglobionts. *Chetoneura*, the only genus of fungus gnats known to develop in caves, has only two described species in the oriental region, but unidentified non-glowing sticky worms are common in many caves of the oriental region, and may uncover a larger taxonomic diversity.

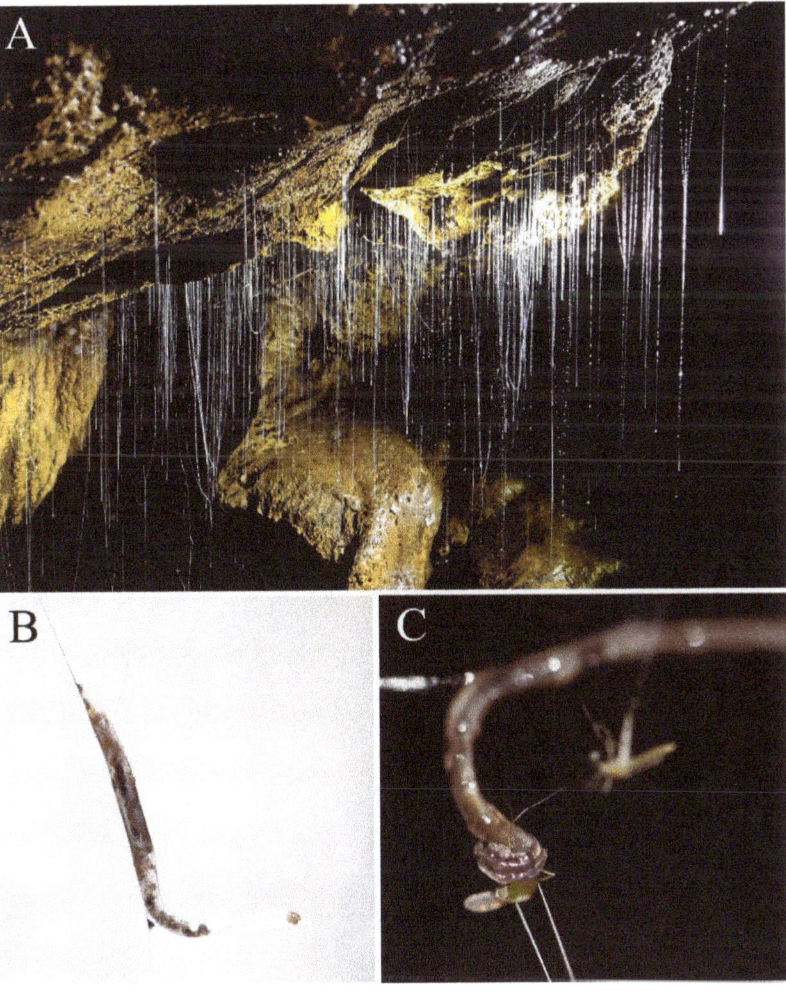

Figure 8. Non-glowing sticky worms (larvae of Keroplatidae) in Ganxiao Dong. (**A**) a cluster of larvae and threads; (**B**) a larva; (**C**) a larva feeding on a mosquito.

3.2.11. Ground Beetles

Three species of troglobiotic ground beetles (Carabidae) are living in Ganxiao Dong, i.e., *Pilosaphaenops hybridiformis* (Figure 9A), *Sinaphaenops wangorum* (Figure 9B), and *Libotrechus nishikawai* (Figure 9C) (Tian 2010 and unpublished). All of them belong to the subfamily Trechinae, and are anophthalmic. The last one is not strongly modified, whereas the other two are aphaenopsian, with elongated body and appendages, and modified mouthparts. They are predators, feeding on springtails (pers. obs.) or even probably eggs of other invertebrates, and are present in most caves of the region. All of them are narrowly endemic [10].

Figure 9. Ground beetles in Ganxiao Dong. (**A**) *Pilosaphaenops hybridiformis* (Uéno, 2002); (**B**) *Sinaphaenops wangorum* Uéno & Ran, 1998; (**C**) *Libotrechus nishikawai* Uéno, 1998.

3.2.12. Round Fungus Beetles

One species of round fungus beetles, *Micronemadus pusillimus* (Kraatz, 1877) (Leiodidae), was found in Ganxiao Dong. This species is similar in morphology to its surface relatives, and is considered as a troglophile. It is widespread, found in Mashan Dong from the same karst [2] and reported also in Malaysia, Indonesia, Japan [21].

3.2.13. Vertebrates

Although they are not cited in Table 1, bats are nevertheless very common mammals in Chinese caves, at the basis of the food webs, because of their guano and carcasses which represent food sources for micro-organisms and saprophagous invertebrates. In Ganxiao Dong only *Rhinolophus pearsonii* (Figure 10A) was observed. The species is widespread in southern continental Asia, listed in the IUCN Red List of Threatened Species as Least Concerned. It generally lives in groups from a dozen to several dozens, hanging on rock walls or cave ceilings. It is a food source of the parasite *Sinospelaeobdella*, a jawed land leech which is present in that cave, though it was not observed on the bat itself.

Figure 10. Other vertebrates in Ganxiao Dong. (**A**) *Rhinolophus pearsonii* Horsfield, 1851; (**B**) *Elaphe moellendorffi* (Boettger, 1886); (**C**) Fossil of *Stegodon* sp.

The beautiful Colubridae *Elaphe moellendorffi* (Boettger, 1886) (Figure 10B), frequent in Chinese caves, may prey upon bats, but its impact on their presence or abundance is unknown.

Bones of a fossil *Stegodon* were also found in the innermost part of the eastern branch of the cave (Figure 10C). The genus *Stegodon* is a very large size extinct mammal genus of the family Stegodontidae, order Proboscidea (elephants).

4. Discussion

The species richness of Ganxiao Dong is relatively low compared with several other temperate or tropical hot spots, but slightly above that of most of species-rich caves listed for continental Southeast Asia [22]. This, however, does not reflect a real biological pattern for two reasons. The first one is that the study of cave invertebrates in China is much more recent than in Europe or northern America, as illustrated by accumulation curves of taxa descriptions [19], and even more recent than that of several caves of Southeast Asia. The second reason is that multitaxa sampling in caves of China began only very recently compared to most regions which include subterranean diversity hotspots. The first multitaxa inventory of a Chinese cave is probably that of Feihu Dong in Hunan province, done in 1995 [23], but this kind of investigation really started in 2005 with a World Bank GEF project that included a karst biodiversity component focused on the Mulun karst.

Not surprisingly, the 26 invertebrate species listed above are therefore only a part of the cave-adapted animals occurring in Ganxiao Dong. Many cave-restricted invertebrates, such as mites, pseudoscorpions, diplurans or ant-loving beetles (Pselaphinae), that are known from other nearby caves in the Maolan-Mulun karst, have not been observed in Ganxiao Dong during our surveys. Moreover, aquatic fauna, which is, in many caves of the world, richer than terrestrial fauna, has not been sampled (with the exception of isopods).

At present, 150 cave species have been found in the Huanjiang World Heritage Site of South China Karst [3] and more than 40 caves have been biologically surveyed to various extents in the Mulun karst [24,25]. Apart from Ganxiao Dong, several caves in Mulun also have a promising species richness. For example, 21 troglobionts have been found in Mashan Dong, as well as 13 troglobionts and 3 troglophiles in Dongzai Dong [22]. Outside the Mulun karst, Ji Dong has 14 troglobionts [24], including three sympatric species of troglobitic ground beetles.

Candidate caves for "hotspot" label are also from other karstic regions of China, in particular Feihu Fong and Shuanghe Dong, both situated more to the north. In the cave of Feihu Dong from the Huoyan karst (northwestern Hunan), 21 troglobiotic species or morphospecies have been recorded so far [23]. In Shuanghe Dong, the longest cave system in China which is over 300 km at present (Jean Bottazzi, pers. comm.), over 50 species were found from several caves, but none of them held more than 20 cave species (unpublished data).

In short, the cave Ganxiao Dong, together with Mashan Dong and Feihu Dong, harbours the richest cave fauna in China according, but several other caves, less intensively sampled, are likely to reach similar levels of richness. The sampling gaps underlined above, combined with the fast pace of discoveries of new taxa in the region, lead us to foresee a significant increase in species richness for the near future in South China caves and karsts.

All the species inventories mentioned above have the same limitation, i.e., undersampling of aquatic fauna and several terrestrial groups. Taking these biases into account, Ganxiao Dong and Feihu Dong would often compare favourably with most other tropical or temperate cave systems [22,23,26].

The number of troglobiotic species measures the frequency of adaptations to cave life in a fauna. It does not inform about the impact of cave life on troglobiont biology and morphology. In this last respect, cave-obligate terrestrial species of southern China are at the front line, with exceptionally high level of troglomorphy in several major groups of cave invertebrates: millipedes, crickets and beetles [8,27,28].

Author Contributions: Conceptualization, M.T., W.L. and S.H.; methodology, W.L., S.H. and H.W.; software, S.H. and W.L.; validation, L.D., A.B., M.T., W.L. and S.H.; formal analysis, S.H., L.D., M.T., W.L. and A.B.; investigation, M.T., G.W., H.W., S.H., L.D., A.B. and W.L.; resources, M.T., G.W. and H.W.; data curation, M.T., L.D., A.B., W.L. and S.H.; writing—original draft preparation, S.H., W.L. and M.T.; writing—review and editing, L.D., S.H., A.B., M.T. and W.L.; visualization, S.H., L.D., A.B., M.T. and W.L.; supervision, M.T., L.D. and W.L.; project administration, M.T., W.L. and G.W.; funding acquisition, M.T., L.D., W.L. and G.W. All authors have read and agreed to the published version of the manuscript.

Funding: This work was funded by the Protection Centre of Huanjiang World Natural Heritage and the National Natural Science Foundation of China, grant numbers 41871039 and 31801956. Field sampling was supported until 2012 by a World Bank GEF linked to the project "Guangxi integrated forestry development and conservation".

Institutional Review Board Statement: Not applicable.

Informed Consent Statement: Not applicable.

Data Availability Statement: The data presented in this study is available in the cited references.

Acknowledgments: We would like to thank the following colleagues who provided help to identify the cave animals: Guchun Zhou (Gannan Normal College, Ganzhou) and Huiming Chen (Institute of Biology, Guizhou Academy of Sciences, Guiyang) for spiders, Libiao Zhang (Institute of Zoology, Guangdong Academy of Sciences, Guangzhou) for the bat, Feng Zhang (Nanjing Agricultural University) for springtails, Stefano Taiti (Istituto per lo Studio degli Ecosistemi, Firenze) and Zhihong Xue (Guangdong Agricultural Academy of Sciences, Guangzhou) for isopods, Zhixiao Liu (Jishou University, Jishou) for the leech, Haomiao Zhang (Kunming Institute of Zoology, Chinese Academy of Sciences) for the dragonfly and Qidi Zhu (Hebei University, Baoding) for crickets.

Conflicts of Interest: The authors declare no conflict of interest.

References

1. Deharveng, L.; Brehier, F.; Bedos, A.; Tian, M.Y.; Li, Y.B.; Zhang, F.; Qin, W.G.; Tan, X.F. Mulun and surrounding karsts (Guangxi) host the richest cave fauna of China. *Subterr. Biol.* **2008**, *6*, 75–79.
2. Tian, M.Y.; Deharveng, L.; Bedos, A.; Li, Y.B.; Xue, Z.H.; Feng, B.; Wei, G.F. Advances of cave biodiversity survey: Result based mainly on invertebrates. In Proceedings of the 17th National Conference of Speleology, Tongshan, China, 1–3 November 2011; pp. 149–163.
3. Tian, M.Y.; Wei, G.F. (Eds.) *Life in Dark World: Cavernicolous Creatures in Huanjiang World Heritage Site of South China Karst*; Guangxi Science & Technology Publishing House: Nanning, China, 2017; p. 136.
4. Liu, Z.Q.; Xiong, K.N.; Li, G.C.; Xiao, S.Z.; Wang, L.Y.; Wang, H.S.; Luo, D. Geomorphologic value and contributions of the Huanjiang Karst Extension to South China Karst World Heritage. *Carsologica Sin.* **2014**, *33*, 64–76.
5. Hu, F.; Dum, H.; Zeng, F.P.; Song, T.Q.; Peng, W.X.; Zhang, F. Dynamics of soil nutrient content and microbial diversity following vegetation restoration in a typical karst peak-cluster depression landscape. *Acta Ecol. Sin.* **2018**, *38*, 2170–2179.
6. Loksa, I. Einige neue Diplopoden-und Chilopodenarten aus chinesischen Höhlen. *Acta Zool. Acad. Sci. Hung.* **1960**, *6*, 137–148.
7. Liu, W.X.; Golovatch, S.I.; Wesener, T.; Tian, M.Y. Convergent evolution of unique morphological adaptations to a subterranean environment in cave millipedes (Diplopoda) *PLoS ONE* **2017**, *12*, e0170717. [CrossRef] [PubMed]
8. Golovatch, S.I.; Liu, W.X. Diversity, distribution patterns, and fauno-genesis of the millipedes (Diplopoda) of mainland China. *ZooKeys* **2020**, *930*, 153–198. [CrossRef] [PubMed]
9. Taiti, S.; Xue, Z.H. The cavernicolous genus *Trogloniscus* nomen novum, with descriptions of four new species from Southern China (Crustacea, Oniscidea, Styloniscidae). *Trop. Zool.* **2012**, *25*, 183–209. [CrossRef]
10. Tian, M.Y.; Huang, S.B.; Wang, X.H.; Tang, M.R. Contributions to the knowledge of subterranean trechine beetles in southern China's karsts: Five new genera (Insecta, Coleoptera, Carabidae, Trechinae). *ZooKeys* **2016**, *564*, 121–156. [CrossRef] [PubMed]
11. Uéno, S.-I.; Wang, F.X. Discovery of a highly specialized cave trechine (Carabidae: Trechinae) in southwest China. *Elytra* **1991**, *19*, 127–135.
12. Uéno, S.I.; Ran, J.C. Notes on *Sinaphaenops* (Coleoptera: Trechinae), with description of two new species. *Elytra* **1998**, *26*, 51–59.
13. Chen, J.J.; Tang, M.R.; Yang, P.J.; Tian, M.Y. Contribution to the knowledge of the aphaenopsian genus *Sinaphaenops* Uéno et Wang, 1991 (Coleoptera: Carabidae: Trechini). *Zootaxa* **2017**, *4227*, 106–118. [CrossRef] [PubMed]
14. Uéno, S.I. Two new genera and species of anophthalmic trechine beetles (Coleoptera, Trechinae) from limestone caves of southeastern Guizhou, South China. *Elytra* **1998**, *26*, 37–50.
15. Deuve, T. Nouveaux Trechidae cavernicoles chinois, découverts dans les confins karstiques du Sichuan, du Hubei et du Yunnan (Coleoptera, Adephaga). *Rev. Française D'entomologie* **1999**, *21*, 151–161.
16. Schiner, J.R. Fauna der Adelsberger-, Lueger- und Magdalenen Grotte. In *Die Grotten und Höhlen von Adelsberg, Lueg, Planina und Laas*; Schmidl, A., Ed.; Braumuller: Vienna, Austria, 1854; pp. 231–272.
17. Sket, B. Can we agree on an ecological classification of subterranean animals? *J. Nat. Hist.* **2008**, *42*, 1549–1563. [CrossRef]
18. Huang, T.; Liu, Z.; Gong, X.; Wu, T.; Liu, H.; Deng, J.; Zhang, Y.; Peng, Q.; Zhang, L.; Liu, Z. Vampire in the darkness: A new genus and species of land leech exclusively bloodsucking cave-dwelling bats from China (Hirudinda: Arhynchobdellida: Haemadipsidae). *Zootaxa* **2019**, *4560*, 257–272. [CrossRef] [PubMed]
19. Deharveng, L.; Bedos, A. Diversity of terrestrial invertebrates in subterranean habitats. In *Cave Ecology*; Moldovan, O.T., Kovác, L., Halse, S., Eds.; Springer: Cham, Switzerland, 2018; pp. 107–172.
20. Li, X.Z.; Niu, C.Y.; Huang, Q.Y.; Lei, C.L.; Stanley, D.W. Life cycle of *Chetoneura shennonggongensis* (Diptera: Keroplatidae: Keroplatinae) from Jiangxi Province, China. *Insect Sci.* **2009**, *16*, 351–359. [CrossRef]
21. Nishikawa, M. New records of *Micronemadus pusillimus* (Coleoptera, Cholevidae) from Malaysia and Indonesia. *Kanagawa-Chuho* **1989**, *90*, 158–160.
22. Deharveng, L.; Bedos, A. Biodiversity in the tropics. In *Encyclopedia of Caves*, 3rd ed.; White, W.B., Culver, D.C., Pipan, T., Eds.; Academic Press: Waltham, MA, USA, 2019; pp. 146–162.
23. Deharveng, L.; Bedos, A. The cave fauna of southeast Asia. Origin, evolution, and ecology. In *Subterranean Ecosystems*; Wilken, H., Culver, D.C., Humphreys, W.F., Eds.; Elsevier: Amsterdam, The Netherlands, 2000; pp. 603–632.

24. Deharveng, L.; Tian, M.Y. Guangxi Integrated Forestry Development and Biodiversity Conservation Project. In *Final Report on Subterranean Biodiversity of Guangxi Karsts and Natural Reserves Prepared for the Provincial Forestry Bureau of Guangxi*; Provincial Forestry Bureau of Guangxi: Nanning, China, 2012; Annexes 34; p. 44.
25. Chen, J.J. Survey of Cave Biodiversity in Huanjiang World Heritage Site of South China Karst. Master's Thesis, South China Agricultural University, Guangzhou, China, 2017.
26. Culver, D.C.; Sket, B. Hotspots of subterranean biodiversity in caves and wells. *J. Cave Karst Stud.* **2000**, *62*, 11–17.
27. Latella, L. Biodiversity: China. In *Encyclopedia of Caves*, 3rd ed.; White, W.B., Culver, D.C., Pipan, T., Eds.; Academic Press: Waltham, MA, USA, 2019; pp. 127–135.
28. Faille, A. Beetles. In *Encyclopedia of Caves*, 3rd ed.; White, W.B., Culver, D.C., Pipan, T., Eds.; Academic Press: Waltham, MA, USA, 2019; pp. 102–108.

diversity

Article

Mammoth Cave: A Hotspot of Subterranean Biodiversity in the United States

Matthew L. Niemiller [1],*, Kurt Helf [2] and Rickard S. Toomey [3]

[1] Department of Biological Sciences, The University of Alabama in Huntsville, 301 Sparkman Dr NW, Huntsville, AL 35899, USA

[2] Cumberland Piedmont Network, National Park Service, Mammoth Cave National Park, 61 Maintenance Rd., Mammoth Cave, KY 42259, USA; kurt_helf@nps.gov

[3] Division of Science and Resources Management, Mammoth Cave National Park, P.O. Box 7, Mammoth Cave, KY 42259, USA; rick_toomey@nps.gov

* Correspondence: matthew.niemiller@uah.edu or cavemander17@gmail.com

Abstract: The Mammoth Cave System in the Interior Low Plateau karst region in central Kentucky, USA is a global hotspot of cave-limited biodiversity, particularly terrestrial species. We searched the literature, museum accessions, and database records to compile an updated list of troglobiotic and stygobiotic species for the Mammoth Cave System and compare our list with previously published checklists. Our list of cave-limited fauna totals 49 species, with 32 troglobionts and 17 stygobionts. Seven species are endemic to the Mammoth Cave System and other small caves in Mammoth Cave National Park. The Mammoth Cave System is the type locality for 33 cave-limited species. The exceptional diversity at Mammoth Cave is likely related to several factors, such as the high dispersal potential of cave fauna associated with expansive karst exposures, high surface productivity, and a long history of exploration and study. Nearly 80% of the cave-limited fauna is of conservation concern, many of which are at an elevated risk of extinction because of small ranges, few occurrences, and several potential threats.

Keywords: checklist; karst; species richness; stygobiont; troglobiont

Citation: Niemiller, M.L.; Helf, K.; Toomey, R.S. Mammoth Cave: A Hotspot of Subterranean Biodiversity in the United States. *Diversity* **2021**, *13*, 373. https://doi.org/10.3390/d13080373

Academic Editors: Tanja Pipan, David C. Culver and Louis Deharveng

Received: 10 July 2021
Accepted: 3 August 2021
Published: 12 August 2021

Publisher's Note: MDPI stays neutral with regard to jurisdictional claims in published maps and institutional affiliations.

1. Introduction

The Mammoth Cave System in central Kentucky, USA is the most extensive cave system in the world with over 663 km (412 miles) of mapped passaged, including 27 entrances and 10 significant caves that have been connected since explorations began in the late 1700s: Colossal, Crystal (=Floyd Collins' Crystal), Donkey, Hoover, Mammoth, Morrison, Proctor, Roppel, Salts, and Unknown caves. Colossal, Crystal, Salts, and Unknown caves comprise the 206 km (128 mile) Flint Ridge Cave system (Figure 1). Mammoth Cave National Park was created in 1941 and includes two-thirds of the Mammoth Cave System [1]. The Mammoth Cave System was recognized as a UNESCO World Heritage Site in 1981 because of its uniqueness as the world's longest cave system as well as its extensive geological, mineral, and biological resources. The region was recognized as the core of an UNESCO Biosphere Reserve—Mammoth Cave Biosphere Region—in 1990.

The Mammoth Cave System is developed in three major limestone layers at the northwestern extent of the Pennyroyal Plateau, an expansive flat karst plain within the Interior Low Plateau physiographic province. The limestone layers include, from youngest to oldest, the Girkin Formation (40 m thick), Ste. Genevieve Limestone (35 m thick), and St. Louis Limestone (53–60 m thick) [2–5]. The Girkin Formation is capped by resistant sandstone and shale of the Big Clifty Formation that form the Mammoth Cave, Flint, Joppa, and Toohey Ridges. Most of the cave system is developed in the Ste. Genevieve Limestone and the upper 40 m of St. Louis Limestone [5]. The limestone strata gently slope from the southeast to the northwest. The Pennyroyal Plateau is exposed at the surface to the southeast, while insoluble strata of the Chester Upland, including the Big Clifty Formation, form a rugged hilly terrain that overlies the cave system to the northwest. The Green River, a tributary of the Ohio River, has cut into the Pennyroyal Plateau about 60 m such that most of the Mammoth Cave watershed now occurs underground [6]. The karst watershed of Mammoth Cave includes seven groundwater basins (Pike Spring, Great Onyx, Echo River, Double Sink, River Styx, Floating Mill Hollow, and Turnhole Bend); in addition, flood overflow occurs into an eighth basin (Sand Cave). These basins encompass 317 km^2 and ultimately drain at springs at base level into the Green River [7,8].

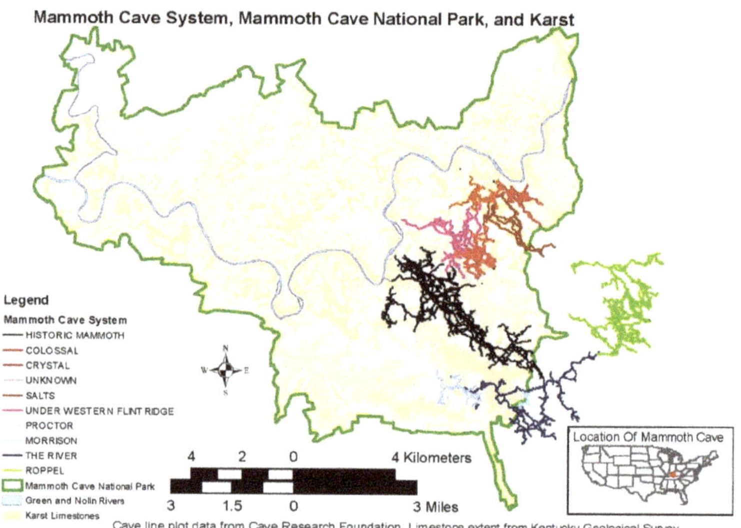

Figure 1. Map showing the location of Mammoth Cave National Park (MCNP) and the extent of the Mammoth Cave System in and adjacent to MCNP. The major segments of the Mammoth Cave System are shown as line plots in various colors. The different segments explored from different entrances (27 total). Line plot data from Cave Research Foundation. MCNP also contains over 500 smaller caves developed in various karstified limestones that are not attached to the Mammoth Cave System. These are grouped on the map, but include the St. Louis, Ste. Genevieve, Haney, Glen Dean and Girkin Formations. These smaller caves contain a variety of habitats from epikarst to base-level streams.

The Mammoth Cave System is characterized by a complex network of vadose and phreatic passages with at least five primary horizontal levels of passages (four fossil stream levels and the modern base level) representing distinct stages of development in association with past periods of water table stability and intervening periods of downcutting of the Green River valley through the resistant caprock into the soluble limestone layers below [1,6]. The evolution of the Mammoth Cave system is linked to the incision history of the Green River, drainage reorganizations, and significant climatic changes from the Pliocene through the Pleistocene, with the oldest upper-level passages dating to 3.2 Mya and the lower levels developing over the past 2 Mya [9].

Mammoth Cave has long been a focal region of study for North American subterranean biodiversity and for advancing our foundational knowledge of the ecology and evolution of cave fauna. Studies of the biodiversity in the Mammoth Cave System have an extensive history dating back to the 1820s (see [10]) when Constantin S. Rafinesque first visited Mammoth Cave [11]. Darwin [12] even mentions cave life from the Mammoth Cave region in *On the Origin of Species by Means of Natural Selection*. Much of our early knowledge of the North American cave fauna was derived from visits and studies by biologists to Mammoth Cave in the 1800s, such as DeKay [13], Wyman [14–19], Tellkampf [20–22], Agassiz [23–25], Von Motschulsky [26,27], Call [28], and Packard [29–36] (reviewed in [10] and [37]).

Additional significant early publications on the fauna and ecology of Mammoth Cave include Putnam [38], Eigenmann [39–42], Bolivar and Jeannel [43], Bailey [44], Buchanan [45], Park [46], Dearolf [47], Hubricht [48–53], Jeannel and Henrot [54], and Barr [55–60]. Barr [10] provided the first comprehensive review of the fauna of the Mammoth Cave system. More recently, Poulson [61,62] and Helf and Olson [63] provided reviews of terrestrial and aquatic ecosystems in Mammoth Cave. Culver and Hobbs [37] comprehensively reviewed the obligate cave fauna of the Mammoth Cave system and compared the fauna with other global hotspots of terrestrial cave biodiversity. Toomey et al. [1] presented a general review of the Mammoth Cave system that included a checklist of cave obligate fauna.

Herein we present an updated list of terrestrial and aquatic cave obligate fauna (i.e., troglobionts and stygobionts, respectively) of the Mammoth Cave system. Our goal is not to duplicate recently published checklists by Culver and Hobbs [37] and Toomey et al. [1] but rather complement these works by including a comprehensive bibliography on the cave obligate fauna of Mammoth Cave. In addition, we compare our list with past checklists from Mammoth Cave and comment on the exceptional biodiversity of this North American and global hotspot of subterranean biodiversity.

2. Materials and Methods

We conducted a search of the scientific literature to compile an updated list of troglobiont and stygobiont species for the Mammoth Cave System. For an overview of taxa that are not cave-limited, we refer readers to Barr [10], Culver and Hobbs [37], Helf and Olson [63], and Poulson [62]. Scientific literature sources included journal articles, book chapters, books, conference proceedings, theses and dissertations, and government reports. Searches of literature sources included keyword queries of ISI Web of Science, Google Scholar, and Zoological Record. In addition, we also searched biodiversity databases including the Global Biodiversity Information Facility (GBIF; Available online: https://gbif.org (accessed on 28 June 2021)), VertNet (Available online: http://www.vertnet.org (accessed on 28 June 2021)), Symbiota Collections of Arthropods Network (SCAN; Available online: https://scan-bugs.org/portal/(accessed on 28 June 2021)), and InvertEBase (Available online: http://www.invertebase.org/portal/index.php (accessed on 28 June 2021)). The list of cave obligate fauna includes the scientific name, authority, and conservation status of each species. Taxonomic nomenclature followed primarily the Integrated Taxonomic Information System (ITIS; Available online: http://itis.gov (accessed on 28 June 2021)). For conservation status, we include the International Union for Conservation of Nature (IUCN) Red List of Threatened Species (Available online: http://www.iucnredlist.org (accessed on 28 June 2021)) and NatureServe (Available online: http://www.natureserve.org (accessed on 28 June 2021)) conservation statuses when available. The status of a species according to the United States list of threatened and endangered species under the U.S. Endangered Species Act is included (Available online: http://www.fws.gov/endangered (accessed on 28 June 2021)), as well its status (endangered, threatened, or of greatest conservation need) under the latest Kentucky State Wildlife Action Plan (Available online: https://fw.ky.gov/WAP/Pages/default.aspx (accessed on 28 June 2021)).

3. Results

Packard [36] summarized the North America cave fauna, which at that time was primarily limited to the fauna of Mammoth Cave. He reported 31 permanent cave species, 18 of which we recognize as cave-limited species today, including 12 troglobionts and six stygobionts (Table 1). Barr [10] reported 44 cave-limited species (28 troglobionts and 16 stygobionts). More recently, Culver and Hobbs [37] listed 48 species (32 troglobionts and 16 stygobionts, 11 of which (nine troglobionts and two stygobionts) are endemic to the Mammoth Cave System, while Toomey et al. [1] reported 50 cave-limited species (32 troglobionts and 18 stygobionts). The authors also included two springtails not yet identified to species (*Willemia* sp. and *Onychiurus* sp.) on their list of cave-limited taxa, which were also reported by Barr [10].

Table 1. Troglobionts and stygobionts of the Mammoth Cave System, Kentucky, USA. NatureServe conservation ranks include Secure (G5), Apparently Secure (G4), Vulnerable (G3), Imperiled (G2), Critically Imperiled (G1), Possibly Extinct (GH), Presumed Extinct (GX), Unranked (GNR), and Unrankable (GU). IUCN Red List categories include Least Concern (LE), Near Threatened (NT), Vulnerable (VU), Endangered (EN), Critically Endangered (CR), Extinct in the Wild (EW), and Extinct (EX). Kentucky State Nature Preserves Commission statuses include Endangered (E), Threatened (T), Special Concern (S), Historic (H), and Extirpated (X). Federal conservation status under the U.S. Endangered Species Act includes Listed Endangered (LE) and Listed Threatened (LT).

Taxon	Authority	Cons. Status	Packard [36]	Barr [10]	Culver and Hobbs [37]	Toomey et al. [1]	This Study
TROGLOBIONTS			**13 Species**	**28 Species**	**32 Species**	**32 Species**	**32 Species**
Phylum Arthropoda							
Class Arachnida							
Order Araneae							
Family Linyphiidae							
Anthrobia monmouthia [T]	Tellkampf, 1844	G5	X	X	X	X	X
Bathyphantes weyeri	(Emerton, 1875)	G4		X	X	X	X
Phanetta subterranea	(Emerton, 1875)	G5		X	X	X	X
Porhomma cavernicola	(Keyserling, 1886)	G5		X	X	X	X
Family Zoropsidae							
Liocranoides unicolor [T]	Keyserling, 1881	GU	X				X
Order Opiliones							
Family Phalangodidae							
Phalangodes armata [T]	Tellkampf, 1844	G3	X	X	X	X	X
Order Pseudoscorpiones							
Family Chernetidae							
Hesperochernes mirabilis	(Banks, 1895)	G5		X	X	X	X
Family Chthoniidae							
Kleptochthonius cerberus [T,E]	Malcolm and Chamberlin, 1961	G1		X	X	X	X
Kleptochthonius hageni [T]	Muchmore, 1963	G1	X	X	X	X	X
Tyrannochthonius hypogeus [T,E]	Muchmore, 1996	G1			X		X
Order Acari							
Family Belbidae							
Dameus bulbipedata [T,E]	(Packard, 1888)	G1	X		X	X	X
Family Cocceupodidae							
Linopodes mammouthia [T]	Banks, 1897	GNR		X	X	X	X
Family Galumnidae							
Galumna alata	(Hermann, 1804)	G1			X	X	X
Family Laelapidae							
Laelaps cavernicola [T]	Packard, 1888	GNR	X	X	X	X	X
Family Macrochelidae							
Macrocheles troglodytes [T]	(Packard, 1888)	G1			X	X	X
Family Rhagidiidae							
Traegaardhia holsingeri [T]	(Zacharda, 1980)	GNR	X	X	X	X	X
Class Collembola							

187

Table 1. *Cont.*

Taxon	Authority	Cons. Status	Packard [36]	Barr [10]	Culver and Hobbs [37]	Toomey et al. [1]	This Study
TROGLOBIONTS			13 Species	28 Species	32 Species	32 Species	32 Species
Order Entomobryomorpha							
Family Entomobryidae							
Pseudosinella espanita [T,E]	Christiansen and Bellinger, 1996	G1			X	X	X
Order Symphypleona							
Family Arrhopalitidae							
Pygmarrhopalites altus [T,E]	(Christiansen, 1966)	G2		X	X	X	X
Class Diplura							
Order Rhabdura							
Family Campodeidae							
Litocampa cookei [T]	(Packard, 1871)	G5	X	X	X	X	X
Class Diplopoda							
Order Chordeumatida							
Family Trichopetalidae							
Scoterpes copei [T]	(Packard, 1871)	G3	X	X	X	X	X
Order Polydesmida							
Family Macrosternodesmidae							
Chaetaspis fragilis [T]	(Loomis, 1943)	GNR		X	X	X	X
Class Insecta							
Order Coleoptera							
Family Carabidae							
Neaphaenops tellkampfi [T]	(Erichson, 1844)	G3	X	X	X	X	X
Pseudanophthalmus audax	(Horn, 1883)	G1		X	X	X	X
Pseudanophthalmus inexpectatus [T,E]	Barr, 1959	G1		X	X	X	X
Pseudanophthalmus menetriesi [T]	(Motschulsky, 1862)	G3	X	X	X	X	X
Pseudanophthalmus pubescens	(Horn, 1868)	G3		X	X	X	X
Pseudanophthalmus striatus [T]	(Motschulsky, 1862)	G2	X	X	X	X	X
Family Leiodidae							
Ptomaphagus hirtus [T]	(Tellkampf, 1844)	G4	X	X	X	X	X
Family Staphylinidae							
Batrisodes henroti	Park, 1956	G2		X	X	X	X
Order Diptera							
Family Phoridae							
Megaselia cavernicola	(Brues, 1906)	GNR				X	
Family Sphaeroceridae							
Spelobia tenebrarum	(Aldrich, 1897)	G5			X	X	X
Order Psocodea							
Family Psyllipsocidae							
Psyllipsocus ramburii	Selys-Longchamps, 1872	GNR		X			

Table 1. *Cont.*

Taxon	Authority	Cons. Status	Packard [36]	Barr [10]	Culver and Hobbs [37]	Toomey et al. [1]	This Study
TROGLOBIONTS			13 Species	28 Species	32 Species	32 Species	32 Species
Phylum Mollusca							
Class Gastropoda							
Order Basommatophora							
Family Carychiidae							
Carychium stygium [T]	Call, 1897	G3		X	X	X	X
Order Stylommatophora							
Family Helicodiscidae							
Helicodiscus hadenoecus	Hubricht, 1962	G3		X			
Helicodiscus punctatellus [T]	Morrison, 1942	G1		X	X	X	
Family Zonitidae							
Glyphyalinia specus	Hubricht, 1965	G4		X	X	X	X
STYGOBIONTS			6 species	16 species	16 species	18 species	17 species
Phylum Platyhelminthes							
Class Turbellaria							
Order Tricladida							
Family Kenkiidae							
Sphalloplana buchanani [T]	(Hyman, 1937)	G1		X	X	X	X
Sphalloplana percoeca [T]	(Packard, 1879)	G5	X	X	X	X	X
Phylum Arthropoda							
Class Malacostraca							
Order Amphipoda							
Crangonyctidae							
Crangonyx barri [T]	Zhang and Holsinger, 2003	G5		X	X	X	X
Stygobromus exilis	Hubricht, 1943	G5		X	X	X	X
Stygobromus vitreus [T]	Cope, 1872	G4; S	X	X	X	X	X
Order Decapoda							
Family Atyidae							
Palaemonias ganteri [T]	Hay, 1901	G1; VU; E; LE		X	X	X	X
Family Cambaridae							
Orconectes pellucidus [T]	(Tellkampf, 1844)	G4; LC; S	X	X	X	X	X
Order Isopoda							
Family Asellidae							
Caecidotea bicrenata	Lewis and Bowman, 1981	G5				X	X
Caecidotea stygia [T]	Packard, 1871	G5	X	X	X	X	X
Class Maxillopoda							
Order Cyclopoida							

188

Table 1. *Cont.*

Taxon	Authority	Cons. Status	Packard [36]	Barr [10]	Culver and Hobbs [37]	Toomey et al. [1]	This Study
TROGLOBIONTS			13 Species	28 Species	32 Species	32 Species	32 Species
Family Cyclopidae							
Megacyclops donnaldsoni	(Chappuis, 1929)	G3		X	X	X	X
Order Harpacticoida							
Family Canthocamptidae							
Atthevella pilosa [T]	Chappuis, 1929	GNR		X	X	X	
Bryocamptus morrisoni	(Chappuis, 1928)	G3		X	X	X	X
Order Siphonostomatoida							
Family Lernaeopodidae							
Cauloxenus stygius	Cope, 1872	G1				X	X
Class Ostracoda							
Order Podocopida							
Family Entocytheridae							
Sagittocythere barri	(Hart and Hobbs, 1961)	G5		X	X	X	X
Sagittocythere stygia [T,E]	Hart and Hart, 1966	G1		X	X	X	X
Phylum Mollusca							
Class Gastropoda							
Order Neotaenioglossa							
Family Hydrobiidae							
Antroselates spiralis [T]	Hubricht, 1963	G3		X	X	X	X
Phylum Chordata							
Class Actinopterygii							
Order Percopsiformes							
Family Amblyopsidae							
Amblyopsis spelaea [T]	DeKay, 1842	G2; NT; S	X	X	X	X	X
Typhlichthys subterraneus	Girard, 1859	G4; NT; S	X	X	X	X	X

[T] Type locality in Mammoth Cave National Park; [E] Mammoth Cave National Park endemic.

Our list of cave-limited fauna includes 49 species, with 32 troglobionts and 17 stygobionts (Table 1; Figure 2). Both Culver and Hobbs [37] and Toomey et al. [1] included the snail *Helicodiscus punctatellus* and copepod *Atteyella pilosa* in their respective lists of cave-limited taxa. *Helicodiscus punctatellus* is known from surface collections [64]. *Atteyella pilosa* is a facultative associate of several species of surface and cave-limited crayfishes and is also known from surface collections [65]. Culver and Hobbs [37] did not include the isopod *Caecidotea bicrenata*, which was included in our list and that of Toomey et al. [1]. Lewis [66] reported several collections of *C. bicrenata* from the Mammoth Cave System where it predominately occurs in lower-level aquatic habitats. Toomey et al. [1] included the phorid fly *Megaselia cavernicola* in their list of cave-limited taxa. *Megaselia cavernicola* is a widely occurring species in caves on eastern North America that lacks obvious troglomorphic characters, is known from surface collections [67], and has been treated as a troglophile (i.e., non-obligate) by most past authors (e.g., [68,69]).

Figure 2. Representative cave-limited fauna from the Mammoth Cave System, Kentucky, USA: (**A**) *Scoterpes copei* (photo by Rickard A. Olson); (**B**) *Neaphaenops tellkampfi* feeding on the egg of the *Hadenoecus subterraneus* (photo by Rickard A. Olson); (**C**) *Hesperochernes mirabilis* with *Macrocera nobilis* larva (photo by Rickard A. Olson); (**D**) *Amblyopsis spelaea* (photo by Dante B. Fenolio); (**E**) *Phalangodes armata* (photo by Rickard A. Olson); (**F**) *Palaemonias ganteri* (photo by Rickard A. Olson); (**G**)—*Orconectes* pellucidus (photo by Dante B. Fenolio); (**H**) *Litocampa cookei* (photo by Rickard A. Olson); (**I**)—*Sphalloplana buchanani* (photo by Rickard A. Olson).

Mammoth Cave is the type locality for 33 cave-limited species (Table 1). Seven species are endemic to the Mammoth Cave system and other smaller caves in Mammoth Cave National Park (Table 1).

3.1. Terrestrial Fauna

Two troglobiotic snails have been documented in the Mammoth Cave System. *Carychium stygium* is found in association with cricket guano and is the most common of the two species [37]. Weigand et al. [70,71] suggest *C. stygium* may be an ecotype of the troglophile *C. exile*, as *C. stygium* shows limited mitochondrial COI sequence divergence from and is nested within a clade containing *C. clappi* and *C. exile*. However, this inference is based on a single locus and only two populations of *C. stygium* were included in analyses. Alternative hypotheses such as incomplete lineage sorting and mitochondrial introgression cannot be ruled out at present and warrant study. Regardless, these studies suggest that it is likely that *C. stygium* has recently colonized caves. *Glyphyalinia specus* is a wide-ranging snail known from 27 occurrences in five states [72]. Significant publications include Call [28], Hubricht [49,50,52,53], Barr [10], Poulson et al. [73], Dourson [74], Poulson [62], and Gladstone et al. [72].

Troglobiotic spiders documented in the Mammoth Cave System include four linyphiids and one zoropsid. All four linyphiids have broad distributions in caves of the eastern United States [75]. *Bathyphantes weyeri* is predominantly known from caves but has rarely been collected from surface habitats in Canada [75–77]. Holsinger et al. [78] hypothesized that the species may be troglobiotic in the southern parts of its range and troglophilic in the northern areas. Moreover, *B. weyeri* may represent a species complex. Most authors, including herein, still treat this species as a troglobiont [1,37,76,78,79]. *Liocranoides unicolor* was described by Keyserling [80] from Mammoth Cave. This species is pale in coloration but does not possess other troglomorphic characters [81]. Significant publications include Packard [29,32,33,36], Emerton [82], Hubbard [83], Keyserling [80], Call [28], Mcindoo [84], Berland [85], Bailey [44], Barr [10], Poulson and Culver [86], Poulson [62,87], Platnick [81], and Miller [75].

A single troglobiotic opilionid (*Phalangodes armata*) is known from several areas in the Mammoth Cave System. Significant publications include Tellkampf [20,21], Packard [29,36], Hubbard [83], Call [28], Bailey [44], Goodnight and Goodnight [88], Barr [10], Poulson and Culver [86], Hedin and Thomas [89], and Poulson [62].

Four troglobiotic pseudoscorpions occur in the Mammoth Cave System. *Hesperochernes mirabilis* is a widely distributed species most abundant near entrances. It is often observed in and near rodent (*Neotoma* and *Peromyscus* sp.) nests, which may facilitate phoretic dispersal. The other three species are thought to be associated with deep cave habitats. *Kleptochthonius cerberus* was described from White's Cave in Mammoth Cave National Park [90] and has to date, only been found there. *Kleptochthonius hageni* was described from Mammoth Dome in Mammoth Cave [91]. *Kleptochthonius cereberus* is thought to be endemic to Mammoth Cave National Park. *Kleptochthonius hageni* is reported to occur in the Mammoth Cave System and possibly some nearby caves not on the park (C.D.R. Stephen, pers. comm.). *Tyrannochthonius hypogeus* is a small, eyeless species with attenuated appendages first collected from log litter in Bruce Hollow [92]. Muchmore [92] considered this species to be cave adapted and associated with the Mammoth Cave fauna. Notable publications include Hubbard [83], Packard [36], Banks [93], Malcolm and Chamberlin [90], Muchmore [91,92], and Barr [10].

The troglobiotic mite fauna is particularly diverse with six species but has been little studied since their descriptions [37]. Notable publications include Packard [36], Call [28], Vitzthum [94], Bailey [44], Holsinger [95], Barr [10], and Zacharda [96].

Two troglobiotic millipedes have been documented in the Mammoth Cave System. *Scoterpes copei* is a common trichopetalid distributed throughout the cave system where it can be found in moist habitats with organic matter (rotting wood, debris, and cricket guano). *Chaetaspis fragilis* is a small polydesmid infrequently encountered in the Mammoth Cave

System but more common in White Cave, Mammoth Cave National Park [10]. Significant publications include Packard [29,36], Cope [97], Hubbard [83], Loomis [98], Barr [10], Poulson and Culver [86], Poulson et al. [73], Shear [99], and Poulson [62].

Although more than 10 species of collembolans (i.e., springtails) have been documented in the Mammoth Cave System [10], just two taxa are considered troglobionts and both are endemic to the cave system. *Pygmarrhopalites altus* was described by Christiansen [100] from Eyeless Fish Trail in the Unkown Cave section of Mammoth Cave. *Pseudosinella espanita* was described by Christiansen and Bellinger [101] from Styx River near Charon's Cascade in Mammoth Cave. Notably absent from the fauna of the Mammoth Cave System are *P. hirsuta* and *Sinella cavernarum*, which have broad distributions that include the Western Pennyroyal Karst of nearby Barren County, Kentucky [102]. Barr [10] reported two undescribed collembolans as potential troglobionts from Mammoth Cave: *Willemia* sp. have been collected from rotting boards in the Roaring River section. This genus includes several edaphic species, but no troglobionts are known to date and it is unlikely that this taxon represents a true troglobiont. *Onychiurus* sp. also have been collected from Mammoth Cave. Four described species in this genus are considered troglobionts in caves of the eastern United States. Additional study is needed on the collembolans of the Mammoth Cave System. Significant publications include Packard [36], Call [28], Christiansen [97,103–105], Barr [10], Poulson and Culver [86], Christiansen and Bellinger [101], and Poulson [62].

A single troglobiotic dipluran occurs in the Mammoth Cave System. *Litocampa cookei* has the largest distribution of any troglobiotic dipluran in the United States [106] but may represent a cryptic species complex. It was described from Mammoth Cave [29]. Notable publications include Packard [29,30,36], Hubbard [83], Silvestri [107,108], Conde [109], Barr [10], Poulson and Culver [86], Ferguson [106,110], and Poulson [62].

The troglobiotic beetle fauna is the most well known and studied of all taxonomic groups in the Mammoth Cave System. Eight species have been documented, namely six carabids, one leiodid, and one staphylinid species. *Neaphaenops tellkampfii* is the largest troglobiotic carabid species in Mammoth Cave and is also the first troglobiotic trechine beetle discovered in North America [111]. It was described from Mammoth Cave [112]. This species is found in silty habitats, where it feeds mostly on the eggs of the cave cricket *Hadenoecus subterraneus* [111,113]. Five species in the genus *Pseudanophthalmus* occur in a variety of habitats throughout the Mammoth Cave System. Three species were described from Mammoth Cave and one species (*P. inexpectatus*) is endemic to MCNP. All six species are blind and wingless. In some locations in the cave system, all six carabid species can be found but appear to have different microhabitat preferences and can be readily distinguished morphologically [59,60,113]. *Ptomaphagus hirtus* is an abundant small carrion beetle that is becoming an important model for studying the genetics of circadian rhythms [114,115]. *Batrisodes henroti* is a small rove beetle that has been infrequently collected in the Mammoth Cave System. Relevant publications include Erichson [112], Tellkampf [20,21], Von Motschulsky [26,27], Horn [116,117], Packard [29,31,34,36], Hubbard [83], Jeannel [118–121], Valentine [122,123], Hatch [124], Jeannel and Henrot [54], Park [125–127], Barr [10,55–60,128–131], Poulson and Culver [86], Barr and Kuehne [132], Peck [133–137], Kane et al. [138], Norton et al. [139], Kane and Poulson [140], Laing et al. [141], Giuseffi et al. [142], Kane and Ryan [143], Barr and Holsinger [144], Kane and Brunner [145], Poulson et al. [73], Friedrich et al. [115], Friedrich [114], Helf [146], Poulson [62], and Leray et al. [147].

The only other troglobiotic insect documented from Mammoth Cave is the dipteran *Spelobia tenebrarum*, a widely distributed species in caves of eastern North America [148]. Notable publications include Barr [10], Marshall and Peck [148], and Poulson [62].

3.2. Aquatic Fauna

Two cave flatworms occur in and were described from the Mammoth Cave System. *Sphalloplana percoeca* occurs primarily in epikarst-fed drip pools in upper-level passages, while *S. buchanani* is associated with stream gravels [37]. Significant publications on cave flatworms include Packard [29,36], de Beauchamp [149], Buchanan [45], Hyman [150], Barr [10], Carpenter [151,152], Barr and Kuehne [132], Kenk [153], Lewis [66], Pearson and Boston [154], and Helf and Olson [63].

A single groundwater snail has been documented in the Mammoth Cave System. *Antroselates spiralis* occurs in base-level streams in cave system. It was described from Echo River Spring, a major drain of the Mammoth Cave System. Notable publications include Hubricht [51], Barr [10], Barr and Kuehne [132], Hershler and Hubricht [155], Lewis [66], Pearson and Boston [154], and Helf and Olson [63].

The copepods of the Mammoth Cave System have not been well studied [37]. Three stygobionts have been documented—*Megacyclops donnaldsoni*, *Bryocamptus morrisoni*, and *Cauloxenus stygius*. *Cauloxenus stygius* is an ectoparasite of the cavefish *Amblyopsis spelaea* [156]. Notable publications include Cope [97], Kofoid [157], Chappuis [158], Barr [10], Barr and Kuehne [132], Whitman [159], Lewis [160], Niemiller and Poulson [156], and Helf and Olson [63].

Two ostracods are ectocommensals primarily of the stygobiotic crayfish *Orconectes pellucidus*—*Sagittocythere barri* and *S. stygia*. *Sagittocythere stygia* was described from River Styx in Mammoth Cave. Significant publications include Kofoid [157], Klie [161], Hart and Hobbs [162], Hart and Hart [163], Barr [10], Barr and Kuehne [132], Hart and Hart [164], and Helf and Olson [63].

Isopods are represented by two aquatic stygobionts—*Caecidotea stygia* and *C. bicrenata*. *Caecidotea stygia* was described from Mammoth Cave by Packard [29] and is more abundant in upper to mid-levels of the cave system, whereas *C. bicrenata* is more common in low to mid-levels [66]. Significant publications include Packard [29,35,36], Hubbard [79], Garman [165], Hay [166], Giovannoli [167], Dearolf [47], Chappuis [168], Barr [10], Barr and Kuehne [132], Lewis and Bowman [169], Lewis [66], Helf and Olson [63], and Helf et al. [170].

Three species of stygobiotic amphipods have been reported from the Mammoth Cave System. *Stygobromus vitreus* is more common in upper levels of the cave system, while *S. exilis* is more common in low to mid-levels [66,132]. *Stygobromus vitreus* was described from Richardson Spring within Mammoth Cave. Cathedral Domes in Mammoth Cave is the type locality of *Crangonyx barri*, an inhabitant of small cave streams and drip pools [171]. Signification publications include Cope [97], Packard [36], Giovannoli [167], Hubricht [48], Barr [10], Barr and Kuehne [132], Holsinger [172], Lewis [66], Zhang [173], Zhang and Holsinger [171], Helf and Olson [63], and Helf et al. [170].

Two stygobiotic decapods occur in the Mammoth Cave System. *Palaemonias ganteri* is a federally endangered atyid shrimp found in slow-flowing base-level streams of eleven groundwater basins in the Mammoth Cave System ([174]; updated by R. Toomey with new data). Significant publications on *P. ganteri* include Hay [166], Fage [175], Giovanolli [167], Barr [10], Barr and Kuehne [132], Hobbs et al. [176], Holsinger and Leitheuser [177–179], Lisowski [180,181], Lisowski and Poulson [182], Leitheuser and Holsinger [183], Leitheuser et al. [184,185], Lewis [66], USFWS [174,186], Pearson and Boston [154], Pearson and Jones [187], Cooper and Cooper [188], Helf and Olson [63], and Stump [189]. *Orconectes pellucidus* was described from Mammoth Cave and is the only stygobiotic crayfish in the cave system. While *O. pellucidus* is a ubiquitous stygobiont in the Mammoth Cave System, it is more abundant in mid- and base-level streams and pools. Notable publications include Tellkampf [20,21], Hagen [190,191], Packard [29,36], Cope [97], Garman [192], Fage [175], Bailey [44], Park et al. [193], Rhoades [194], Hobbs and Barr [195,196], Brown [197], Wolfe and Cornwell [198], Barr [10], Hobbs et al. [176], Pearson and Boston [154], Pearson and Jones [187], Compson [199], Taylor and Schuster [200], Helf and Olson [63], and Helf et al. [170].

The only cave-limited vertebrates known from the Mammoth Cave System are the amblyopsid cavefishes *Amblyopsis spelaea* and *Typhlichthys subterraneus*. *Amblyopsis spelaea* was described from River Styx in Mammoth Cave by Dekay [13] and represents the first cave-adapted fish formally described [156,201]. Mammoth Cave is one of only a handful of cave systems globally with two or more syntopic cavefish species [156,201]. *Typhlichthys subterraneus* are more abundant in upstream sections of streams that drain vertical shafts, whereas *A. spelaea* are more common in deeper pools at base level [61,156]; both are top predators. It remains unclear whether *A. spelaea* outcompetes *T. subterraneus* in base-level habitats. Significant publications on cavefishes of Mammoth Cave include Davidson [202], DeKay [13], Wyman [14–19], Thompson [203], Tellkampf [21,22], Agassiz [23–25], Girard [204], Putnam [38], Packard [36], Eigenmann [39–42], Bailey [44], Woods and Inger [205], Poulson [61,206–208], Barr and Kuehne [209], Rosen [210], Barr [10], Poulson and White [211], Barr and Kuehne [132], Clay [212], Swofford et al. [213], Lisowski and Poulson [182], Swofford [214], Burr and Warren [215], Lewis [66,160,216], Keith [217], Branson [218], Pearson and Boston [154], Pearson and Jones [187], Romero [219], Romero and Bennis [220], Compson [199], Proudlove [201], Niemiller and Poulson [156], Niemiller [221], Niemiller and Fitzpatrick [222], Niemiller et al. [223], Helf and Olson [63], Helf et al. [170], and Hart et al. [224].

4. Discussion

The Mammoth Cave obligate cave fauna is exceptionally rich with 49 troglobionts and stygobionts, making it one of the most diverse systems globally [37,225,226]. The terrestrial fauna is particularly diverse—tied for the third richest cave system in the world behind the Postojna Planina Cave System (36 species) in Slovenia and Cueva de Felipe Revention (34 species) in the Canary Islands [226]. With respect to stygobiotic fauna, the Mammoth Cave System ranks second in North America behind San Marcos Artesian Well in San Marcos, Texas (55 taxa, 39 described and 16 undescribed; [227]).

Several hypotheses have been proposed [10,61,127,224,228] to explain the high species richness in the Mammoth Cave System (recently reviewed in [37]). First, high species richness in the Mammoth Cave System may reflect the long history of more intensive sampling and study compared to other cave systems in the region [37]. While sampling intensity and bias may partially explain the high species richness at Mammoth Cave, several other biogeographical hypotheses warrant mention. The Mammoth Cave System is developed within a thick, continuous karst exposure over a large area in the Interior Low Plateau, which supports larger and more stable population sizes, more complex communities, and greater dispersal potential [113,129,130]. Moreover, the Mammoth Cave System is located at an intersection of hypothesized dispersal routes for cave-limited species from other karst areas, such as the Pennyroyal Karst Plain, Cumberland Saddle, and Bluegrass Region, and its cave fauna includes not only endemic species but also taxa also found in these adjacent regions [10,37,66,130]. The Mammoth Cave System lies within a hypothesized ridge of high troglobiont diversity found in temperate North America and Europe identified by Culver et al. [228]. This ridge corresponds to a general region of high surface primary productivity, which provides higher levels of allochthonous input into cave systems [228]. Mammoth Cave is noted for having high levels of allochthonous productivity but also chemoautotrophic productivity [37,63,229]. However, whether chemosynthesis subsidizes troglobiont communities or contributes significantly to the high troglobiont diversity found in the Mammoth Cave System remains speculative, as it is not well supported by empirical evidence.

The obligate fauna of the Mammoth Cave System is diverse and includes 39 cave-limited species (18 troglobionts an 11 stygobionts) of conservation concern, highlighted by the federally endangered cave shrimp *Palaemonias ganteri*. Most of these species are at an elevated risk of extinction due to their limited distributions and/or are known from few occurrences. For example, the cave pseudoscorpion *Tyrannochthonius hypogeus* is known from just two specimens collected from a single locality [92]. Cave-limited fauna face many threats, such as habitat loss and degradation, groundwater overexploitation and contamination, and climate change [230,231].

Although much of the Mammoth Cave System lies within the boundaries of Mammoth Cave National Park, the cave system is not immune to direct and indirect threats to its biodiversity, particularly those stressors that originate from outside of the park, such as industrial and tourism development, oil and gas drilling, runoff from agriculture, residential areas, and highways, and emergent diseases [63,232–236]. For example, sewage from the town of Park City was previously known to drain into the headwaters of the Echo River basin potentially impacted the stygobiotic fauna [130], including *Typhlichthys subterraneus, Amblyopsis spelaea, Palaemonias ganteri, Orconectes pellucidus*, and *Antroselates spiralis*. A hydrocarbon spill along Interstate 65 was responsible for a significant die-off of aquatic cave life [232,236]. Flow reversals and back-flooding from the Green River into cave springs also may transport sediment, potential contaminants, pathogens, and invasive aquatic species into base level streams in the Mammoth Cave System [237–239].

Great potential still exists to discover new taxa and add to the list of obligate species at Mammoth Cave. Two potentially cave-limited springtails that we do not include in our checklist (*Willemia* sp. and *Onychiurus* sp.) are known from Mammoth Cave and have not been identified to species [1,10]. Terrestrial woodlice are notably absent from the troglobiotic fauna of Mammoth Cave and may be discovered in the future. Seven troglobiotic trichoniscids (Isopoda, family Trichoniscidae) are known caves of the Interior Low Plateau and Appalachians karst regions [240], including *Miktoniscus barri* known from several caves of Indiana and Kentucky [241]. A troglophilic species, *Miktoniscus mammothensis*, occurs in cave and surface habitats at MCNP [242]. Other taxonomic groups have not been particularly well studied in the Mammoth Cave System, such as flatworms, copepods, springtails, and mites. More intensive work on these groups may uncover additional taxa. With more than 651 km of passage, much of the Mammoth Cave System has not been comprehensively bioinventoried, and some habitats, such as epikarst, have been disproportionately under-sampled and may harbor undescribed taxa [37]. In addition, over 500 other caves occur in MCNP, including several biologically rich sites, such as White and Great Onyx caves. These cave systems also may harbor undocumented diversity. Finally, few genetic studies to date have incorporated samples from the Mammoth Cave System. Comprehensive sampling within the Mammoth Cave System has the potential to uncover cryptic diversity in some taxonomic groups, which is an increasingly common discovery of genetic and phylogenetic studies in cave-limited taxa [223,243–245].

Author Contributions: Conceptualization, M.L.N.; methodology and analysis, M.L.N.; data acquisition, M.L.N., K.H. and R.S.T.; original draft preparation, M.L.N.; review and editing, M.L.N., K.H. and R.S.T. All authors have read and agreed to the published version of the manuscript.

Funding: M.L.N. was supported by the National Science Foundation (award No. 2047939).

Institutional Review Board Statement: Not applicable.

Informed Consent Statement: Not applicable.

Data Availability Statement: No new data were created or analyzed in this manuscript. Data sharing is not applicable for this manuscript.

Acknowledgments: We thank Dante B. Fenolio and Rickard A. Olson for providing photographs. The views presented herein are those of the authors and do not necessarily represent those of the National Park Service.

Conflicts of Interest: The authors declare no conflict of interest.

References

1. Toomey, R.S.; Hobbs, H.H., III; Olson, R.A. An orientation to Mammoth Cave and this volume. In *Mammoth Cave: A Human and Natural History, Cave and Karst Systems of the World*; Springer: Berlin/Heidelberg, Germany, 2017; pp. 1–28.
2. Miotke, F.D.; Palmer, A.N. *Genetic Relationship between Cave Land-Forms in the Mammoth Cave National Park Area*; Boehler Verlag: Wuerzburg, Germany, 1972.
3. Palmer, A.N. *A Geological Guide to Mammoth Cave National Park*; Zephyrus Press: Teaneck, NJ, USA, 1981.
4. Palmer, A.N. *Cave Geology*; Cave Books: Dayton, OH, USA, 2007.

5. Palmer, A.N. Geology of Mammoth Cave. In *Mammoth Cave: A Human and Natural History, Cave and Karst Systems of the World*; Springer: Berlin/Heidelberg, Germany, 2017; pp. 97–110.

6. Palmer, A.N. Geologic history of Mammoth Cave. In *Mammoth Cave: A Human and Natural History, Cave and Karst Systems of the World*; Springer: Berlin/Heidelberg, Germany, 2017; pp. 111–121.

7. Quinlan, J.F.; Ewers, R.O. Hydrogeology of the Mammoth Cave region, Kentucky. In *GSA Cincinnati 1981 Field Trip Guidebook 3*; American Geological Institute: Falls Church, VA, USA, 1981; pp. 457–506.

8. Glennon, A.; Groves, C. An examination of perennial stream drainage patterns within the Mammoth Cave watershed, Kentucky. *J. Cave Karst Stud.* **2002**, *64*, 82–91.

9. Granger, D.E.; Fabel, D.; Palmer, A.N. Pliocene-Pleistocene incision of the Green River, Kentucky, determined from radioactive decay of cosmogenic ^{26}Al and ^{10}Be in Mammoth Cave sediments. *Geo. Soc. Am. Bull.* **2001**, *113*, 825–836. [CrossRef]

10. Barr, T.C., Jr. Ecological studies in the Mammoth Cave System of Kentucky I: The biota. *Int. J. Speleol.* **1968**, *3*, 147–204. [CrossRef]

11. Rafinesque, C. The caves of Kentucky. *Atl. J.* **1832**, *1*, 27–30.

12. Darwin, C. *On the Origin of Species by Means of Natural Slection, or the Preservation of Favoured Races in the Struggle for Life*, 1st ed.; John Murray: London, UK, 1859.

13. DeKay, J.E. *Zoology of New York or the New York Fauna; Comprising Detailed Descriptions of All the Animals Hitherto Observed within the State of New York, with Brief Notices of Those Occasionally found Near Its Borders, and Accompanied by Appropriate Illustrations. Part IV, Fishes*; W.& A. White & J. Visscher: Albany, NY, USA, 1842.

14. Wyman, J. Description of a blind-fish from a cave in Kentucky. *Am. J. Sci. Arts* **1843**, *45*, 94–96.

15. Wyman, J. Description of a blind-fish from a cave in Kentucky. *Ann. Mag. Nat. Hist.* **1843**, *12*, 298–299.

16. Wyman, J. Account of dissections of the blind fishes (*Amblyopsis spelaeus*) from the Mammoth Cave, Kentucky. *Proc. Boston Soc. Nat. Hist.* **1851**, *3*, 349–375.

17. Wyman, J. The eyes and organs of hearing in *Amblyopsis spelaeus*. *Proc. Boston Soc. Nat. Hist.* **1854**, *4*, 149–151.

18. Wyman, J. On the eye and the organ of hearing in the blind fishes (*Amblyopsis spelaeus* DeKay) of the Mammoth Cave. *Proc. Boston Soc. Nat. Hist.* **1854**, *4*, 395–396.

19. Wyman, J. Notes and drawings of the rudimentary eyes, brain and tactile organs of *Amblyopsis spelaeus*, in Putnam (1872). *Am. Nat.* **1872**, *6*, 6–30.

20. Tellkampf, T.G. Ueber den blinden Fisch der Mammuth-Hohle in Kentucky, mit Bemerkungen ueber einige undere in dieser Hohle lebenden Thiere. *Arch. Anat. Phys. Wiss. Med.* **1844**, *4*, 381–394.

21. Tellkampf, T.G. Beschreibung einiger neuer in der Mammuth-Höhle in Kentucky aufgefundener Gattungen von Gliedertieren. *Arch. Nat.* **1844**, *10*, 318–322. [CrossRef]

22. Tellkampf, T.G. Memoirs on the blind-fishes and some other animals living in the Mammoth Cave in Kentucky. *N. Y. J. Med. Collat. Sci.* **1845**, *5*, 84–93.

23. Agassiz, J.L.R. Plan for an investigation of the embryology, anatomy and effect of light on the blind-fish of the Mammoth Cave, *Amblyopsis spelaeus*. *Proc. Am. Acad. Arts Sci.* **1847**, *1*, 1–180.

24. Agassiz, J.L.R. Observations of the blind fish of the Mammoth Cave. *Am. J. Sci. Arts* **1851**, *11*, 127–128.

25. Agassiz, J.L.R. Recent researches of Prof. Agassiz. *Am. J. Sci. Arts* **1853**, *16*, 134.

26. von Mutschulsky, T.V. *Etudes Entomologiques, 3e Annee*; Helsingfors, Finland, 1854.

27. von Mutschulsky, T.V. *Etudes Entomologiques, 11e Annee*; Dresden, Germany, 1862.

28. Call, R.E. Some notes on the flora and fauna of Mammoth Cave, Kentucky. *Am. Nat.* **1897**, *31*, 377–392. [CrossRef]

29. Packard, A.S. On the crustaceans and insects of the Mammoth Cave. *Am. Nat.* **1871**, *5*, 744–761.

30. Packard, A.S. Occurrence of Japan in the United States. *Am. Nat.* **1874**, *8*, 501–502.

31. Packard, A.S. Larvae of *Anophthalmus* and *Adelops*. *Am. Nat.* **1874**, *8*, 562–563.

32. Packard, A.S. The invertebrate cave fauna of Kentucky and adjoining states, Araneina. *Am. Nat.* **1875**, *9*, 271–278. [CrossRef]

33. Packard, A.S. Cave-inhabiting spiders. *Am. Nat.* **1875**, *9*, 663–664.

34. Packard, A.S. The cave-beetles of Kentucky. *Am. Nat.* **1876**, *10*, 282–287. [CrossRef]

35. Packard, A.S. On the structure of the brain of *Asellus* and the eyeless form *Caecidotea*. *Am. Nat.* **1885**, *19*, 85–86.

36. Packard, A.S. The cave fauna of North America, with remarks on the anatomy of the brain and origin of the blind species. *Mem. Nat. Acad. Sci. USA* **1888**, *4*, 1–156.

37. Culver, D.C.; Hobbs, H.H., III. Biodiversity of Mammoth Cave. In *Mammoth Cave: A Human and Natural History, Cave and Karst Systems of the World*; Hobbs, H.H., III, Olson, R.A., Winkler, E.G., Culver, D.C., Eds.; Springer: Berlin/Heidelberg, Germany, 2017; pp. 227–234.

38. Putnam, F.W. The blind fishes of the Mammoth Cave and their allies. *Am. Nat.* **1872**, *6*, 6–30. [CrossRef]

39. Eigenmann, C.H. The Amblyopsidae, the blind fish of America. *Rep. Br. Assoc. Adv. Sci.* **1897**, *1897*, 685–686.

40. Eigenmann, C.H. The blind fishes of North America. *Pop. Sci. Mon.* **1899**, *56*, 473–486.

41. Eigenmann, C.H. Divergence and convergence in fishes. *Biol. Lect. Mar. Biol. Lab. Woods Hole* **1905**, *8*, 59–66. [CrossRef]

42. Eigenmann, C.H. *Cave Vertebrates of America: A Study in Degenerative Evolution*; Carnegie Institution of Washington: Washington, DC, USA, 1909.

43. Bolivar, C.; Jeannel, R. Campagne Speleologique dans I'Amerique du Nord en 1928. Biospeleologie LVI. *Arch. Zool. Exp. Gen.* **1931**, *71*, 293–499.

44. Bailey, V. Cave life of Kentucky, mainly in the Mammoth Cave region. *Am. Midl. Nat.* **1933**, *14*, 385–635. [CrossRef]
45. Buchanan, J.W. Notes on an American cave flatworm, *Sphalloplana percaeca* (Packard). *Ecology* **1936**, *17*, 194–241. [CrossRef]
46. Park, O. Key to the more common adult animals of Mammoth and adjacent caves. In *A Laboratory Introduction to Animal Ecology and Taxonomy*; Park, O., Allee, W.C., Shelford, V.E., Eds.; University of Chicago Press: Chicago, IL, USA, 1939; pp. 118–124.
47. Dearolf, K. Report of a biological reconnaissance of the New Discovery in Mammoth Cave, Kentucky, January 7–11, 1941. *Bul. Nat. Speleol. Soc.* **1942**, *4*, 48–52.
48. Hubricht, L. Studies of the Nearctic freshwater Amphipoda, III. Notes on the freshwater Amphipoda of eastern United States, with description of ten new species. *Am. Midl. Nat.* **1943**, *29*, 683–712. [CrossRef]
49. Hubricht, L. The cave snail, *Carychium stygium* Call. *Trans. Ky. Acad. Sci.* **1960**, *21*, 35–38.
50. Hubricht, L. New species of *Helicodiscus* from the eastern United States. *Nautilus* **1962**, *75*, 102–107.
51. Hubricht, L. New species of Hydrobiidae. *Nautilus* **1963**, *76*, 138–140.
52. Hubricht, L. Land snails from the caves of Kentucky, Tennessee, and Alabama. *Nat. Speleol. Soc. Bull.* **1964**, *26*, 33–36.
53. Hubricht, L. Four new land snails from the southeastern United States. *Nautilus* **1965**, *79*, 4–7.
54. Jeannel, R.; Henrot, H. Les coleopteres cavernicoles de la region des Appalaches. I. La region des Appalaches (Jeannel), and II. Enumeration des grottes explorees (Henrot). *Notes Biospel.* **1949**, *4*, 11–36.
55. Barr, T.C., Jr. New cave beetles (Carabidae) from Tennessee and Kentucky. *J. Tenn. Acad. Sci.* **1959**, *34*, 5–30.
56. Barr, T.C., Jr. The male of *Pseudanophthalmus audax* (Coleoptera: Carabidae). *Trans. Ky. Acad. Sci.* **1959**, *20*, 1–3.
57. Barr, T.C., Jr. The blind beetles of Mammoth Cave, Kentucky. *Am. Midl. Nat.* **1962**, *68*, 278–284. [CrossRef]
58. Barr, T.C., Jr. Studies on the cavernicole *Ptomaphagus* of the United States (Coleoptera, Catopidae). *Psyche* **1963**, *70*, 50–58. [CrossRef]
59. Barr, T.C., Jr. Cave Carabidae (Coleoptera) of Mammoth Cave. *Pysche* **1966**, *73*, 284–287.
60. Barr, T.C., Jr. Cave Carabidae (Coleoptera) of Mammoth Cave, Part II. *Pysche* **1967**, *74*, 24–25.
61. Poulson, T.L. The Mammoth Cave ecosystem. In *The Natural History of Biospeleology*; Camacho, A., Ed.; Museo Nacional de Ciencias Naturales: Madrid, Spain, 1992; pp. 569–611.
62. Poulson, T.L. Terrestrial cave ecology of the Mammoth Cave region. In *Mammoth Cave: A Human and Natural History, Cave and Karst Systems of the World*; Hobbs, H.H., III, Olson, R.A., Winkler, E.G., Culver, D.C., Eds.; Springer: Berlin/Heidelberg, Germany, 2017; pp. 199–207.
63. Helf, K.L.; Olson, R.A. Subsurface aquatic ecology of Mammoth Cave. In *Mammoth Cave: A Human and Natural History, Cave and Karst Systems of the World*; Hobbs, H.H., III, Olson, R.A., Winkler, E.G., Culver, D.C., Eds.; Springer: Berlin/Heidelberg, Germany, 2017; pp. 209–226.
64. Coney, C.C.; Tarpley, W.A.; Bohannan, R. Ecological studies of land snails in the Hiawassee River Basin of Tennessee, U.S.A. *Malacol. Rev.* **1982**, *15*, 69–106.
65. Bowman, T.E.; Prins, R.; Morris, B.F. Notes on the harpacticoid copepods *Attheyella pilosa* and *A. carolinensis*, associates of crayfishes in the eastern United States. *Proc. Biol. Soc. Wash.* **1968**, *81*, 571–586.
66. Lewis, J.J. The Systematics, Zoogeography and Life History of the Troglobitic Isopods of the Interior Plateaus of the Eastern United States. Ph.D. Dissertation, University of Louisville, Louisville, KY, USA, 1988.
67. Borgmeier, T. Revision of the North American Phorid flies. Part III. The species of the genus *Megaselia*, subgenus *Megaselia* (Diptera, Phoridae): Studia Entomologica. *Petropolis* **1966**, *8*, 1–160.
68. Lewis, J.J. *Bioinventory of Caves of the Cumberland Escarpment Area of Tennessee, Final Report*; Tennessee Chapter of the Nature Conservancy: Nashville, TN, USA, 2005.
69. Zigler, K.S.; Niemiller, M.L.; Stephen, C.D.R.; Ayala, B.N.; Milne, M.A.; Gladstone, N.S.; Engel, A.S.; Jensen, J.B.; Camp, C.D.; Ozier, J.C.; et al. Biodiversity from caves and other subterranean habitats of Georgia, USA. *J. Cave Karst Stud.* **2020**, *82*, 125–167. [CrossRef]
70. Weigand, A.M.; Jochum, A.; Pfenninger, M.; Steinke, D.; Klussmann-Kolb, A. A new approach to an old conundrum—DNA barcoding sheds new light on phenotypic plasticity and morphological stasis in microsnails (Gastropoda, Pulmonata, Carychiidae). *Mol. Ecol. Res.* **2011**, *11*, 255–265. [CrossRef] [PubMed]
71. Weigand, A.M.; Jochum, A.; Slapnik, R.; Schnitzler, J.; Zarza, E.; Klussmann-Kolb, A. Evolution of microgastropods (Ellobioidea, Carychiidae): Integrating taxonomic, phylogenetic and evolutionary hypotheses. *BMC Evol. Biol.* **2013**, *13*, 18. [CrossRef]
72. Gladstone, N.S.; Carter, E.T.; McKinney, M.L.; Niemiller, M.L. Status and distribution of the cave-obligate land snails in the Appalachians and Interior Low Plateau of the eastern United States. *Am. Malacol. Bull.* **2018**, *36*, 62–78. [CrossRef]
73. Poulson, T.L.; Lavoie, K.H.; Helf, K. Long-term effects of weather on the cricket (*Hadenoecus subterraneus*, Orthoptera, Rhaphidophoridae) guano community in Mammoth Cave National Park. *Am. Midl. Nat.* **1995**, *134*, 226–236. [CrossRef]
74. Dourson, D.C. *Kentucky's Land Snails and Their Ecological Communities*; Goatslug Publications: Bakersville, NC, USA, 2010.
75. Miller, J.A. A redescription of *Porrhomma cavernicola* Keyserling (Araneae, Linyphiidae) with notes on Appalachian troglobites. *J. Arachnol.* **2005**, *33*, 426–439. [CrossRef]
76. Holsinger, J.R.; Culver, D.C. The invertebrate cave fauna of Virginia and a part of eastern Tennessee: Zoogeography and ecology. *Brimleyana* **1988**, *14*, 1–162.
77. Paquin, P.; Dupérré, N. Guide d'identifcation des Araignées (Araneae) du Québec. *Fabreries Supplément* **2003**, *11*, 1–251.

78. Holsinger, J.R.; Culver, D.C.; Hubbard, D.A., Jr.; Orndorff, W.D.; Hobson, C.S. The invertebrate cave fauna of Virginia. *Banisteria* **2013**, *42*, 9–56.
79. Peck, S.B. A summary of diversity and distribution of the obligate cave-inhabiting faunas of the United States and Canada. *J. Caves Karst Stud.* **1998**, *60*, 18–26.
80. Keyserling, E.G. Neue Spinnen aus Amerika, III. Verh. zool.-bot. *Ges. Wien* **1881**, *31*, 269–314.
81. Platnick, N.I. A revision of the Appalachian spider genus *Liocranoides* (Araneae: Tengellidae). *Amer. Mus. Nov.* **1999**, *3285*, 1–13.
82. Emerton, J.H. Notes on spiders from caves in Kentucky and Indiana. *Am. Nat.* **1875**, *9*, 278–281. [CrossRef]
83. Hubbard, H.G. Two days collecting in the Mammoth Cave, with contributions to a study of its fauna. *Am. Entomol.* **1880**, *3*, 34–40.
84. McIndoo, N.E. Notes on some arachnids from Ohio valley caves. *The Biol. Bull.* **1911**, *20*, 183–186. [CrossRef]
85. Berland, L. *Arachnides araneides* (Biospeologica LVI) . *Arch Zool. Exp. Gen.* **1931**, *71*, 383–388.
86. Poulson, T.L.; Culver, D.C. Diversity in terrestrial cave communities. *Ecology* **1969**, *50*, 153–158. [CrossRef]
87. Poulson, T.L. Variations in life history of Linyphiid cave spiders. In Proceedings of the 8th International Congress of Speleology, Tampa, FL, USA, 18–24 July 1981; Volume 1, pp. 60–62.
88. Goodnight, C.J.; Goodnight, M.L. Speciation among cave opilionids of the United States. *Am. Midl. Nat.* **1960**, *64*, 34–38. [CrossRef]
89. Hedin, M.; Thomas, S.M. Molecular systematics of eastern North American Phalangodidae (Arachnida: Opiliones: Laniatores), demonstrating convergent morphological evolution in caves. *Mol. Phylogenet. Evol.* **2010**, *54*, 107–121. [CrossRef]
90. Malcolm, D.R.; Chamberlin, J.C. The pseudoscorpion genus *Kleptochthonius* Chamberlin (Chelonethida, Chthoniidae). *Am. Mus. Nov.* **1961**, *2063*, 1–35.
91. Muchmore, W.B. Redescription of some cavernicolous pseudoscorpions (Arachnida, Chelonethida) in the collection of the Museum of Comparative Zoology. *Breviora* **1963**, *188*, 1–16.
92. Muchmore, W.B. The genus *Tyrannochthonius* in the eastern United States (Pseudoscorpionida: Chthoniidae). Part II. More recently described species. *Insecta Mundi* **1996**, *10*, 153–168.
93. Banks, N. Notes on the Pseudoscorpionida. *J. N. Y. Entomol. Soc.* **1895**, *3*, 1–13.
94. Vitzthum, H. Die unterirdische Acarofauna. *Z. Naturw.* **1925**, *62*, 125–186.
95. Holsinger, J.R. Redescriptions of two poorly known species of cavernicolous rhagidiid mites (Acarina, Trombidiformes) from Virginia and Kentucky. *Acarologia* **1965**, *7*, 654–662.
96. Zacharda, M. Soil mites of the family Rhagidiidae (Actinedida: Eupodoidea). Morphology, systematics, ecology. *Acta Univ. Carol. Biol.* **1980**, *1978*, 489–785.
97. Cope, E.D. On the Wyandotte Cave and its fauna. *Am. Nat.* **1872**, *6*, 406–422. [CrossRef]
98. Loomis, H.F. New cave and epigean millipedes of the United States, with notes on some established species. *Bull. Mus. Comp. Zool.* **1943**, *92*, 373–410.
99. Shear, W.A. The milliped family Trichopetalidae, Part 2: The genera *Trichopetalum, Zygonopus* and *Scoterpes* (Diplopoda: Chordeumatida, Cleidogonoidea). *Zootaxa* **2010**, *2385*, 1–62. [CrossRef]
100. Christiansen, K.A. The genus *Arrhopalites* (Collembola, Sminthuridae) in the United States and Canada. *Int. J. Speleol.* **1966**, *2*, 43–73. [CrossRef]
101. Christiansen, K.; Bellinger, P. Cave *Pseudosinella* and *Oncopodura* new to science. *J. Cave Karst Stud.* **1996**, *58*, 38–53.
102. Harker, D.F.; Barr, T.C. Western Kentucky coal field: Preliminary investigations of natural features and cultural resources. In *Caves and Associated Fauna of the Western Kentucky Coal Field*; Technical Report; Kentucky Nature Preserves Commission: Frankfort, KY, USA, 1980; Volume 2.
103. Christiansen, K.A. The genus *Pseudosinella* (Collembola, Entomobryidae) in caves of the United States. *Psyche* **1960**, *67*, 1–25. [CrossRef]
104. Christiansen, K.A. A preliminary survey of the knowledge of North American cave Collembola. *Am. Midi. Nat.* **1960**, *64*, 39–44. [CrossRef]
105. Christiansen, K.A. The genus *Sinella* Brook (Collembola, Entomobryidae) in Nearctic caves. *Ann. Entomol. Soc. Am.* **1960**, *53*, 481–491. [CrossRef]
106. Ferguson, L.M. Systematics, Evolution, and Zoogeography of the Cavernicolous Campodeids of the Genus Litocampa (Diplura: Campodeidae) in the United States. Ph.D. Dissertation, Virginia Polytechnic Institute and State University, Blacksburg, VA, USA, 1981.
107. Silvestri, F. Campodeidae (Biospeologica LVI). *Arch. Zool. Expo Gen.* **1934**, *76*, 379–383.
108. Silvestri, F. On some Tapygidae in the Museum of Comparative Zoology. *Psyche* **1947**, *51*, 209–229. [CrossRef]
109. Conde, B. Campodeides cavernicoles de la region des Appalaches. *Notes Biospeol.* **1949**, *4*, 125–139.
110. Ferguson, L.M. Taxonomy, Distribution, and Evolution of Cavernicolous Campodeids in Virginia (Diplura: Campodeidae). Master's Thesis, Virginia Polytechnic Institute and State University, Blacksburg, VA, USA, 1974.
111. Barr, T.C., Jr. The taxonomy, distribution, and affinities of *Neaphanops*, with notes on associated species of *Pseudanophthalmus* (Coleoptera, Carabidae). *Am. Mus. Nov.* **1979**, *2682*, 1–20.
112. Erichson, W.F. Footnote, p. 384; In Ueber den blinden Fisch der Mammuth-Hohle in Kentucky, mit Bemerkungen uber einige andere in diser Hohle lebenden Thiere; Tellkampf, T. *Mullers Arch Anat. Physiol.* **1844**, *4*, 384–394.

113. Barr, T.C., Jr. Pattern and process in speciation of trechine beetles in eastern North America (Coleoptera: Carabidae: Trechinae). In *Taxonomy, Phylogeny, and Biogeography of Beetles and Ants*; Ball, G.E., Ed.; Series Entomologia 33; Junk: Dordrecht, The Netherlands, 1985; pp. 350–407.
114. Friedrich, M. Biological clocks and visual systems in cave-adapted animals at the dawn of speleogenomics. *Integr. Comp. Biol.* **2013**, *53*, 50–67. [CrossRef]
115. Friedrich, M.; Chen, R.; Daines, B.; Bao, R.; Caravas, J.; Rai, P.K.; Zagmajster, M.; Peck, S.B. Phototransduction and clock gene expression in the troglobiont beetle *Ptomaphagus hirtus* of Mammoth cave. *J. Exp. Biol.* **2011**, *214*, 3532–3541. [CrossRef]
116. Horn, G.H. Catalogue of Coleoptera from south-west Virginia. *Trans. Am. Entomol. Soc.* **1868**, *2*, 123–128.
117. Horn, G.H. Miscellaneous notes and short studies of North American Coleoptera. *Trans. Am. Entomol. Soc.* **1883**, *10*, 269–293. [CrossRef]
118. Jeannel, R. Notes sur les Trechini. *Bull. Soc. Entomol. Fr.* **1920**, *1920*, 150–155.
119. Jeannel, R. Monographie des Trechinae, 3e livraison. *L'Abeille* **1928**, *35*, 1–808.
120. Jeannel, R. Revision des Trechinae de l'Amerique du nord. *Arch. Zool. Expo Gen.* **1931**, *71*, 403–499.
121. Jeannel, R. Les coleopteres cavernicoles de la region des Appalaches. III. Etude systematique. *Notes Biospeol.* **1949**, *4*, 37–104.
122. Valentine, J.M. New cavernicole Carabidae of the subfamily Trechinae Jeannel. *J. Elisha Mitchell Sci. Soc.* **1931**, *46*, 247–258.
123. Valentine, J.M. A classification of the genus *Pseudanophthalmus* Jeannel (fam. Carabidae) with descriptions of new species and notes on distribution. *J. Elisha Mitchell Sci. Soc.* **1932**, *48*, 261–280.
124. Hatch, M.H. Studies on the Leptodiridae (Catopidae) with descriptions of new species. *J. N. Y. Entomol. Soc.* **1933**, *41*, 187–239.
125. Park, O. New or little-known species of pselaphid beetles from southeastern United States. *J. Tenn. Acad. Sci.* **1956**, *31*, 54–100.
126. Park, O. New or little-known species of pselaphid beetles, chiefly from southeastern United States. *J. Tenn. Acad. Sci.* **1958**, *33*, 39–74.
127. Park, O. Cavernicolous pselaphid beetles of the United States. *Amer. Midl. Nat.* **1960**, *64*, 66–104. [CrossRef]
128. Barr, T.C., Jr. Non-troglobitic Carabidae (Coleoptera) from caves in the United States. *Coleopt. Bull.* **1964**, *18*, 1–4.
129. Barr, T.C., Jr. Cave ecology and the evolution of troglobites. *Evol. Biol.* **1968**, *2*, 35–102.
130. Barr, T.C., Jr. Ecology and evolution of cave faunas. In Proceedings of the Third Annual Symposium on the Natural History of the Lower Tennessee and Cumberland River Valleys; Hamilton, S.W., Finley, M.T., Eds.; Center for Field Biology, Austin Peay State University: Clarksville, TN, USA, 1990; pp. 1–19.
131. Barr, T.C., Jr. *A Classification and Checklist of the Genus Pseudanophthalmus Jeannel (Coleoptera: Carabidae: Trechinae)*; Special Publication 11; Virginia Museum of Natural History: Martinsville, VA, USA, 2004; p. 52.
132. Barr, T.C., Jr.; Kuehne, R.A. Ecological studies in the Mammoth Cave System of Kentucky, II: The ecosystem. *Ann. Spéléologie* **1971**, *26*, 47–96.
133. Peck, S.B. A systematic revision and evolutionary biology of the *Ptomaphagus adelops*. *Bull. Mus. Comp. Zool.* **1973**, *145*, 29–162.
134. Peck, S.B. The life cycle of a Kentucky cave beetle, *Ptomaphagus hirtus*, (Coleoptera; Leiodidae; Catopinae). *Int. J. Speleol.* **1975**, *7*, 7–17. [CrossRef]
135. Peck, S.B. Experimental hybridizations between populations of cavernicolous *Ptomaphagus* beetles (Coleoptera: Leiodidae: Cholevinae). *Canad. Entomol.* **1983**, *115*, 445–452. [CrossRef]
136. Peck, S.B. The distribution and evolution of cavernicolous *Ptomaphagus* beetles in the southeastern United States (Coleoptera; Leiodidae; Cholevinae) with new species and records. *Canad. J. Zool.* **1984**, *62*, 730–740. [CrossRef]
137. Peck, S.B. Evolution of adult morphology and life-history characters in cavernicolous *Ptomaphagus* beetles. *Evolution* **1986**, *40*, 1021–1030. [CrossRef]
138. Kane, T.C.; Norton, R.M.; Poulson, T.L. The ecology of a predaceous troglobitic beetle, *Neaphaenops tellkampfii* (Coleoptera: Carabidae, Trechinae). I. Seasonality of food input and early life history stages. *Int. J. Speleol.* **1975**, *7*, 45–54. [CrossRef]
139. Norton, R.M.; Kane, T.C.; Poulson, T.L. The ecology of a predaceous troglobitic beetle, *Neaphaenops tellkampfii* (Coleoptera: Carabidae, Trechinae). II. Adult seasonality, feeding and recruitment. *Int. J. Speleol.* **1975**, *7*, 55–64. [CrossRef]
140. Kane, T.C.; Poulson, T.L. Foraging by cave beetles: Spatial and temporal heterogeneity of prey. *Ecology* **1976**, *57*, 793–800. [CrossRef]
141. Laing, C.; Carmody, G.R.; Peck, S.B. Population genetics and evolutionary biology of the cave beetle *Ptomaphagus hirtus*. *Evolution* **1976**, *30*, 484–498. [CrossRef]
142. Giuseffi, S.; Kane, T.C.; Duggleby, W.F. Genetic variability in the Kentucky cave beetle *Neaphaenops tellkampfii* (Coleoptera: Carabidae). *Evolution* **1978**, *32*, 679–681. [CrossRef] [PubMed]
143. Kane, T.C.; Ryan, T. Population ecology of carabid cave beetles. *Oecologia* **1983**, *60*, 46–55. [CrossRef] [PubMed]
144. Barr, T.C., Jr.; Holsinger, J.R. Speciation in cave faunas. *Ann. Rev. Ecol. Syst.* **1985**, *16*, 313–337. [CrossRef]
145. Kane, T.C.; Brunner, G.D. Geographic variation in the cave beetle, *Neaphaenops tellkampfi* (Coleoptera: Carabidae). *Psyche* **1986**, *93*, 231–251. [CrossRef]
146. Helf, K.L. Foraging Ecology of the Cave Cricket *Hadenoecus subterraneus*: Effects of Climate, Ontogeny, and Predation. Ph.D. Dissertation, University of Illinois at Chicago, Chicago, IL, USA, 2003.
147. Leray, V.L.; Caravas, J.; Friedrich, M.; Zigler, K.S. Mitochondrial sequence data indicate "Vicariance by Erosion" as a mechanism of species diversification in North American *Ptomaphagus* (Coleoptera, Leiodidae, Cholevinae) cave beetles. *Subterr. Biol.* **2019**, *29*, 35–57. [CrossRef]

148. Marshall, S.A.; Peck, S.B. Distribution of cave-dwelling Sphaeroceridae (Diptera) of eastern North America. *Proc. Entomol. Soc. Ontario* **1984**, *115*, 37–41.
149. De Beachamp, P. Turbellaries triclades (Biospeologica LVI). *Arch. Zool. Expo Gen.* **1931**, *71*, 317–331.
150. Hyman, L.H. Studies on the morphology, taxonomy, and distribution of North American triclad Turbellaria. VIII. Some cave planarians of the United States. *Trans. Am. Micr. Soc.* **1937**, *56*, 457–477. [CrossRef]
151. Carpenter, J.H. Systematics and Ecology of Cave Planarians of the United States. Ph.D. Thesis, University of Kentucky, Lexington, KY, USA, 1970.
152. Carpenter, J.H. Observations on the biology of cave planarians of the United States. *Int. J. Speleol.* **1982**, *12*, 9–26. [CrossRef]
153. Kenk, R. Freshwater triclads (Turbellaria) of North America, IX. The genus *Sphalloplana. Smithson. Contrib. Zool.* **1977**, *246*, 1–38. [CrossRef]
154. Pearson, W.D.; Boston, C.H. *Distribution and Status of the Northern Cavefish*; Amblyopsis Spelaea, Final Report; Nongame and Endangered Wildlife Program, Indiana Department of Natural Resources: Indianapolis, IN, USA, 1995.
155. Hershler, R.; Hubricht, L. Notes on *Antroselates* Hubricht, 1963 and *Antrobia* Hubricht, 1971 (Gastropoda: Hydrobiidae). *Proc. Biol. Soc. Wash.* **1988**, *101*, 730–740.
156. Niemiller, M.L.; Poulson, T.L. Subterranean fishes of North America: Amblyopsidae. In *The Biology of Subterranean Fishes*; Trajano, E., Bichuette, M.E., Kappor, B.G., Eds.; Science Publishers: Enfield, NH, USA, 2010; pp. 169–280.
157. Kofoid, C.A. The plankton of Echo River, Mammoth Cave. *Trans. Am. Micr. Soc.* **1899**, *21*, 113–126. [CrossRef]
158. Chappuis, P.A. Crustaces copepodes (Biospeologica LVI). *Arch. Zool. Expo Gen.* **1931**, *71*, 345–360.
159. Whitman, R.L. *Meiofaunal Sampling at Mammoth Cave National Park. Draft Report*; National Park Service, Indiana Dunes National Lakeshore: Chesterton, IN, USA, 1989.
160. Lewis, J.J. *Conservation Assessment for Northern Cavefish Copepod (Cauloxenus stygius). Report*; USDA Forest Service: Eastern Region, Ghana, 2002.
161. Klie, W. *Crustaces ostracodes* (Biospeologica LVI). *Arch. Zool. Expo Gen.* **1931**, *71*, 333–344.
162. Hart, C.W., Jr.; Hobbs, H.H., Jr. Eight new troglobitic ostracods of the genus *Entocythere* (Crustacea, Ostracoda) from the eastern United States. *Proc. Acad. Nat. Sci. USA* **1961**, *113*, 173–185.
163. Hart, C.W., Jr.; Hart, D.G. Four new entocytherid ostracods from Kentucky, with notes on the troglobitic *Sagittocythere barri. Nolulae Nat.* **1966**, *388*, 1–10.
164. Hart, D.G.; Hart, C.W., Jr. The ostracod family Entocytheridae. *Acad. Nat. Sci. Phila. Monogr.* **1974**, *18*, 1–239.
165. Garman, H. The origin of the cave fauna of Kentucky, with a description of a new blind beetle. *Science* **1892**, *20*, 240–241. [CrossRef] [PubMed]
166. Hay, W.P. Observations on the crustacean fauna of the region about Mammoth Cave, Kentucky. *Proc. U.S. Nat. Mus.* **1902**, *25*, 223–236. [CrossRef]
167. Giovanolli, L. Invertebrate life of Mammoth and other neighboring caves. Pp. 600–623 In: V. Bailey. Cave Life of Kentucky. *Am. Midl. Nat.* **1933**, *14*, 385–635.
168. Chappuis, P.A. Biospeologica 71: Campagne speologique de C. Bolivar et R. Jeannel dans PAmerique du Nord (1928), 13, Asellides. *Arch. Zool. Expo Gen.* **1950**, *87*, 177–182.
169. Lewis, J.J.; Bowman, T.E. The subterranean asellids (*Caecidotea*) of Illinois (Crustacea: Isopoda: Asellidae). *Smithson. Contrib. Zool.* **1981**, *335*, 1–66. [CrossRef]
170. Helf, K.L.; Moore, W.; Wells, B. *Monitoring Cave Aquatic Biota at Selected Parks in the Cumberland Piedmont Network: Protocol Narrative—Version 1.0. Natural Resource Report NPS/CUPN/NRR—2018/1705*; National Park Service: Fort Collins, CO, USA, 2018.
171. Zhang, J.; Holsinger, J.R. Systematics of the freshwater amphipod genus *Crangonyx* (Crangonyctidae) in North America. *Va. Mus. Nat. Hist. Memoir.* **2003**, *6*, 1–274.
172. Holsinger, J.R. *Freshwater Amphipod Crustaceans (Gammaridae) of North America, Biota of Freshwater Ecosystems, Identification Manual No. 5*; U.S. Environmental Protection Agency: Washington, DC, USA, 1972.
173. Zhang, J. Systematics of the Freshwater Amphipod Genus *Crangonyx* (Crangonyctidae) in North America. Ph.D. Dissertation, Old Dominion University, Norfolk, VA, USA, 1997.
174. United States Fish & Wildlife Service. *Kentucky Cave Shrimp (Palaemonias ganteri), 5-Year Review: Summary and Evaluation*; Southeast Region, Kentucky Ecological Services Field Office: Frankfort, KY, USA, 2016.
175. Fage, L. Crustaces amphipodes et decapodes. In: Biospeologica, LVI: Campagne speologique de C. Bolivar et R. Jeannel dans l'Amerique du Nord (1928). *Arch. Zool. Exp. Gen.* **1931**, *71*, 361–374.
176. Hobbs, H.H., Jr.; Hobbs, H.H., III; Daniel, M.A. A review of the troglobitic decapod crustaceans of the Americas. *Smithson. Contrib. Zool.* **1977**, *244*, 1–83. [CrossRef]
177. Holsinger, J.R.; Leitheuser, A.T. *Ecological Analysis of the Kentucky Cave Shrimp, Palaemonias ganteri Hay, Mammoth Cave National Park (Phase I), Final Report*; Old Dominion Research University Foundation: Norfolk, VA, USA, 1982.
178. Holsinger, J.R.; Leitheuser, A.T. *Ecological Analysis of the Kentucky Cave Shrimp, Palaemonias ganteri Hay, Mammoth Cave National Park (Phase II), Final Report*; Old Dominion Research University Foundation: Norfolk, VA, USA, 1982.
179. Holsinger, J.R.; Leitheuser, A.T. *Ecological analysis of the Kentucky Cave Shrimp, Palaemonias ganteri Hay, Mammoth Cave National Park (Phase III), Final Report*; Old Dominion Research University Foundation: Norfolk, VA, USA, 1983.

180. Lisowski, E.A. The endangered Kentucky blind cave shrimp. In *Proceedings of the National Cave Management Symposium, Carlsbad, New Mexico and Mammoth Cave National Park, Kentucky*; Wilson, R.C., Lewis, J.J., Eds.; Pygmy Dwarf Press: Oregon City, OR, USA, 1982; pp. 138–142.
181. Lisowski, E.A. Distribution, habitat, and behavior of the Kentucky cave shrimp *Palaemonias ganteri* Hay. *J. Crustacean Biol.* **1983**, *3*, 88–92. [CrossRef]
182. Lisowski, E.A.; Poulson, T.L. Impacts of Lock and Dam Six on base level ecosystems in Mammoth Cave. In *Cave Research Foundation 1979 Annual Report*; The Cave Research Foundation: Cave, KY, USA, 1981; pp. 48–54.
183. Leitheuser, A.T.; Holsinger, J.R. *Ecological Analysis of the Kentucky Cave Shrimp, Palaemonias ganteri Hay, Mammoth Cave National Park (Phase IV), Final Report*; Old Dominion University Research Foundation: Norfolk, VA, USA, 1983.
184. Leitheuser, A.T.; Holsinger, J.R.; Olson, R.; Pace, N.R.; Whitman, R.L.; White, T. *Ecological Analysis of the Kentucky Cave Shrimp, Palaemonias ganteri Hay, at Mammoth Cave National Park (Phase V), Final Report*; Old Dominion University Research Foundation: Norfolk, VA, USA, 1985.
185. Leitheuser, A.T.; Whitman, R.L.; Gochee, A.V.; Holsinger, J.R. *Ecological Analysis of the Kentucky Cave Shrimp, Palaemonias ganteri Hay, at Mammoth Cave National Park (Phase VI), Final Report*; Old Dominion University Research Foundation: Norfolk, VA, USA, 1986.
186. United States Fish & Wildlife Service. *Kentucky Cave Shrimp Recovery Plan*; United States Fish & Wildlife Service: Atlanta, GA, USA, 1988.
187. Pearson, W.D.; Jones, T.G. *A Final Report Based on a Faunal Inventory of Subterranean Streams and Development of a Cave Aquatic Biological Monitoring Program Using a Modified Index of Biotic Integrity, Final Report*; National Park Service, Mammoth Cave National Park: Brownsville, KY, USA, 1998.
188. Cooper, J.E.; Cooper, M.R. Observations on the biology of the endangered stygobiotic shrimp *Palaemonias alabamae*, with notes on *P. ganteri* (Decapoda: Atyidae). *Subterr. Biol.* **2011**, *8*, 9–20. [CrossRef]
189. Stump, A.J. The Use of Environmental DNA for the Detection of *Palaemonias ganteri* (Hay, 1901), a Federally Endangered Species. Master's Thesis, Eastern Kentucky University, Richmond, KY, USA, 2019.
190. Hagen, H.H. Monograph of the North American Astacidae. *Illus. Cat. Mus. Com. Zool.* **1870**, *3*, 1–109.
191. Hagen, H.H. The blind crayfish. *Am. Nat.* **1872**, *6*, 494.
192. Garman, H. A little-known cave crayfish. *Trans. Ky. Acad. Sci.* **1924**, *1*, 87–94.
193. Park, O.; Roberts, T.W.; Harris, S.J. Preliminary analysis of activity of the cave crayfish, *Cambarus pellucidus*. *Am. Nat.* **1941**, *75*, 154–171. [CrossRef]
194. Rhoades, R. The crayfishes of Kentucky, with notes on variation, distribution and descriptions of new species and subspecies. *Am. Midl. Nat.* **1944**, *31*, 111–149. [CrossRef]
195. Hobbs, H.H., Jr.; Barr, T.C., Jr. Origins and affinities of the troglobitic crayfishes of North America (Decapoda: Astacidae), I: Genus Cambarus. *Am. Midl. Nat.* **1960**, *64*, 12–33. [CrossRef]
196. Hobbs, H.H., Jr.; Barr, T.C., Jr. Origins and affinities of the troglobitic crayfishes of North America (Decapoda: Astacidae), II: Genus *Orconectes*. *Smithson. Contrib. Zool.* **1972**, *105*, 1–84. [CrossRef]
197. Brown, F.A. Diurnal rhythm in cave crayfish. *Nature* **1961**, *191*, 929–930. [CrossRef]
198. Wolfe, D.A.; Cornwell, D.G. Carotenoids of cavernicolous crayfish. *Science* **1964**, *144*, 1467–1469. [CrossRef]
199. Compson, Z.G. An Isotopic Examination of Cave, Spring and Epigean Trophic Structures in Mammoth Cave National Park. Master's Thesis, Western Kentucky University, Bowling Green, KY, USA, 2004.
200. Taylor, C.A.; Schuster, G.A. *The Crayfishes of Kentucky*; Illinois Natural History: Champaign, IL, USA, 2004.
201. Proudlove, G.S. *Subterranean Fishes of the World*; International Society for Subterranean Biology: Moulis, France, 2006.
202. Davidson, R. *An Excursion to the Mammoth Cave and the Barrens of Kentucky, with Some Notices of the Early Settlement of the State*; A.T. Skillman and Son, Lexington; Thomas Cowperthwait and Co.: Philadelphia, PA, USA, 1840.
203. Thompson, W. Notice of the blindfish, crayfish, and insects from Mammoth Cave, Kentucky. *Ann. Mag. Nat. Hist.* **1844**, *13*, 3.
204. Girard, C.F. Ichthyological notices. *Proc. Acad. Nat. Sci. USA* **1860**, *1859*, 56–68.
205. Woods, L.P.; Inger, R.F. The cave, spring, and swamp fishes of the family Amblyopsidae of central and eastern United States. *Am. Midl. Nat.* **1957**, *58*, 232–256. [CrossRef]
206. Poulson, T.L. Cave Adaptation in Amblyopsid Fishes. Ph.D. Dissertation, Department of Zoology, University of Michigan,, Ann Arbor, MI, USA, 1961.
207. Poulson, T.L. Cave adaptation in amblyopsid fishes. *Am. Midl. Nat.* **1963**, *70*, 257–290. [CrossRef]
208. Poulson, T.L. *Cave Research Foundation Annual Report*; Aquatic Cave Communities: Washington, DC, USA, 1968; pp. 16–18.
209. Barr, T.C., Jr.; Kuehne, R.A. The cavefish, *Amblyopsis spelaea*, in northern Kentucky. *Copeia* **1962**, *1962*, 662. [CrossRef]
210. Rosen, D.E. Comments on the relationships of the North American cave fishes of the family Amblyopsidae. *Am. Mus. Nov.* **1962**, *2109*, 1–35.
211. Poulson, T.L.; White, W.B. The cave environment. *Science* **1969**, *165*, 971–981. [CrossRef] [PubMed]
212. Clay, W.M. *The Fishes of Kentucky*; Kentucky Department of Fish and Wildlife Resources: Frankfurt, KY, USA, 1975.
213. Swofford, D.L.; Branson, B.A.; Sievert, G. Genetic differentiation of cavefish populations (Amblyopsidae). *Isozyme Bull.* **1980**, *13*, 109–110.
214. Swofford, D.L. Genetic Variability, Population Differentiation, and Biochemical Relationships in the Family Amblyopsidae. Master's Thesis, Eastern Kentucky University, Richmond, KY, USA, 1982.

215. Burr, B.M.; Warren, M.L., Jr. *A Distributional Atlas of Kentucky Fishes, Vol. 4*; Kentucky State Nature Preserves Commission Scientific and Technical Series: Frankfort, KY, USA, 1986.
216. Lewis, J.J. *Conservation Assessment for Southern Cavefish (Typhlichthys subterraneus) Report*; USDA Forest Service: Eastern Region, Ghana, 2002.
217. Keith, J.H. Distribution of Northern cavefish, *Amblyopsis spelaea* DeKay, in Indiana and Kentucky and recommendations for its protection. *Nat. Areas J.* **1988**, *8*, 69–79.
218. Branson, B.A. The Mammoth Cave blindfish. *Trop. Fish Hobbyist* **1991**, *40*, 39–40.
219. Romero, A. Threatened fishes of the world: *Typhlichthys subterraneus* Girard, 1860 (Amblyopsidae). *Environ. Biol. Fishes* **1998**, *53*, 74. [CrossRef]
220. Romero, A.; Bennis, L. Threatened fishes of the world: *Amblyopsis spelaea* DeKay, 1842 (Amblyopsidae). *Environ. Biol. Fishes* **1998**, *51*, 421–428. [CrossRef]
221. Niemiller, M.L. Evolution, Speciation, and Conservation of Amblyopsid Cavefishes. Ph.D. Dissertation, University of Tennessee, Knoxville, TN, USA, 2011.
222. Niemiller, M.L.; Fitzpatrick, B.M. Status and life history of the amblyopsid cavefishes in Kentucky. *Ky. Dept. Fish Wildl. Resour.* **2012**, *5*, 9–15.
223. Niemiller, M.L.; Near, T.J.; Fitzpatrick, B.M. Delimiting species using multilocus data: Diagnosing cryptic diversity in the southern cavefish, *Typhlichthys subterraneus* (Teleostei: Amblyopsidae). *Evolution* **2012**, *66*, 846–866. [CrossRef]
224. Hart, P.B.; Niemiller, M.L.; Burress, E.D.; Armbruster, J.W.; Ludt, W.B.; Chakrabarty, P. Cave-adapted evolution in the North American amblyopsid fishes inferred using phylogenomics and geometric morphometrics. *Evolution* **2020**, *74*, 936–949. [CrossRef]
225. Culver, D.C.; Sket, B. Hotspots of subterranean biodiversity in caves and wells. *J. Cave Karst Stu.* **2000**, *62*, 11–17.
226. Culver, D.C.; Pipan, T. Subterranean ecosystems. In *Encyclopedia of Biodiversity*, 2nd ed.; Levin, S.A., Ed.; Elsevier: Amsterdam, The Netherlands, 2013.
227. Hutchins, B.T.; Gibson, J.R.; Diaz, P.H.; Schwartz, B.F. Stygobiont diversity in the San Marcos Artesian Well and Edwards Aquifer groundwater ecosystem, Texas, USA. *Diversity* **2021**, *13*, 234. [CrossRef]
228. Culver, D.C.; Deharveng, L.; Bedos, A.; Lewis, J.J.; Madden, M.; Reddell, J.R.; Sket, B.; Trontelj, P.; White, D. The mid-latitude biodiversity ridge in terrestrial cave fauna. *Ecography* **2006**, *29*, 120–128. [CrossRef]
229. Olson, R. Potential effects of hydrogen sulfide and hydrocarbon seeps on Mammoth Cave ecosystems. In *Mammoth Cave National Park's 10th Research Symposium*; Mammoth Cave National Park: Brownsville, KY, USA, 2013; pp. 25–30.
230. Niemiller, M.L.; Bichuette, E.; Taylor, S.J. Conservation of cave fauna in Europe and the Americas. In *Ecological Studies: Cave Ecology*; Moldovan, O.T., Kovac, L., Halse, S., Eds.; Springer: Dordrecht, The Netherlands, 2018; pp. 451–478.
231. Mammola, S.; Cardoso, P.; Culver, D.C.; Deharveng, L.; Ferreira, R.L.; Fišer, C.; Galassi, D.M.P.; Griebler, C.; Halse, S.; Humphreys, W.F.; et al. Scientists' warning on the conservation of subterranean ecosystems. *BioScience* **2019**, *69*, 641–650. [CrossRef]
232. Brucker, R. Conservation at Mammoth Cave. In *Cave Research Foundation 1979 Annual Report*; Mammoth Cave National Park: Brownsville, KY, USA, 1979; pp. 40–41.
233. Pfaff, R.M.; Glennon, J.A.; Groves, C.G.; Anderson, M.; Fry, J.; Meiman, J. Landuse and water quality threats to the Mammoth Cave karst aquifer, Kentucky. In Proceedings of the 12th National Cave and Karst Management Symposium, Chattanooga, TN, USA, 19–22 October 1999.
234. Meiman, J.; Hopper, H.L.; Brucker, R.W. Management issues and threats to the longest cave. In Proceedings of the 15th National Cave and Karst Management Symposium, Tucson, AZ, USA, 16–19 October 2001.
235. Toomey, R.; Thomas, S.; Gillespie, J.; Carson, V.; Trimboli, S.R. White-nose Syndrome at Mammoth Cave National Park: Actions before and after its detection. In Proceedings of the 10th Mammoth Cave Research Symposia, Mammoth Cave, KY, USA, 14–15 February 2013; p. 13.
236. Olson, R.A. Environmental issues relevant to the Mammoth Cave area. In *Mammoth Cave: A Human and Natural History, Cave and Karst Systems of the World*; Hobbs, H.H., III, Olson, R.A., Winkler, E.G., Culver, D.C., Eds.; Springer: Berlin/Heidelberg, Germany, 2017; pp. 265–275.
237. Ruhl, M. Flow Reversal Events Increase the Abundance of Nontroglobitic Fish in the Subterranean Rivers of Mammoth Cave National Park. Master's Thesis, Western Kentucky University, Bowling Green, KY, USA, 2005.
238. Trimboli, S.R.; Weber, K.; Ryan, S.; Toomey, R.S. An overview of the reverse flow patterns of River Styx in Mammoth Cave, Kentucky: 2009–2012. In Proceedings of the 11th Mammoth Cave Research Symposia, Mammoth Cave, KY, USA, 18–20 April 2016.
239. Trimboli, S.R.; Toomey, R.S. Temperature and reverse-flow patterns of the River Styx, Mammoth Cave, Kentucky. *J. Cave Karst Stud.* **2019**, *81*, 174–187. Available online: https://digitalcommons.wku.edu/mc_reserch_symp/11th_Research_Symposium_2016/Day_three/3 (accessed on 28 June 2021). [CrossRef]
240. Niemiller, M.L.; Taylor, S.J.; Slay, M.E.; Hobbs, H.H., III. Biodiversity in the United States and Canada. In *Encyclopedia of Caves*, 3rd ed.; Culver, D.C., White, W.B., Pipan, T., Eds.; Academic Press: Cambridge, MA, USA, 2019; pp. 163–177.
241. Lewis, J.J.; Lewis, S.L. Cave fauna study for the Interstate 66 EIS (Somerset to London, Kentucky). In Proceedings of the 2005 National Cave and Karst Management Symposium, Albany, NY, USA, 31 October–4 November 2005; pp. 15–20.
242. Muchmore, W. New terrestrial isopods of the genus *Miktoniscus* from eastern United States (Crustacea: Isopoda: Oniscoidea). *Ohio J. Sci.* **1964**, *64*, 51–57.

243. Zakšek, V.; Sket, B.; Gottstein, S.; Franjević, D.; Trontelj, P. The limits of cryptic diversity in groundwater: Phylogeography of the cave shrimp *Troglocaris anophthalmus* (Crustacea: Decapoda: Atyidae). *Mol. Ecol.* **2009**, *18*, 931–946. [CrossRef] [PubMed]
244. Ethridge, J.Z.; Gibson, J.R.; Nice, C.C. Cryptic diversity within and amongst spring-associated *Stygobromus* amphipods (Amphipoda: Crangonyctidae). *Zool. J. Linn. Soc.* **2013**, *167*, 227–242. [CrossRef]
245. Devitt, T.J.; Wright, A.M.; Cannatella, D.C.; Hillis, D.M. Species delimitation in endangered groundwater salamanders: Implications for aquifer management and biodiversity conservation. *Proc. Natl. Acad. Sci. USA* **2019**, *116*, 2624–2633. [CrossRef] [PubMed]

Article

The Towakkalak System, A Hotspot of Subterranean Biodiversity in Sulawesi, Indonesia

Louis Deharveng [1,*], Cahyo Rahmadi [2], Yayuk Rahayuningsih Suhardjono [2] and Anne Bedos [1]

[1] Institut de Systématique, Évolution, Biodiversité (ISYEB), UMR7205 CNRS, Muséum National d'Histoire Naturelle, Sorbonne Université, EPHE, 45 Rue Buffon, 75005 Paris, France; bedosanne@yahoo.fr
[2] Museum Zoologicum Bogoriense, Research Center for Biology, Indonesian Institute of Sciences, Cibinong Science Center, Cibinong 16911, Indonesia; cahyo.rahmadi@lipi.go.id (C.R.); yayukrs@indo.net.id (Y.R.S.)
* Correspondence: dehar.louis@wanadoo.fr

Abstract: The Towakkalak System located in the Maros karst of South Sulawesi is currently the richest of Southeast Asia in obligate subterranean species. It comprises several caves and shafts that give access to the subterranean Towakkalak river as well as many unconnected fossil caves, stream sinks, and springs located within its footprint. The total length of the caves linked to the active system is 24,319 m and comprises two of the longest caves of Indonesia, Gua Salukkan Kallang and Gua Tanette. Studies of its fauna began in 1985. There are 10 stygobionts and 26 troglobionts that are known from the system. The smaller adjacent system of Saripa has 6 stygobionts and 18 troglobionts, of which 1 and 3, respectively, are absent from Towakkalak. Like all tropical cave inventories, our dataset has limits due to identification uncertainties, gaps in habitat (waters, guano) and taxonomic coverage (micro-crustaceans, mites), sampling methods (pitfall trapping, Karaman–Chappuis), and problems of ecological assignment. A number of additional species are therefore expected to be found in the future. The Towakkalak and Saripa cave systems are included in the Bantimurung-Bulusaraung National Park and are under efficient protection, but parts of the Maros karst outside the park are under serious threat, mainly from quarrying.

Keywords: cave biology; stygobionts; troglobionts; hotspot cave; sampling biases; Southeast Asia

check for **updates**

Citation: Deharveng, L.; Rahmadi, C.; Suhardjono, Y.R.; Bedos, A. The Towakkalak System, A Hotspot of Subterranean Biodiversity in Sulawesi, Indonesia. *Diversity* **2021**, *13*, 392. https://doi.org/10.3390/d13080392

Academic Editor: Luc Legal

Received: 27 July 2021
Accepted: 10 August 2021
Published: 20 August 2021

Publisher's Note: MDPI stays neutral with regard to jurisdictional claims in published maps and institutional affiliations.

1. Introduction and Context

Cave-restricted species were long considered to be exceptional or rare in tropical caves. They are actually widespread in all tropical regions of the world [1,2]. In Indonesia, the presence of such cave-restricted species has been documented since 1985 in the karst of Maros in South Sulawesi, which has been pointed to as one of the richest tropical caves of the world in this respect [2–4]. In the present paper, we give a commented and updated list of the cave fauna of the Towakkalak and Saripa cave systems, which have the highest species richness within the Maros karst [5]. We confirm Towakkalak as the richest hotspot of the tropical world for subterranean biodiversity. We briefly circumvent the shortfalls that affect our cave species inventory as well as the few ones available in other regions of tropical Asia. We characterize the Towakkalak system fauna within the Maros karst and in a broader context. We skim through the threats that exist to the Towakkalak and Maros cave biodiversity, focusing on growing concerns about the limestone quarrying that affects large parts of the karst that are not currently included in the Bantimurung-Bulusaraung National Park.

1.1. Geographical Setting

The Maros karst is located in the province of Sulawesi Selatan (South Sulawesi), close to the city of Makassar, located between the latitudes 4.650 °S and 5.088 °S, and is within an altitudinal range from sea level to about 700 m. It is an increasingly popular region in Indonesia, as it combines a series of exceptional features in several domains: fine tower karst landscape, unusual geological formations, huge and beautiful caves, an abundance of prehistoric artefacts, the oldest rock painting on earth, and a rich fauna. Accessible by road in one hour from the big city of Makassar, the Maros karst is under very strong human pressure, mostly due to limestone exploitation, which has destroyed large parts of its unique landscape since the 1980s. The Bantimurung-Bulusaraung National Park (Babul) was created in 2004, and today, it protects the core of this invaluable heritage, though the northern part of the karst is not included.

The Maros region has a tropical climate, but it departs from the surrounding areas of Sulawesi by a contrasted seasonal climate. Over a 30 year period (1931–1960), annual rainfall exceeded 3000 mm per year in Maros. Monthly rainfall was 50 mm per month or less during the dry season (July to September), and was more than 400 mm per month from December to March, with a peak in January (761 mm) [6]. In January 1986, rainfall reached 1067 mm, and the whole plain from Samanggi to Maros was under water. According to [7], during the 1985–2014 period, the rainfall peak in January was 827.40 mm, and the lowest was in August (46.4 mm), with average monthly rainfall of 337.02 mm.

1.2. Archeology

The Maros karst is a major site for the prehistory of Sunda islands. Since the pioneer works of [8], a large amount of buried remains and artefacts have been discovered in several caves on the western border of the karst, especially in the Leang Leang valley and in the nearby karst of Mallawa, that characterize the Toalean culture. Since the 1950s, a number of prehistoric paintings have also been found in several caves of the region, generating a strong interest in knowledge of Southeast Asia prehistory [9,10]. More recently, Maros made headlines again in the archeological world when the dating of several cave paintings revealed ages of up to 45,000 years, leading them to become the oldest rock art in the world [11,12]. Many caves of the Maros karst harbor Pleistocene parietal art, rivaling the 'ice age' cave art of western Europe [13].

1.3. Geology

The geology of the Maros karst has been synthetized by [14]. The karst is composed of Tonasa limestone from the upper Eocene to mid-Miocene period. The Tonasa formation, which is 3000 m thick, is mostly composed of massive coralline limestone, bioclastic limestone, and calcarenite. These limestones are discontinuously overlaid by the volcano-sedimentary rocks of the Camba formation, dated from the mid- to upper Miocene period. The Camba formation, with its overlying limestone, subsists in the lowlands south of the Maros karst. It also subsists east of the karst at a higher altitude, suggesting that the two streamlets that sink there, far from the main karst block, probably circulate kilometers under the Camba formation to join the Towakkalak system and its resurgence.

The mid- to late-Miocene volcanic rocks (basalt, trachytes, and diorites) are visible in the karst in several areas as laccoliths, sills, and dykes. Sills are volcanic rocks interbedded horizontally between limestone layers and have a thickness of 10–30 m (Figure 1D). The small K11 spring near Gua Salukkan Kallang emerges at the foot of a sill. But the most unusual feature of the Maros karst is the presence of deep (up to 150 m), narrow (4 to 50 m), long (up to 4 km), straight, or weakly curved corridors in several areas (Figure 1A), which occur in the footprint of the Towakkalak system in particular. These corridors are predominantly generated by volcanic dykes that are more rapidly eroded than the surrounding Tonasa limestone. Interestingly, the subterranean passages that have been explored so far rarely follow corridor dykes. Conversely, in several caves, especially in the case of Gua Salukkan Kallang, narrow to large dykes (1 dm to several meters wide) have been intersected by subterranean galleries, giving large volcanic pebbles in the subterranean streams. These cave passages are therefore posterior to the late Miocene period, though no volcanic filling has been found in the explored caves, which could allow more precise dating [14].

Figure 1. Karst landscape of the Maros karst; (**A**) corridors generated by volcanic dykes in the Towakkalak area (Google Earth); (**B**) Batuputte giant shafts (Google Earth); (**C**) limestone cliffs NW of Bantimurung (Leang Leang valley); (**D**) volcanic sill topped with limestone near Pangea. Reproduced with permission from Louis Deharveng (**C,D**).

1.4. Geomorphology

'Such gorges, chasms, and precipices as here abound, I have nowhere seen in the Archipelago. A sloping surface is scarcely anywhere to be found, huge walls and rugged masses of rock terminating all the mountains and inclosing the valleys. In many parts there are vertical or even overhanging precipices five or six hundred feet high, yet completely clothed with a tapestry of vegetation'. This is how Alfred Russel Wallace described the Maros karst in 1890, where he spent weeks searching for insects [15].

The Maros karst is pinpointed, together with the Sierra de los Organos in Cuba, as a model of a tropical tower karst and is repeatedly documented and interpreted by karstologists [6,16]. It is actually mostly constituted by massive cliff-bound karst units, with tower morphology that is only developed in their upper part. Delimited on its western side by subvertical cliffs arising abruptly up to 200 m above the alluvial plain almost at sea level (Figure 1C), it forms large limestone plateaus dissected by a few deep valleys, up to an elevation of 700 m on the slopes of the Bulusaraung extinct volcano. Together with the nearby karsts of Mallawa to the east and Barru to the north, it covers about 700 km^2 and is mostly concentrated around Bulusaraung.

The Maros karst offers an amazing variety of surface landforms. The corridors discussed above are its most remarkable feature, which have only been reported elsewhere from the Bau karst in Sarawak, though at a smaller extent [14,17]. They often give access to the core of the karst massif, while the terrain between them is practically impassable. More classical karstic features are noticeable, including vestigial karstic outcrops sparse in the alluvial plain, isolated outcrops of various morphology, flat subsurface limestone with water circulation up to 4 meters under the level of the plain, hemispherical hills, deep poljes and depressions, rocky gorges, impassable lapiez, and tsingy-like formations. The deep karst involves big karstic springs, river sinks, numerous foothill caves, long, clean and beautiful underground rivers, mega shafts (Figure 1B) comparable in size to those of Papua New Guinea, and a variety of speleothems.

1.5. Caves

A well-documented synthesis on the Maros karst and caves was published by [18]. A first expedition was organized by Denis Wellens in 1984, but its results were not published [19]. The discovery of big caves began in 1985, with the first expedition of the Association Pyrénéenne de Spéléologie to Maros, where more than 8 km in Gua Salukkan Kallang were explored and mapped [20], while an Italian team explored and mapped 3500 m of Leang Assuloang [21]. Several other expeditions followed, sometimes in collaboration with local Makassar cavers. Expedition reports provide localization and description of explored caves as well as hydrogeological and biological original data [22–30]. All of these caving expeditions were associated with biological surveys and were done with the participation of Indonesian biologists and researchers from Indonesian Institute of Sciences (LIPI) and the Museum Zoologicum Bogoriense. A few independent expeditions were also conducted by Italian and Indonesian cavers, in particular, the Acintyacunyata Speleological Club from Yogyakarta [31] and Korpala UNHAS from Makassar, while several biological sampling campaigns in caves and springs have also been conducted by LIPI researchers.

A total of 219 caves (68 km surveyed) are documented in the APS reports from 1985 to 2001 for the Maros karst [30], to which about 20 caves mapped by the Acintyacunyata Speleological Club from Yogyakarta [31], several tens of caves surveyed by various Indonesian biologists [32] as well as a large number of archaeological caves should be added.

There are five of these caves that have a depth of 190 m or more. Lubang Leaputte is a huge shaft with a 263 m depth and a 100 × 70 m entrance (Figure 2A); Lubang Kapa Kapasa has a 210 m depth and a vertical pit of 205 m; Lubang Beru has a 207 m depth; Gua Salukkan Kallang has a 205 m depth; and Lubang Tomanangna has a 190 m depth with a pit of 170 m and a section of 30 × 20 m [33–35].

There are eight caves that are more than 1 km long. A total of four of them belong to the two systems of our study: the Towakkalak system, which contains Gua Salukkan Kallang (12263 m surveyed), Gua Tanette (9692 surveyed m), and Lubang Kabut (1095 m) [35], and the Saripa system, which contains Leang Saripa (2336 m) [34,36]. There have been four other caves that have been explored and mapped outside of these systems: Leang Assuloang (10048 m surveyed) [37]; Gua Londron (5893 m) [38]; Gua Mimpi (1395 m surveyed) [33]; Gua Kacici (1058 m) [39].

Figure 2. (**A**) Lubang Leaputte; (**B**) Gua Salukkan Kallang underground river; (**C**) Bantimurung waterfall, downstream Towakkalak spring; (**D**) Fossil gallery in Gua Tanette; (**E**) underground river in Gua Tanette. Reproduced with permission from Didier Rigal.

2. Material and Methods

In this paper, we focus on the stygobionts and troglobionts, defined independently of their morphology as species only known from caves [1], and also consider the most characteristic stygophiles and troglophiles. Species included in these last ecological categories were all eutroglophiles. The authors of the species names are given in Tables 1 and 2 or in the text for any cited species that are not listed in these tables. The following local terms for landscape features are used: Gua (Indonesian), Leang (Makassar): cave; Gunung (Indonesian), Bulu (Makassar): mountain; Lubang (Indonesian): shaft; Sungai (Indonesian): river.

In this paper, we deal with the biodiversity of the large underground system of Towakkalak and the small adjacent system of Saripa as point of comparison, which are both located in the southern part of the Maros karst.

2.1. The Towakkalak System (Figures 2B–E and 3)

A detailed description of the system can be found in [40,41], and information on the system's hydrology and hydrogeology can be found in [42,43]. The Towakkalak system (SKT or 'Gua Salukkan Kallang–Towakkalak' system in [4]) is a large underground river that can be accessed by several caves between the Sungai Gallang sinks to the big emergence of Towakkalak at Bantimurung, which is located 7 km away in a straight line. There are two caves that represent the main access to the system: Gua Salukkan Kallang and Gua Tanette.

The Towakkalak system includes the second and fourth longest caves in Indonesia, Gua Salukkan Kallang (12,263 m long, 205 m deep, 4 entrances: K1, K2, K3, K4) and Gua Tanette (9692 m, 25 m deep, a single entrance). Cave passages are typically large galleries of a regular section, often more than 15 m in diameter, where the underground Towakkalak river flows and is navigable in two sections by several kilometers. The river can be accessed through two other smaller caves located between two larger and previously cited caves: Lubang Batu Neraka (749 m long, 85 m deep) and Lubang Kabut (1095 m long, 74 m deep) [35,41]. The upstream sump of the system in Gua Salukkan Kallang was dived in 2001 on a length of more than 100 m [36]. Several other caves, unconnected to the Towakkalak underground river, are likely to belong to the same system.

The Towakkalak underground river emerges at about 40 m of altitude as a large spring which provides water to the Bantimurung waterfall, which is 700 m downstream and is a highly praised tourism spot in South Sulawesi (Figure 2C). The Towakkalak resurgence drains a hydrogeological basin of at least 57 km^2, including 23.5 km^2 of outcropping limestone in its lower part (highest elevation probably 500 m), and 33.5 km^2 as volcanic rocks and sediments in its upper part (highest elevation 1353 m at Bulusaraung). Its discharge was measured in the dry season (July) to be about 0.6 m^3/s [42], and it probably reaches several tens times this value in the rainiest months. Towakkalak is also fed by several stream sinks, those of the Salukkan Kallang and Pangni streamlets (260 m and 420 m of altitude) and the sinks of the Sungai Gallang at 420 m of altitude in particular. The Salukkan Kallang stream sinks at the contact point between volcanic-alluvia and limestone. Lubang Pangni is located more than 3 km northeast of the Salukkan Kallang sink in a very small limestone isolated outcrop; its 100-meter depth suggests a long water circulation under the volcanic and alluvia surfaces. The Sungai Gallang, which sinks in a very tiny and isolated outcrop of limestone, is supposed to circulate for 2.7–2.9 km under a volcanic ridge towards the upstream sump of Gua Salukkan Kallang. In fact, half of the supposed catchment area of the Towakkalak spring is a black box, as the extent and continuity of the limestone underneath the volcanic terrains and the recent alluvial terrain are unknown.

The limits of the Towakkalak catchment in its limestone part are also uncertain but for another reason. About 700 m downstream of the Towakkalak spring and 30–40 m below its elevation are two other springs: Jamala spring and Baharuddin spring. Dye injected into Gua Salukkan Kallang subterranean stream 4.8 km to the east strongly colored the Towakkalak spring, but none of these other springs [42]. The Baharuddin spring, the smallest (135 l/s in dry season), has no evident link with any of the surrounding caves. The Jamala spring is about two times bigger than Towakkalak (1.3 m^3/s) [42]. The origin of its water is not known, and not a single cave can be attached to this spring, while at least 50 caves are located in the footprint of the Towakkalak catchment. Moreover, a long fossil gallery of Gua Tanette, the second largest cave of the Towakkalak system, heads to the west and would pass above the most likely predictable passages of Jamala according to the sump map drawn by divers [36]. Therefore, the northern limits of the Towakkalak system are currently impossible to trace.

2.2. The Saripa System

Biologically, the best documented cave of the Maros karst is Leang Saripa and its spring (Figures 3 and 4). This small hydrogeological system, which is of easier access than the Towakkalak system, is adjacent to it in a same large limestone unit. Its cave fauna is slightly less rich, differing from that of Towakkalak by several taxa. The Saripa system includes the spring of Saripa, which flows out at the plain level from a small cave that is about 50 m long, and Leang Saripa, which is located about 30 m above it.

Many other caves open along the alluvial plain of the Patunuang river down to Bantimurung and are probably connected to the Saripa system during the rainy season when the plain is flooded. Leang Saripa is unusual for its complex system of galleries, contrasting with most caves of the Maros karst. In 2000, about 1700 m of passages, arranged in two or three levels, were explored and mapped [31,33,34]. In 2001, the main sump of the system was dived, giving access to 700 additional meters of big galleries, where a large number of bats were present [36]. Saripa has a large array of aquatic and terrestrial habitats that are favorable to stygobiotic and troglobiotic fauna.

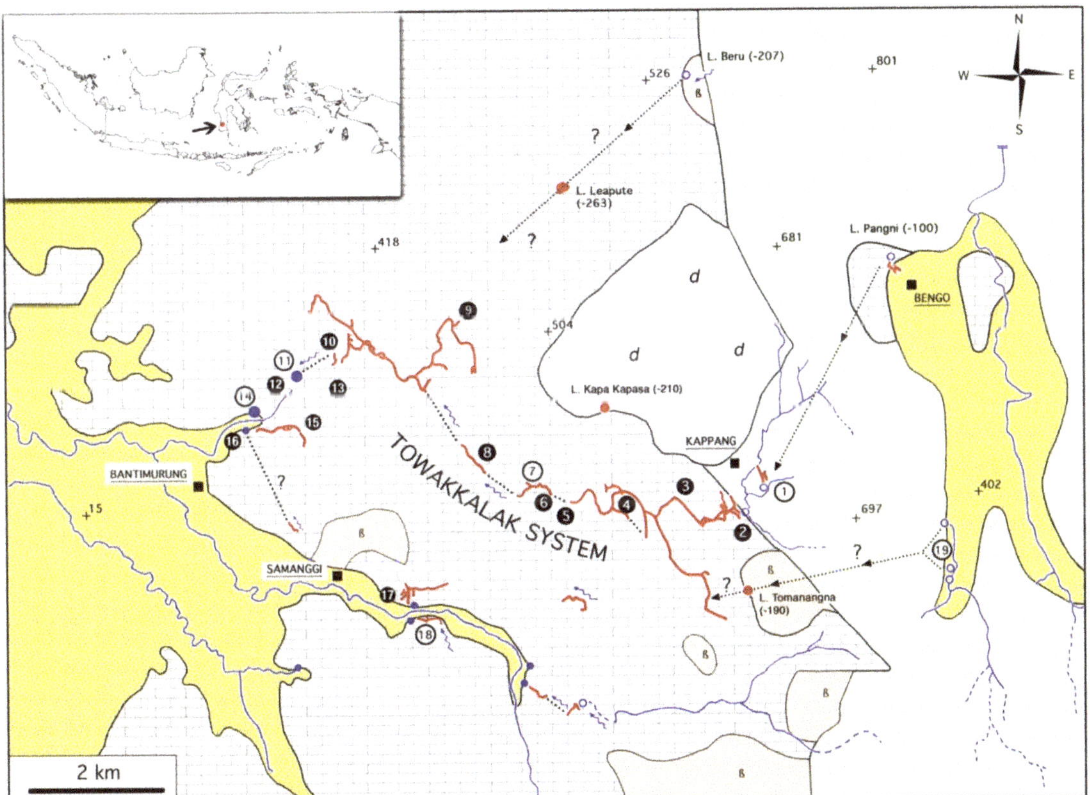

Figure 3. Hydrogeological synthesis of the Maros karst east of Bantimurung, including Towakkalak and Saripa systems [35] updated. Upper left: map of Indonesia with the localization of the studied region in Sulawesi. Legend. Limestone, area filled with rectangles; recent alluvial deposit, area filled with dots, yellow background; basalt (ß), orange; diorite (d), grey; volcanic sediments of the Camba formation, pale green. Square, village; cross, elevation point; empty blue circle, stream sink; filled blue circle, karstic spring; blue line, surface river; blue arrow, underground river; filled red circle, big shaft; red line, explored cave; dotted line, underground water passage, proven by coloration or hypothesized. Encircled numbers, caves and springs (white on black, sampled caves; black on white, large unsampled caves): 1, Gua Burung Salangan Geram; 2, Gua Salukkan Kallang; 3, K11 spring; 4, Gua K9; 5, Gua Alolu; 6, Gua Broukiss; 7, Lubang Kabut; 8, Lubang Batu Neraka; 9, Gua Tanette (entrance); 10, Gua Wattanang (a) and Gua Uri (b); 11, Towakkalak spring; 12, Gua Lumpur (a) and Gua Bantimurung (b); 13, Gua B2 and Gua B3; 14, Jamala spring; 15, Gua Mimpi and Gua Istani Toakala; 16, Gua Baharuddin; 17, Leang Saripa and Saripa spring; 18, Gua Restauran; 19, sinks of the Sungai Gallang.

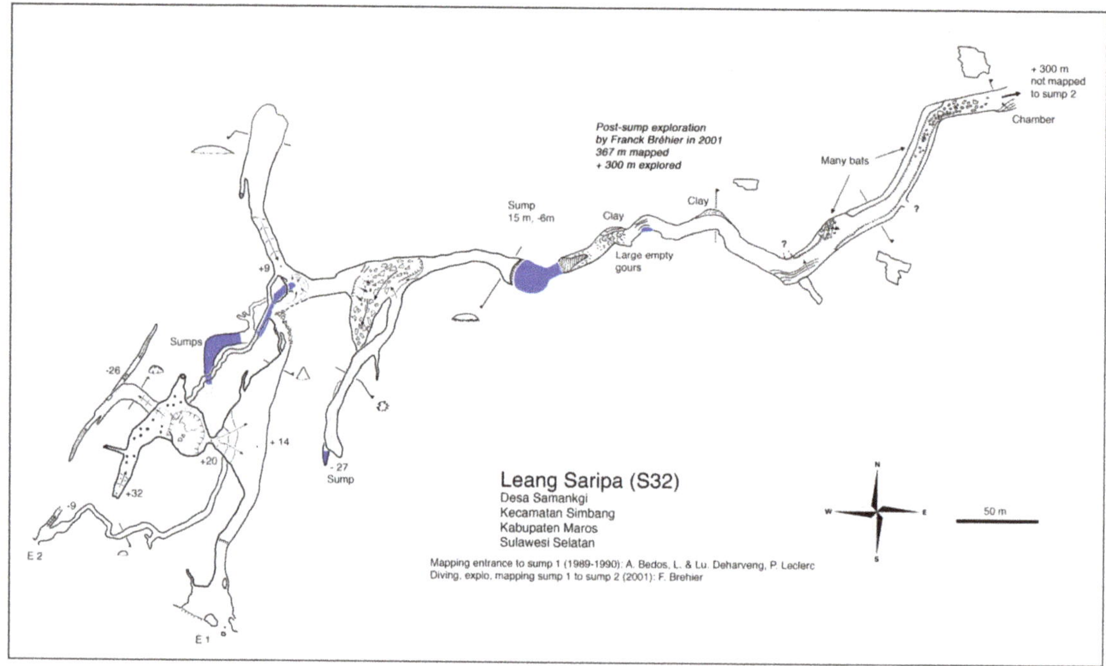

Figure 4. Map of Leang Saripa, synthesis after [31,34,36].

2.3. Sampling

Aside taxonomic descriptions, a number of papers deal with the cave fauna of the Maros karst, in particular that of Towakkalak and Saripa systems, which have been more studied than others [4,5,23,44–47]. The fauna was mostly collected by the authors of this paper by hand picking, netting, and substrate extraction in the Berlese funnels, over the course of several field trips from 1985 to 2008 and in collaboration with cavers of the Association Pyrénéenne de Spéléologie (Toulouse, France) and the Indonesian caving clubs of Yogyakarta and Makassar. Sampled aquatic habitats include freshwater pools, percolating water, endogenous stream, exogenous streams (often eutrophic), and more rarely, phreatic water. Sampled terrestrial habitats were the oligotrophic and 'SCAT' habitats (where the main food supply is scattered swiftlet and bat feces) [48], periodically, inundated habitats with fine silt deposits, flood detritus, bat guano accumulation, parietal and twilight habitats, and soil within cave. Milieu Souterrain Superficiel-type habitats have not been detected, while deep soil habitats outside of the cave host a diverse and rich interstitial fauna not considered here, where troglophilic species are sometimes present.

3. Results

The Towakkalak and Saripa systems, hereafter referred to as Towakkalak and Saripa, are the best studied and richest cave systems of the Maros karst. Their stygobiotic, troglobiotic, main stygophilic, and main troglophilic species are listed in Tables 1 and 2. In 2000, Deharveng and Bedos listed 7 stygobionts and 21 troglobionts in Towakkalak (under the name Gua Salukkan Kallang system) [4]. The updated list includes 10 stygobionts and 26 troglobionts due to changes in the ecological status of several species and the integration of some overlooked species. In addition, one stygobiont and three troglobionts that were absent in Towakkalak but were present in the adjacent Saripa system were added.

3.1. Aquatic Fauna

As shown in Tables 1 and 2 below.

Tricladida: Only four species of flatworms are cited from Southeast Asian caves, three of which are members of the genus *Dugesia* Girard, 1850 [49]. *Dugesia leclerci* is the only eyeless stygobiotic species from Indonesia. It is described and only known from Gua Tanette [50]. A blind flatworm, tentatively reported here as *D. leclerci*, is present at a very high density in the lake (sump 1) of Leang Saripa. The ecological status of *D. uenorum*, another cave flatworm described and only known from Lubang Pangni, a shaft probably linked to Towakkalak, is pendent, as it is oculated and weakly pigmented [50].

Table 1. Stygobiotic and troglobiotic species of the Towakkalak and Saripa systems. Abbreviations—Ecology: SB, stygobiont (aquatic); TB, troglobiont (terrestrial). Distribution (dist): end, restricted to the Maros karst; end *, restricted to a single system. Towakkalak: see Figure 3 for cave numbering; 10a, Gua Wattanang; 12a, Gua Lumpur; 12b, Gua Bantimurung; Pa, Lubang Pangni. Saripa: L, Leang Saripa; S, Saripa Spring; ?, uncertain identification. Number of species in the genus and taxonomic validity of the species (gen val): +, corrected numbers from various sources. n.a., not applicable.

Taxonomic Group	Taxon	Ecology	Dist	Towakkalak	Saripa	Gen Val
Tricladida: Dugesiidae	*Dugesia leclerci* Kawakatsu and Mitchell, 1995	SB	end	9	L?,S?	121 [51]
Tricladida: Dugesiidae	*Dugesia uenorum* Kawakatsu and Mitchell, 1995	SB?	end *	Pa		121 [51]
Amphipoda: Bogidiellidae	Bogidiellidae sp.	SB	end	9		n.a.
Isopoda: Cirolanidae	*Cirolana marosina* Botosaneanu, 2003	SB	end	12a	S	146 [51]
Caridea: Atyidae	*Caridina leclerci* Cai and Ng, 2009	SB	end *	2,3,9		344 [51]
Caridea: Atyidae	*Marosina brevirostris* Cai and Ng, 2005	SB	end *	2		2 [51]
Caridea: Atyidae	*Marosina longirostris* Cai and Ng, 2005	SB	end	2,8,9	L	2 [51]
Caridea: Atyidae	*Parisia deharvengi* Cai and Ng, 2009	SB	end *	9		9 [51]
Brachyura: Hymenosomatidae	*Cancrocaeca xenomorpha* Ng, 1991	SB	end	8,9	L,S	1 [51]
Brachyura: Gecarcinucidae	*Parathelphusa sorella* Chia and Ng, 2006	SB	end *		L	49 [51]
Pisces: Eleotridae	*Bostrychus microphthalmus* Hoese and Kottelat, 2005	SB	end	9	L	9 [51]
Acari: Leeuwenhoekiidae	Leeuwenhoekiidae sp. (cf.)	TB?	n.a.		L,S	n.a.
Amblypygi: Charinidae	*Sarax* sp.	TB	end?	2		17 [52]+
Araneae: Ctenidae	*Amauropelma* cf. sp.	TB	end	4,9,12a	L	24 [53]
Araneae: Psilodercidae	*Psiloderces leclerci* Deeleman-Rheinhold, 1995	TB?	end	2,4,13		38 [53]
Araneae: Ochyroceratidae	*Speocera caeca* Deeleman-Rheinhold, 1995	TB	end	2,4,13	L	84 [53]
Araneae: Pholcidae	*Spermophora maros* Huber, 2005	TB	end *	2,9		45 [53]
Araneae: Pholcidae	*Uthina mimpi* Huber, Caspar and Eberle, 2019	TB	end	2,15		17 [53]
Opiliones	Opiliones sp.	TB	end	2,9,12b,15		n.a.
Palpigradi: Eukoeneniidae	*Eukoenenia maros* Condé, 1992	TB	end *	9,13		93 [54]
Palpigradi: Prokoeneniidae	*Prokoenenia celebica* Condé, 1994	TB	end *	2		6 [54]
Pseudoscorpiones	Pseudoscorpiones sp.	TB	n.a.	15	L	n.a.
Schizomida	Schizomida spp.	TB	n.a.	2,4,9,10a,15	L	n.a.
Diplopoda: Haplodesmidae	*Eutrichodesmus reductus* Golovatch et al., 2009	TB	end	9	L,S	53 [51]
Diplopoda: Metopidiothrichidae	*Metopidiothrix kalang* Shear, 2002	TB	end *	2		39 [51]
Oniscida: Philosciidae	*Papuaphiloscia* sp.	TB	end	2,9,13	L,S	15 [51]
Oniscida: Armadillidae	*Venezillo* sp.	TB	end	4,10a,12b,13,15	S	137 [51]
Collembola: Neanuridae	*Deuterobella* sp.	TB-TP?	end	9	L,S	4 [55]+
Collembola: Neelidae	*Megalothorax* sp.	TB-TP?	?	2	L	33 [55]+
Collembola: Oncopoduridae	*Oncopodura* sp.	TB	end	9	L	49 [55]+
Collembola: Sminthuridae	*Pararrhopalites* sp.	TB	end?	2,9,12b,15	L	17 [55]+
Collembola: Entomobryidae	*Pseudosinella maros* Deharveng and Suhardjono, 2004	TB	end	2,4,5,9,12a,12b,15	L,S	352 [55]+
Collembola: Entomobryidae	*Sinella* sp.	TB?	?	15		86 [55]+
Diplura: Campodeidae	*Lepidocampa (Lepidocampa) hypogaea* Condé, 1992	TB	end	2,4,6,9,12b,13		17 [56]
Zygentoma: Nicoletiidae	Nicoletiidae	TB?	end		S	n.a.
Blattodea: Nocticolidae	Nocticolidae sp. 1	TB	end	2?,9?,12a,12b?,15?	L?,S?	n.a.
Blattodea: Nocticolidae	Nocticolidae sp. 2	TB	end	2?,9,12a,12b?,15?	L,S?	n.a.
Coleoptera: Carabidae	*Eustra saripaensis* Deuve, 2002	TB	end *		L	28 [57]
Coleoptera: Carabidae	*Mateuellus troglobioticus troglobioticus* Deuve, 1990	TB	end	6,15	L,S	2 [58]
Hemiptera: Cixiidae	Cixiidae sp.	TB?	?	15		n.a.

Table 2. Main stygophilic and troglophilic species of the Towakkalak and Saripa systems. Abbreviations: Ecology: SP, stygophile (aquatic); TP, troglophile (terrestrial); Gu: guanophile (terrestrial); eu, euedaphic (terrestrial). Distribution (dist): cos, cosmopolite; end, restricted to the Maros karst; ipa, Indo-Pacific; pan, pantropical; sul, Sulawesi; wid, widespread. Towakkalak: see Figure 3 for cave numbering; 10b, Gua Uri; 12b, Gua Bantimurung; Pa, Lubang Pangni. Saripa: same as Table 1. Number of species in the genus and taxonomic validity of the species (gen val): same as Table 1.

Taxonomic Group	Taxon	Ecology	Dist	Towakkalak	Saripa	Gen Val
Caridea: Atyidae	*Caridina parvidentata* Roux, 1904	SP	end	9		344 [51]
Caridea: Palaemonidae	*Macrobrachium lar* (Fabricius, 1798)	SP	pan	9		276 [51]
Brachyura: Gecarcinucidae	*Parathelphusa celebensis* (De Man, 1892)	SP	sul	Pa		49 [51]
Brachyura: Gecarcinucidae	*Parathelphusa pareparensis* (De Man, 1892)	SP	end	2,9		49 [51]
Pisces: Gobiidae	*Glossogobius* sp.	SP?	?	9		35 [51]
Gastropoda: Subulinidae	*Allopeas gracile* (Hutton, 1834)	TP	pan	15	L	21 [51]
Gastropoda: Subulinidae	*Paropeas achatinaceum* Pfeiffer, 1846	TP	ipa	15		4 [51]
Amblypygi: Charontidae	*Charon* sp.	TP	?	10b,15	L,S	5 [52]+
Araneae: Sparassidae	*Heteropoda beroni* Jaeger, 2005	TP	end	15	L	189 [53]
Araneae: Ochyroceratidae	*Speocera karkari* (Baert, 1980)	TP	?	13		84 [53]

Table 2. *Cont.*

Taxonomic Group	Taxon	Ecology	Dist	Towakkalak	Saripa	Gen Val
Diplopoda: Glomeridesmidae	*Glomeridesmus* sp.	TP?	?	15		28 [51]
Diplopoda: Cambalopsidae	*Hypocambala helleri* Silvestri, 1897	TP-TB	wid	10b,15	L	15 [51]
Diplura: Campodeidae	*Lepidocampa (L.) weberi borneensis* Silvestri, 1933	TP	wid	2		17 [56]
Collembola: Hypogastruridae	*Acherontiella* sp.	TP(eu)	?	15		20 [55]+
Collembola: Isotomidae	*Isotomodes* sp.	TP(eu)	?	15		36 [55]+
Collembola: Isotomidae	*Folsomides centralis* (Denis, 1931)	TP(Gu)	pan	2,9		70 [55]+
Collembola: Isotomidae	*Folsomides parvulus* Stach, 1922	TP(Gu)	cos	2,5		70 [55]+
Collembola: Isotomidae	*Folsomides pseudoparvulus* Martynova, 1978	TP(Gu)	pan	2		70 [55]+
Collembola: Isotomidae	*Folsomina onychiurina* Denis, 1931	TP(Gu)	pan	2,5,12b		5 [55]+
Collembola: Isotomidae	*Isotomiella nummulifer* Deharveng and Oliveira, 1990	TP	pan	2		55 [55]+
Collembola: Isotomidae	*Isotomiella symetrimucronata* Najt and Thibaud, 1988	TP	pan	2,15		55 [55]+
Collembola: Hypogastruridae	*Willemia* cf. *buddenbrocki* Hüther, 1959	TP	?	2		47 [55]+
Collembola: Hypogastruridae	*Xenylla yucatana* Mills, 1938	TP(Gu)	pan	5		139 [55]+
Orthoptera: Rhaphidophoridae	*Rhaphidophora* sp.	TP	end	2,9,12a,15	L,S	102 [59]+
Coleoptera: Aderidae	Aderidae sp.	TP(Gu)	?	15	L	n.a.
Coleoptera: Histeridae	*Aeletes* sp.	TP	?	2		87 [60]
Coleoptera: Staphylinidae	Staphylinidae spp.	TP(Gu)	?	2,4,12a,12b,15	L	n.a.
Lepidoptera: Tineidae	Tineidae spp.	TP(Gu)	?	5		n.a.

Aquatic Crustacea, Amphipoda, Bogidiellidae: Amphipoda are represented by a single specimen of an unidentified Bogidiellidae collected in Gua Tanette. The rarity of the Maros species is likely due to the under-sampling of interstitial habitats, as the family is represented in several caves of the Sunda Islands by described (*Bogidiella deharvengi* Stock and Botosaneanu, 1989 from Halmahera) and undescribed species (on Muna Island, Kalimantan, Sumatra).

Aquatic Crustacea, Isopoda, Cirolanidae (Figure 5A): There is one species of stygobiotic isopod, *Cirolana marosina*, that is present in the Maros karst. Though derived from a marine stock, it is only known from standing freshwater pools, contrary to several of its subterranean congeners, which live in anchialine habitats [61]. The species is known from four caves of the Maros karst: Gua Assuloang, located more than 15 km from seacoast, Gua Lumpur of the Towakkalak system and the two Saripa caves, more than 25 km from seacoast. A flat alluvial plain almost at sea level lies between these caves and the sea, with mangrove locally present along meandering channels. Interestingly, [61] noticed that the holotype of Assuloang is completely blind, while the Saripa specimens, located more than 10 km SE of Gua Assuloang, have reduced unpigmented eyes, suggesting that a process of incipient speciation might have occurred.

Aquatic Crustacea, Decapoda: The dominant species of stygobiotic and stygophilic mesofauna in the groundwaters of the Towakkalak and Saripa systems are shrimps (Caridea) and crabs (Brachyura) [62].

- Caridea (Figure 5B): The shrimps found in the systems of interest belong to four genera: *Caridina* H. Milne-Edwards, 1837 (with 1 stygobiotic species restricted to Towakkalak system and 1 stygophilic species); *Marosina*, Cai and Ng, 2005 (endemic genus with two stygobiotic species only known from the Maros karst, one of which is limited to Gua Salukkan Kallang), *Parisia* Holthuis, 1956 (endemic genus with one rare troglobiotic species only found in Gua Tanette), and *Macrobrachium lar*, a large size stygophile widespread in Pacific and Indian Ocean islands [62–64]. These shrimps often live as large populations in springs, streams, and puddles. Their diversity in the Maros karst is reminiscent of the radiation of the genus *Caridina* in the lakes of Sulawesi [65], which is not matched elsewhere in tropical caves in Asia and obviously calls for further sampling. The proneness to colonize cave habitats repeats in several clades of the worldwide distributed family Atyidae [66] but is particularly marked in the Maros karst, which has the highest number of troglobiotic species in Australasia and the highest level of troglomorphy with its two *Marosina* species.
- Brachyura (Figure 5C,D): Crabs are frequent in the Maros caves, with four species. A total of three of them belong to the speciose sundaic genus *Parathelphusa* H. Milne-Edwards, 1853, of which *P. sorella*, a cave-obligate species with reduced eyes and that is restricted to Saripa cave, where it is rather common [67]. The most remarkable crab species of Maros is the small *Cancrocaeca xenomorpha*, which is blind and with very long and thin legs [68]. The monospecific genus *Cancrocaeca* Ng, 1991, belongs to a

family of mostly marine species, but the Maros species only lives in freshwater, both in the Towakkalak and Saripa systems, where it occurs sporadically in standing water puddles. A second species of Hymenosomatidae discovered more recently in a cave of the Sangkulirang karst of Kalimantan, *Guaplax denticulata* Naruse, Ng and Guinot, 2008, also lives in freshwater. Both differ from the third cave Hymenosomatidae of the region, *Sulaplax ensifer* Naruse, Ng and Guinot, 2008, from Muna Island in Southeast Sulawesi, which lives in brackish water.

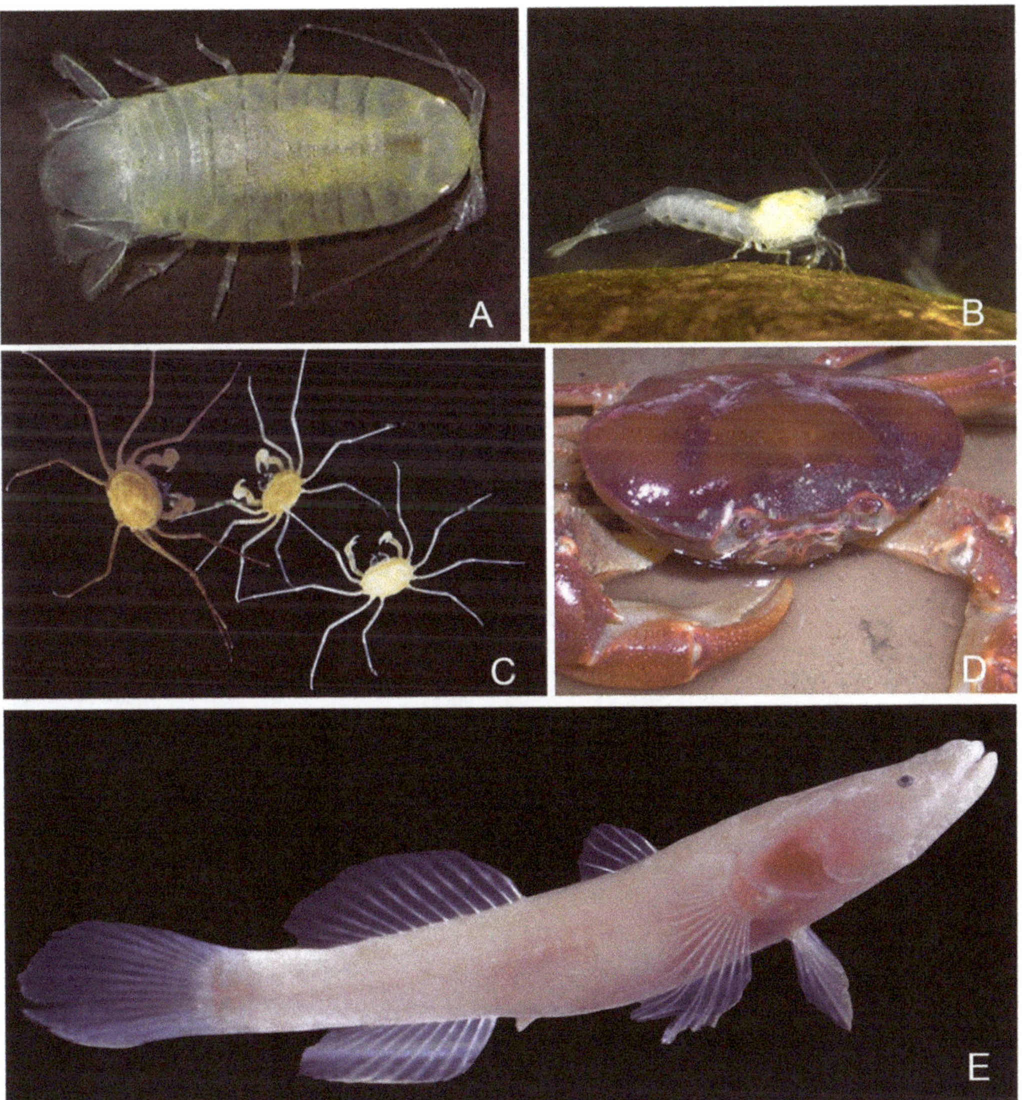

Figure 5. (**A**) *Cirolana marosina* from Leang Saripa; (**B**) *Marosina longirostris* from Leang Saripa; (**C**) *Cancrocaeca xenomorpha* from Lubang Batu Neraka; (**D**) *Parathelphusa sorella* from Leang Saripa; (**E**) *Bostrychus microphthalmus* from Gua Tanette. Reproduced with permission from Louis Deharveng (**A**,**D**); Jean-Yves Rasplus (**B**); Didier Rigal (**C**,**E**).

Aquatic Vertebrata, Pisces (Figure 5E): Only two modified cave fish are known from Indonesia, and both are from Sulawesi. *Bostrychus microphthalmus* from caves of the Maros karst is one of them, which is so far restricted to the Towakkalak and Saripa systems [69,70]. It retains minute eyes covered with skin and is the only *Bostrychus* modified for cave life according to Proudlove [71]. The species is found sporadically in still water pools. In Leang Saripa, the species was collected in 2004 in the sump where several specimens were also observed in August 2020. We also collected a discolored but normal-eyed *Glossogobius* in Gua Tanette, which could be a cave form of an undescribed species found in the surface water of the Maros area [69].

3.2. Terrestrial Fauna

As shown in Tables 1 and 2 above.

Gastropoda: Terrestrial snails of the family Subulinidae are frequent in Maros caves, especially in guano, as well as in most caves of Southeast Asia. They are represented by two large-range species in the systems of interest [72].

Araneae (Figure 6A,B): Spiders are present and are diversified in all of the terrestrial habitats of the Maros caves. A total of nine species, which may be either troglobiotic or troglophilic are listed in Tables 1 and 2.

- *Speocera caeca*: This troglobiont is widely represented in the caves of the two systems. It is the only blind species of the speciose tropical genus *Speocera* Berland, 1914, which includes several other cave species [73].
- The species cf. *Amauropelma* sp.: It is a troglobiotic spider that has been found in several caves of the two systems. By its reduced and unpigmented eyes, pale body color, and rather large size, it is reminiscent of *Amauropelma matakecil* Miller and Rahmadi, 2012, that was recently described from caves of Central Java [74].
- *Psiloderces leclerci*: The speciose Southeast Asia genus *Psiloderces* Simon, 1892, is known by one species in our study, *P. leclerci*, from one cave and one surface site. The latter form is 'much deeper in color' and has shorter legs than that of the cave [75], suggesting a possible separate species status. *Psiloderces leclerci* belongs to the speciose genus *Psiloderces* of the South Asian family Psilodercidae, widespread on Sunda island with a few other cave species.
- *Spermophora maros*: The genus *Spermophora* Hentz, 1841, comprises 45 species that are widely distributed in tropical regions around the world [53] but that are rare in caves. *Spermophora maros* is the only species of the Sunda Islands to have reduced unpigmented eyes, long legs, and whitish coloration [76].
- *Uthina mimpi*: The 17 species of the genus *Uthina* Simon, 1893 [53], mostly live in forest litter. *Uthina mimpi* is, however, a weakly modified species of pale coloration, slightly reduced eyes, and slender legs that seems to be fully troglobiont. In addition, two other *Uthina* species of less clear ecological status, the widespread troglophilic *Uthina luzonica* Simon, 1893, and the Maros endemic *U. sulawesiensis* Yao and Li, 2016, probably exist in our study area [77].

Aside from several unidentified families of troglophilic species and the *Uthina* spp. discussed above, two other identified species are listed in Table 2. *Speocera karkari* is reported from a few surface sites in Southeast Asia. In Sulawesi, it is only known from two caves, Gua Mampu in the Bone karst and Gua B2 of the Towakkalak system. The big *Heteropoda beroni* is a regular inhabitant of the caves of the two systems, quietly resting on cave walls, especially in non-oligotrophic habitats, and often near the cave entrance. A species of this genus that may be *H. beroni* has been found hunting during the night near caves at Bantimurung. It is a major predator of the giant arthropod communities, just like its congeners in other Southeast Asia karsts [4].

Figure 6. (**A**) *Heteropoda beroni* from Leang Saripa; (**B**) cf. *Amauropelma* sp. from Saripa spring cave; (**C**) schizomid from Gua Tanette; (**D**) *Charon* sp. from Saripa spring cave; (**E**) *Eutrichodesmus reductus* from Leang Saripa; (**F**) *Hypocambala helleri* from Leang Saripa. Reproduced with permission from Jean-Yves Rasplus (**A,E**) and Louis Deharveng (**B–D,F**).

Opiliones: There is one unidentified species of Opiliones, possibly an Assamiidae, that is found rather frequently as isolated specimens in the caves of the Towakkalak and Saripa systems. It has small eyes, is pale yellow, and has never been found outside of caves. Similar species are sometimes present in Southeast Asian caves, but most remain unstudied.

Palpigradi: Karsts of South Sulawesi around Maros are unusually rich in Palpigradi, with two families, three genera, and four species, three of which have been described and are only known from the region. There are two species (*Koeneniodes frondiger* Remy, 1950 and *Eukoenenia paulinae* Condé, 1994) that are soil dwellers. The two other species are so far cave restricted. *Eukoenenia maros* is limited to two caves of the Towakkalak system, and

Prokoenenia celebica is known from Gua Salukkan Kallang and Gua Mampu in the Bone karst. Always rare in caves, these minute Arachnida were found as isolated specimens in oligotrophic habitats. As the soil outside caves has been well sampled in Maros, the troglobiotic status of the cave species is reliable. In addition, *E. maros* is considered as troglomorphic in its original description.

Pseudoscorpiones: They are rare, represented by one or two blind unidentified species, which are present in two caves.

Schizomida (Figure 6C): Schizomida of Southeast Asia are severely under-studied compared to those of tropical America or Australia. In Maros, they are frequent in caves and soils, with species differing in particular by the morphology of the male flagellum and the relative length of the appendages. Some morphospecies seem to be cave restricted. Given the proneness to speciation in several lineages of schizomids [78], several new species are expected to occur in Maros as they do in other low elevation karsts of tropical Asia.

Amblypygi (Figure 6D): Whip spiders are, after huntsman spiders of the genus *Heteropoda*, the more common predators of the giant arthropod communities in most caves of Southeast Asia. There are two species of the genera *Sarax* Simon, 1892, and *Charon* Karsch, 1879, that are reported from Maros. The *Sarax* is probably new to science [79]. Both are common in many caves of the region, including the Towakkalak and Saripa system, where they are sometimes encountered rather deep inside caves. They may feed on the *Rhaphidophora* sp. abundant in all caves, probably on juveniles or eggs rather than on adults given the size of this cricket. None of these species has been collected outside caves in Maros so far. The *Sarax* is possibly troglobiont, as suggested by its pale-yellow color and relatively long legs [79]. Several *Charon* described from Southeast Asia were synonymized with *C. grayi* (Gervais, 1842) by Kraepelin [80], but recent publications do not accept these synonymies and consider that several species are likely to occur in the region [81,82].

Diplopoda (Figure 6E,F): *Hypocambala helleri*, a widely distributed saprophagous species, is the most abundant milliped and perhaps arthropod in eutrophic or mesotrophic habitats of the Maros caves. Large populations may be observed on guano, but it is also common in SCAT habitats. The species is considered as a troglophile in the literature but was only found in caves in the Maros karst. A second, less frequent species is the troglobiotic snow white *Eutrichodesmus reductus*, present in several caves of the Maros karst including both systems of interest [83]. It is never found in dense populations nor on guano piles. *Metopidiothrix kalang* has a morphology of a surface species (eyes and pigment present) but has been regularly and only found in Gua Salukkan Kallang so far.

Oniscida (Figure 7A,B): There are two cave restricted terrestrial isopods that are reported from Towakkalak [84] and that are present in Saripa system. The most frequent, undescribed species assigned to the genus *Venezillo* Verhoeff, 1928, which is whitish with reduced eyes, has been found in many caves of the Maros karst, where it lives in oligotrophic to mesotrophic habitats, feeding on organic debris. It has never been found outside of caves. The second species, assigned to the genus *Papuaphiloscia* Vandel, 1970, is an eyeless and transparent species that is also found in several caves of the Towakkalak and Saripa systems. Both genera have cave species in other regions of Southeast Asia.

Collembola [85] (Figure 7C–E): Springtails are at the basis of trophic chains for the terrestrial invertebrate compartment in most caves of the world, especially in oligotrophic habitats. In guano, they are usually second to mites but reach huge density locally. They are preyed upon by beetles, mites, spiders, pseudoscorpions, and probably schizomids. Guano is not as abundant in the Maros karst as it is in other tropical regions of Asia or the Pacific, but three caves of the Towakkalak and Saripa system nevertheless host large bat colonies. The dominant springtails in these habitats are the parthenogenetic pantropical species listed in Table 2. In the oligotrophic or mesotrophic habitats of Maros caves, Collembola are moderately diversified. In the Towakkalak and Saripa systems, a single troglobitic species of the family Entomobryidae, *Pseudosinella maros*, is known which is also present in other caves of the Maros karst. This species is apparently related to blind small surface *Pseudosinella* Schäffer, 1897 that are abundant in the forest soils of the region. *P. maros* mostly differs from them by several troglomorphic characters, i.e., larger body size, proportionally longer appendages, and thinner claws [86]. Similar to its surface relatives in Maros, *P. maros* is eyeless. Though weakly modified, it is more troglomorphic than other tropical cave *Pseudosinella* that have been described so far [87]. Similar undescribed forms of this genus are present in caves of Halmahera [88] and Papua (Fak-Fak) [89]. *Pseudosinella maros* exhibits noticeable variability in claw elongation and antennal S-chaetotaxy. Specimens collected in underground systems of the Maros karst north and south of the type locality (Gua Restauran, close to but independent from Saripa system) might be taxonomically different [86]. This has been corroborated by subsequent barcode analyses, which indicate species-level divergences of 15–20% between populations within and around the systems of interest (unpublished data).

Figure 7. (**A**) *Venezillo* sp. from Leang Saripa; (**B**) *Papuaphiloscia* sp. from Leang Saripa; (**C**) *Pseudosinella maros* from Gua Lumpur; (**D**) *Oncopodura* sp. from Gua Tanette; (**E**) *Pararrhopalites* sp. from Gua Tanette. Reproduced with permission from Louis Deharveng.

Diplura: *Lepidocampa (Lepidocampa) hypogaea* is a clearly troglomorphic species and is widespread in the caves of Towakkalak and Saripa (Table 1). Interestingly, all 11 specimens collected in Gua Baharuddin, a system adjacent to Towakkalak, differ morphologically from those of Towakkalak [90].

Orthoptera, Rhaphidophoridae (Figure 8A): *Diestrammena heinrichi* (Ramme, 1943) is a large Orthoptera described from the Maros region, but we ignore from which habitat it is described. Allegrucci et al. [91] mention a *Rhaphidophora* Serville, 1838, from Leang Saripa, and we assigned the very abundant cricket present in all of the caves of the Maros karst to this last genus, pending a redescription of *D. heinrichi*. This *Rhaphidophora* is the biggest and most common species of the giant arthropod community in the caves of the Towakkalak and Saripa systems. It seems to be saprophagous, but it has not been ascertained that it feeds outside of the cave during night as observed for some other species of the family. The species has normal eyes but a rather pale and uniform coloration.

Figure 8. (**A**) *Rhaphidophora* sp. from Saripa spring cave; (**B**) *Eustra saripaensis* from Leang Saripa; (**C**) Nocticolidae sp. from Leang Saripa; (**D**) *Mateuellus troglobioticus* from Leang Saripa. Reproduced with permission from Anne Bedos (**A**), Jean-Yves Rasplus (**B,C**) and Louis Deharveng (**D**).

Blattodea, Nocticolidae (Figure 8C): Nocticolidae are regular troglobionts in caves in Southeast Asia and the Pacific and are among the most troglomorphic cave arthropods in this area, often combining anophthalmy, marked appendage elongation, and depigmentation. Specimens collected in the Towakkalak and Saripa systems have not been analyzed by specialists, but two eyeless troglomorphic species are present (Fred Stone in litt.), differing by slenderness, degree of troglomorphy, and male tergal structure. They are commonly found in loose groups deep inside caves and mostly in oligotrophic habitats, where they run very quickly when disturbed. Nocticolidae are considered as saprophagous in the literature, probably feeding on degraded organic matter, which is consistent with what we observed in our caves. Interestingly, blind as well as microphthalmic species have been collected in several other caves of the Maros karst.

Coleoptera, Carabidae (Figure 8B,D): Cave obligate beetles are exceptional in tropical cave systems. There are two species of Carabidae (ground beetles) from two different subfamilies that are known in the Towakkalak and Saripa systems, a case that is unique to the Indonesian archipelago: *Mateuellus troglobioticus* (Pterostichinae) and *Eustra saripaensis* (Paussinae). The only species of the genus *Mateuellus* Deuve, 1990, *M. troglobioticus*, is also the only clearly troglobiotic species of the large tribe Abacetini. It exhibits moderately troglomorphic characteristics, i.e., slightly reduced eyes and slightly elongate appendages. The species is a regular inhabitant of the caves of the Maros karst, with a subspecies described from the nearby Tompobulu karst, *M. troglobioticus faillei* Deuve, 2010. Interestingly, another abacetin beetle was recently described from Java caves (*Metabacetus willi* Guéorguiev 2013). Though only found in caves, it is considered to be a troglophile or a trogloxene by its author on the basis of its dark color and inconspicuously modified eyes and leg length. *Eustra saripaensis* is a rare micro-endemic species that is blind and unpigmented, exclusively known from Leang Saripa. It seems to be restricted to a single location, a clay slope in the unflooded part of the cave, where it usually occurs in rotten bamboo or wood. The genus is known by another cave species (*E. pseudomatanga cavernicola* Deuve, 2001) and an edaphic species in Sulawesi. It is also present in a few caves of other Southeast Asian regions [57], where it is always rare.

Lepidoptera: Tineidae are very abundant on guano piles and are represented by several unidentified species.

Swiftlets: Swiftlets are common in the Towakkalak and Saripa systems as well in several caves of the Maros karst. They are an important provider of organic matter in the caves where they nest, though large piles of their guano have not been observed. There have been three species recorded in South Sulawesi—two *Aerodramus* Oberholser, 1906, which echolocate, and *Collocalia esculenta* (Linnaeus, 1758), which does not [92], but their exact identification cannot be ascertained. *Aerodramus* swiftlets were qualified of 'good speciator' [92] for having both extremely good flying, hence dispersal ability, and a strong link to their home caves. Of the many species recognized in Southeast Asia and the Pacific, these two are present in the caves of Sulawesi, but it is not known if they are both present in the Maros karst.

Bats: Bats are present in many Maros caves, where 15 species are recorded [93]. Insectivorous species are much more frequent than frugivorous or nectarivorous ones. The former are encountered, sometimes deep inside caves, as isolated individuals and small groups in many caves of the karst. Rather large colonies with large piles of guano are surprisingly rare in our study area: unidentified species in Gua Alolu and *Emballonura alecto* (Eydoux and Gervais, 1836) in Leang Saripa and in Gua Mimpi. The only very large colony of insectivorous bats, of the species *Chaerephon plicatus* (Buchanan, 1800), is located 13 km NW from Towakkalak in a cave near Salenrang. Fruit bats are recorded from Gua Kelelawar near Kappang as a rather large colony of two species, *Dobsonia exoleta* K. Andersen, 1909, and *Rousettus amplexicaudatus* (E. Geoffroy, 1810). Most records of bats in the Maros karst are actually from other caves outside of our study area, such as Gua Londron, Gua Mattampa Belakang, Gua Togendra, or Gua Peceng [93; Rahmadi pers. comm.].

4. Discussion and Conclusions

4.1. Limitations of The Checklist

The number of species of an underground site makes sense by comparison to others. Comparisons make sense if sampling is comparable in terms of sampled habitats, sampled groups, and sampling efforts. This condition is rarely fulfilled, resulting in false or uncertain species absence (or presence) in lists, which should lead to cautious interpretation and comparison of the results. We provide hereafter an overview of gaps and bias of our dataset, which was summarized in Tables 1 and 2.

4.1.1. Taxonomy

A large proportion of the listed species are undescribed, which is usual in tropical cave species inventories, even the best documented ones [94]. Several groups have been collected but not studied, such as Opiliones, Pseudoscorpiones, or Schizomida, which may uncover several species.

4.1.2. Habitats

Habitats have been unevenly sampled. Aquatic, especially interstitial habitats, guano, and hanging roots, all known to often be species-rich, were not sampled, under-, or minimally sampled, which may account for the low diversity of microcrustacea, guanophiles, and euedaphomorphic species in our caves. Stygobiotic microinvertebrates represent a major part of subterranean biodiversity. They have been sporadically collected but not identified. Guano has only been marginally sampled. Hanging roots have not been sampled in the systems of interest, but in Gua Assuloang, where they were sampled, they hosted the remarkable genus *Celebenna* Hoch and Wessel, 2011 (Cixiidae), which is so far endemic of this cave. The global richness of the fauna has probably been severely truncated for the concerned taxa, a major bias that has to be taken into account when comparing species richness.

4.1.3. Sampling Methods

The collection methods were mostly direct hand-picking and Berlese extraction of substrate cores in terrestrial habitats and netting using large mesh nets in aquatic habitats. Pitfall trapping, baiting, the use of fine mesh nets, or Karaman–Chappuis in sediments, to maintain basic collecting methods, were not or marginally used, and would have probably provided additional species.

4.1.4. Species Ecological Status

Assignment to troglobionts of terrestrial species found in caves on the basis of their anophthalmy and depigmentation is widespread in the literature, while these traits are more common in euedaphic species from deep soil than in troglobionts [95]. Here, we only assigned species that were collected in caves to troglobionts or stygobionts [1], a status that has to be considered provisional, at least for non-troglomorphic species, as it depends on our knowledge of surface fauna, especially soil. To limit this uncertainty hanging over most species lists, we sampled as much as possible inside of and outside of caves.

4.2. Cave Fauna Features of The Towakkalak and Saripa Systems

4.2.1. Species Richness

A total of 26 troglobionts and 10 stygobionts, i.e., 36 obligate cave species, are known for Towakkalak; the numbers are 19, 6, and 25 for Saripa. These values are the highest recorded for any Southeast Asian caves, and probably for any tropical cave. Most species are shared by both systems, but 15 species are only present in Towakkalak, and 4 species are only in Saripa. The much larger size of Towakkalak is probably linked to more diverse habitats that could explain the difference, which might increase in the future, as Towakkalak has been much less studied than Saripa.

4.2.2. Endemism

There are thirty-four cave restricted species that are endemic to the area, having not been detected in the nearest karsts of Bone (east of South Sulawesi), in Muna (Southeast Sulawesi), nor elsewhere in Indonesia. The remaining species are morphospecies of unknown distribution but that are likely to be endemic as well. The figure is completely different for stygophiles and troglophiles, with only 4 probably endemic species of a total of 28. Endemism at the generic level is a marker of stronger geographic isolation. There are four genera that are endemic to the Maros karst. A total of three of them are present in the systems of interest with four species: the minute crab *Cancrocaeca xenomorpha*, two shrimps species of the genus *Marosina,* and the terrestrial beetle *Mateuellus troglobioticus.* The three former species are highly troglomorphic, the latter, only moderately. There are two more genera that are endemic of the Maros karst but were not found in the systems of interest: the stygobiotic beetle *Speonoterus bedosae* Spangler, 1996, from Gua Mangana [42] and *Celebenna thomarosa*, a non-troglomorphic Cixiidae (Homoptera) from Gua Assuloang. The presence of five endemic genera confers an exceptional biological value to the Maros karst. All of the described cave restricted species of Towakkalak and Saripa are endemic to the Maros karst. Other karsts of South Sulawesi are much less known, but none of their cave restricted species have been found in the Maros karst [96].

4.2.3. Shared Diversity Features

The species composition of the Towakkalak and Saripa caves have two common characteristic features: a high diversity of Decapoda among stygobionts and a high diversity of arachnids among troglobionts. Decapoda dominate the aquatic fauna diversity, with 6 stygobionts (5 in Towakkalak, 3 in Saripa) out of a total of 11 (10 in Towakkalak, 6 in Saripa). This dominance seems to be characteristic of Sulawesi caves, which is in line with the radiation undergone by Caridea in the lakes of Southeast Sulawesi (see Caridea above). With 12 troglobiotic or likely troglobiotic species (11 in Towakkalak, 5 in Saripa) of a total of 29 troglobionts (26 in Towakkalak, 18 in Saripa), Arachnida dominate the cave terrestrial fauna in terms of diversity, as is the case in other tropical caves of the region [4], although this figure seems less clear for tropical Brazilian caves [97,98].

4.2.4. Troglomorphy

Patterns of troglomorphy are contrasted among cave species of the systems of interest. In Southeast Asia, for several taxa, the occurrence and degree of troglomorphy seem to roughly decrease with decreasing elevation and with decreasing distance to the equator [2]. Due to the geographical location near equator and low altitude, Maros cave species were therefore not expected to exhibit significant troglomorphic traits. However, they do. The ultimate driver of cave colonization, seasonality, is strong in the Maros karst region and may have led to these modest cave-related morphological modifications that affect several species of Palpigradi, springtails, Diplura, beetles, and fish. On the other hand, Nocticolidae and aquatic taxa generally do not respond to the elevation and latitudinal gradients nor to seasonality and exhibit highly troglomorphic morphologies in many low altitude caves of tropical Asia, as they do in the Towakkalak and Saripa systems.

4.2.5. Guano

As the major food resource for cave fauna in tropical caves, guano habitats deserve some comments. SCAT habitats in Maros caves host, as in most caves of the world, the largest number of cave restricted species [48,99]. On the other hand, guano accumulations have a much richer overall fauna, especially diversified in tropical caves, which remain poorly documented [4]. Remote fossil passages where bats or swiftlets do not venture, such as the large fossil galleries of Gua Tanette, are almost azoic. Surprisingly for a tropical karst of Southeast Asia, guano accumulations are not common in the Maros karst (see above 'bats'). Leang Saripa hosted isolated specimens of insectivorous bats and swiftlets in the first part of the cave but hosted large bat colonies beyond sump 1. Guano accumulation in Maros caves is mostly produced by insectivorous bats, exceptionally by frugivorous ones, and never by swiftlets. As is the case in Vanuatu caves [100], significant faunistic differences between guano of different types were not detected, but the dominant group in this habitat, mites, have not been analyzed. The identified fauna was mostly represented by large populations of troglophiles-guanophiles of wide distribution: *Hypocambala helleri* and several pantropical parthenogenetic Collembola (Table 2). The former species seems to be restricted to caves in the Maros karst, where it is very common, but is reported from surface habitats in other tropical regions. Among Collembola, *Xenylla yucatana*, a pantropical species, is frequent in cave guano and is scarce in surface habitats. The other cited Collembola, mostly pantropical and parthenogenetic, may form dense populations in guano piles but all are more frequent in outside soils. Species of the giant arthropod community as well as, though less frequently, some troglobionts, such as *Mateuellus troglobioticus*, *Venezillo sp.* or Schizomida, are often wandering near guano piles.

4.2.6. Invasives

Big cockroaches are often present at a huge density in low altitude and warm tropical caves of Southeast Asia, especially when they have been disturbed by humans. *Pycnoscelus surinamensis* L., 1767, is reported, for instance, from Gua Mampu in the Bone karst, which is 35 km NE of the Maros karst [101]. However, these giant cockroaches are absent in all of the caves that were surveyed in the Maros karst, including those of Towakkalak and Saripa. In Leang Saripa, the occurrence of rats was noticed in several parts of the cave, indicated by bottles of pitfall traps that were missing or removed from their original places. Some kinds of rodent nest were also found in small holes on the cave walls, mostly composed of plastic garbage.

4.3. Conservation Issues

The cave fauna of the Maros karst, the richest spot of subterranean biodiversity in the tropics, is well protected in the core of the Bantimurung-Bulusaraung National Park, but several peripherical zones are at risk and experience the impact of human pressure that have dramatically increased during the last two decades. Forest logging and land use changes, which may induce important modifications in water flow and circulation, disturbing food supply for underground fauna, are active in these non-protected parts of the karst. Given the terrain roughness, this is not, however, the main threat on cave fauna. Contrary to Gunungsewu in Java, Maros is lucky enough to have agriculture related pollution sources that are mostly downstream karst resurgences, not upstream sinking rivers. Cave over-frequentation linked to the impressive increase in local tourism in Bantimurung and its associated degradations, particularly garbage, is a more recent concern that seriously affects cave habitats in various ways (Gua Mimpi and Leang Saripa). Its direct effects on invertebrate fauna are likely very local. Walls are sometimes extensively tagged, even in the deepest parts of Leang Saripa, but this landscape degradation does not directly impact the fauna. The main biological concern related to over-frequentation is actually that large bat colonies have already abandoned the most visited cave (Gua Mimpi).

However, the major concern for landscape and cave fauna is the multiplication of limestone exploitations during the last two decades, affecting some of the best karst landscape and also several archaeological caves. The extent to which they impact cave invertebrate communities is unknown because they are mostly located in unsurveyed areas north of the National Park boundaries. The only well biologically surveyed outcrops in this area, Mattampa and Lancina northeast of Pangkadjene, host an original and rich cave fauna, including a blind scorpion (*Chaerilus sabinae* Lourenço, 1995) and a large Japygidae that are unknown elsewhere in Maros. This raises serious concerns about the potential impact of these numerous limestone quarries on Maros subterranean biodiversity. Even when the habitat is not destroyed itself, noise and dust linked to quarrying are known to disturb swiftlets, bats, and bat colonies, which are the main food providers for obligate cave fauna and the only one for guanobionts, but we also lack information on the distribution of swiftlets and bats in this part of the karst. They are also, together with climate change, the greatest threat on the preservation of the invaluable prehistoric rock art of the Maros karst [13] as well the main cause of the current disfiguration of the unique landscape of the karst outside of the National Park boundaries.

The upgrade of the Karaenta Nature Reserve to National Park status in 2004 placed the core of the karst, including the Towakkalak system, under strong protection, but the problems have worsened for several limestone blocks outside the Park boundaries, which are currently without guardrail as a response to the impact of quarrying on cave biodiversity. It is hoped that the candidature of the Maros-Pangkep karst as a World Heritage site, a status which it deserves in so many respects, will lead to the reconsideration of the multiplication of limestone exploitations in its northern part from a landscape and biodiversity perspective.

Author Contributions: Prepared the framework of the article, L.D.; all contributed to data compilation, data interpretation, writing, and editing, L.D., C.R., Y.R.S., and A.B. All authors have read and agreed to the published version of the manuscript.

Funding: Funding for field work in 2001–2002 was provided by the ARCBC joint cooperation project between the European Union and ASEAN; by the Museum National d'Histoire Naturelle (Paris) in 2007; and during these few last years by LIPI (DIPA Pusat Penelitian Biologi: 2003-2015) for Maros & Gunungsewu Karst Project for C.R. and Y.R.S.

Institutional Review Board Statement: Not applicable.

Data Availability Statement: Not applicable.

Acknowledgments: This paper is dedicated to the memory of the late Baharuddin, who introduced us to the beautiful karst of Maros, for his unforgettable sense of humor. Our campaigns in the Maros karst were backed by teams of cavers and divers, both Indonesian and French, without whom we would have been unable to sample a number of caves. Thanks to François Brouquisse, Didier Rigal, Franck Brehier, and all those who participated in the Association Pyrénéenne de Spéléologie expeditions to Maros. Philippe Leclerc actively contributed to biological sampling together with several LIPI researchers (late Renny K. Hadiaty, Agustinus Suyanto, Ristiyanti M. Marwoto and Daisy Wowor). Jean-Yves Rasplus and Didier Rigal kindly provided several of their photographs. We are indebted to Roland Barkey from Maros for his kind support from the beginning of the work on the karst, and to many people of Bantimurung and around for their help in the field. Special thanks to the National Park staff, particularly Iqbal Abadi Rasjid, Chaeril, and Dedy Asriady, for their support and kindness during field works. Last but not least, we are grateful to the numerous taxonomists who have efficiently studied our collections.

Conflicts of Interest: The authors declare no conflict of interest.

References

1. Deharveng, L.; Bedos, A. Diversity of Terrestrial Invertebrates in Subterranean Habitats. In *Cave Ecology, Ecological Studies 235;* Moldovan, O.T., Kovác, L., Halse, S., Eds.; Springer Nature: Cham, Switzerland, 2018; pp. 107–172.
2. Deharveng, L.; Bedos, A. Chapter 18—Diversity Patterns in the Tropics. In *Encyclopedia of Caves;* White, W.B., Culver, D.C., Pipan, T., Eds.; Academic Press: Cambridge, MA, USA, 2019; pp. 146–162.
3. Culver, D.C.; Sket, B. Hotspots of subterranean biodiversity in caves and wells. *J. Caves Karst Studies* **2000**, *62*, 11–17.
4. Deharveng, L.; Bedos, A. The cave fauna of Southeast Asia. Origin, evolution and ecology. In *Ecosystems of the World 30. Subterranean Ecosystems;* Wilkens, H., Culver, D.C., Humphreys, W.F., Eds.; Elsevier: Amsterdam, The Netherlands, 2000; pp. 603–632.
5. Suhardjono, Y.R.; Ubaidillah, R. (Eds.) *Fauna Karst dan gua Maros, Sulawesi Selatan;* LIPI Press: Jakarta, Indonesia, 2012; 258p.
6. Balazs, D. Karst Regions in Indonesia. *Karszt Es Barlang* **1968**, *5*, 3–61.
7. Arsyad, M.; Sulistiawaty, U.; Tiwow, V.A. Analysis of Characteristics and Classification of Rainfall in the Maros Karst Region, South Sulawesi. In Proceedings of the International Seminar on Mathematic, Science and Computer Education, Bandung, Indonesia, 15 October 2016.
8. Sarasin, P.; Sarasin, F. *Die Toala-Hoehlen von Lamontjong. Versuch Einer Anthropologie der Insel Celebes;* C.W. Kriedel's Verlag: Wiesbaden, Germany, 1905.
9. Van Heekeren, H.R. Rock-paintings and other prehistoric discoveries near Maros (South West Celebes). *Laporan Tahunan* **1952**, 22–35.
10. Glover, I.C.; Sinha, P. Changes in stone tool use 10,000 years ago: A microwear analysis of flakes with use gloss from Leang Burung 2 and Ulu Leang 1 caves, Sulawesi, Indonesia. *Mod. Quat. Res. S. E. Asia* **1984**, *8*, 137–164.
11. Aubert, M.; Brumm, A.; Ramli, M.; Sutikna, T.; Saptomo, E.W.; Hakim, B.; Morwood, M.J.; van den Bergh, G.D.; Kinsley, L.; Dosseto, A. Pleistocene cave art from Sulawesi, Indonesia. *Nature* **2014**, *514*, 223–227. [CrossRef]
12. Brumm, A.; Oktaviana, A.A.; Burhan, B.; Hakim, B.; Lebe, R.; Zhao, J.X.; Sulistyarto, P.H.; Ririmasse, M.; Adhityatama, S.; Sumantri, I.; et al. Oldest cave art found in Sulawesi. *Sci. Adv.* **2021**, *7*, eabd4648. [CrossRef]
13. Huntley, J.; Aubert, M.; Oktaviana, A.A.; Lebe, R.; Hakim, B.; Burhan, B.; Muhammad Aksa, L.; Made Geria, I.; Ramli, M.; Siagian, L.; et al. The effects of climate change on the Pleistocene rock art of Sulawesi. *Sci. Rep.* **2021**, *11*, 1–10. [CrossRef] [PubMed]
14. Brouquisse, F. 13. Cadre géologique. In *Expédition Thaï-Maros 85, Rapport Spéléologique et Scientifique;* APS: Toulouse, France, 1986; pp. 101–118.
15. Wallace, A.R. *The Malay Archipelago: The Land of the Orang-Utan and the Bird of Paradise: A Narrative of Travel with Studies of Man and Nature;* Macmillan and Co.: London, UK, 1890.
16. Sunartadirdja, M.A.; Lehmann, H. Der tropische Karst von Maros und Nord-Bone in SW-Celebes (Sulawesi). *Z. Geomorph.* **1960**, *4* (Suppl. 2), 49–65.

17. Crabtree, S.; Friederich, H. The Caves of the Bau District, Sarawak. *Cave Sci.* **1982**, *9*, 83–93.
18. Kusch, H. Speläologische Forschungen auf der Insel Sulawesi (Celebes, Indonesien) zwischen 1857 und 1977. *Die Höhle* **1981**, *32*, 91–102.
19. Deharveng, L.; Bedos, A. 10. Les cavités des environs de Bantimurung. In *Expédition Thaï-Maros 85, Rapport Spéléologique et Scientifique*; APS: Toulouse, France, 1986; pp. 81–95.
20. Association Pyrénéenne de Spéléologie. *Expédition Thaï-Maros 85, Rapport Spéléologique et Scientifique*; APS: Toulouse, France, 1986; pp. 1–215.
21. De Vivo, A.; Campion, N.; Menin, A.; Viviani, F. Vecchie storie Indonesiane. *Speleologia* **1992**, *27*, 32–41.
22. Association Pyrénéenne de Spéléologie. *Expédition Thaï-Maros 86, Rapport Spéléologique et Scientifique*; APS: Toulouse, France, 1987; pp. 1–174.
23. Association Pyrénéenne de Spéléologie. *Expéditions de l'A.P.S. en Asie du Sud-Est. Travaux Scientifiques—1*; APS: Toulouse, France, 1988; pp. 1–52.
24. Association Pyrénéenne de Spéléologie. *Expéditions Maros 88-Maros 89, Rapport Spéléologique*; APS: Toulouse, France, 1990; pp. 1–51.
25. Association Pyrénéenne de Spéléologie. *Expédition Indonésie 90, Rapport Spéléologique et Scientifique*; APS: Toulouse, France, 1992; pp. 1–104.
26. Association Pyrénéenne de Spéléologie. *Expédition Maros 94, Rapport Spéléologique et Scientifique*; APS: Toulouse, France, 1997; pp. 1–40.
27. Association Pyrénéenne de Spéléologie. *Indonésie 92, Rapport Spéléologique*; APS: Toulouse, France, 2001; pp. 1–67.
28. Association Pyrénéenne de Spéléologie. *Expédition Maros 99, Rapport Spéléologique*; APS: Toulouse, France, 2002; pp. 1–39.
29. Association Pyrénéenne de Spéléologie. *Maros 2001—Indonésie—Sulawesi Selatan, Rapport Spéléologique*, unpublished, no date. 25p.
30. Brouquisse, F.; Deharveng, L.; Laumanns, M. Indonesia 1985–2001 Expeditions of the Association Pyrénéenne de Spéléologie. *Berl. Höhlenkundliche Ber.* **2015**, *59*, 1–197.
31. Acintyacunyata Speleological Club. *Laporan Ekspedisi Maros 1989—Sulawesi Selatan*, Unpublished, no date. 1–146.
32. Suhardjono, Y.R.; Rahmadi, C.; Nugroho, H.; Wiantoro, S. Bab 2 Karst dan Gua. In *Fauna Karst dan gua Maros, Sulawesi Selatan*; Suhardjono, Y.R., Ubaidillah, R., Eds.; LIPI Press: Jakarta, Indonesia, 2012; pp. 13–52.
33. Bedos, A.; Deharveng, L.; Deharveng, L.; Leclerc, P.; Rigal, D.; Solier, P. 3. Résultats spéléologiques. In *Expéditions Maros 88-Maros 89, Rapport Spéléologique*; APS: Toulouse, France, 1990; pp. 15–48.
34. Brouquisse, F.; Lacas, M.; Rigal, D. 5. Sulawesi: Résultats spéléologiques. In *Expédition Indonésie 90, Rapport Spéléologique et Scientifique*; APS: Toulouse, France, 1992; pp. 37–82.
35. Bedos, A.; Brouquisse, F.; Deharveng, L.; Leclerc, P.; Rigal, D. 4. Grandes Cavités du Karst de Maros. In *Indonésie 92, Rapport Spéléologique*; APS: Toulouse, France, 2001; pp. 39–45.
36. Brehier, F. Reconnaissance de quelques siphons du karst de Maros. *Maros 2001—Indonésie—Sulawesi Selatan, Rapport Spéléologique*, Unpublished. 16–20.
37. Rigal, D. Leang Assuloang. *Maros 2001—Indonésie—Sulawesi Selatan, Rapport Spéléologique*, Unpublished. 7–10.
38. Rigal, D.; Lacas, M. 3. Nouvelles découvertes sur le karst de Maros. In *Indonésie 92, Rapport Spéléologique*; APS: Toulouse, France, 2001; pp. 15–37.
39. Brouquisse, F.; Brouquisse, R. 4. Résultats spéléologiques. In *Expédition Maros 94, Rapport Spéléologique et Scientifique*; APS: Toulouse, France, 1997; pp. 14–28.
40. Brouquisse, F. 9. Le secteur de Kappang et le réseau de Gua Salukkan Kallang. In *Expédition Thaï-Maros 85, Rapport Spéléologique et scientifique*; APS: Toulouse, France, 1986; pp. 68–80.
41. Rigal, D. Gua Salukkan Kallang, karst de Maros, Célèbes Sud, Indonésie. *Spelunca* **1987**, *28*, 32–39.
42. Brouquisse, F.; Rigal, D. 6. Résultats spéléologiques Sulawesi. In *Expédition Thaï–Maros 86, Rapport Spéléologique et Scientifique*; APS: Toulouse, France, 1987; pp. 47–74.
43. Brouquisse, F.; Dalger, D. 8. Hydrogéochimie. In *Expédition Thaï-Maros 86, Rapport Spéléologique et Scientifique*; APS: Toulouse, France, 1987; pp. 85–96.
44. Deharveng, L. 10. Programme zoologique: Bilan général et principaux résultats. In *Expédition Thaï-Maros 86, Rapport Spéléologique et Scientifique*; APS: Toulouse, France, 1987; pp. 111–116.
45. Galletti, I. Nota biospeleologica della spedizione Sulawesi '94. *Speleologia Iblea* **1996**, *4*, 98–101.
46. Deharveng, L.; Bedos, A. Salukkan Kallang, Indonesia: Biospeleology. In *Encyclopedia of Cave and Karst Science*; Gunn, J., Ed.; Fitzroy Dearborn: London, UK, 2004; pp. 631–633.
47. Suhardjono, Y.R. Review of Biospeleology in Sulawesi Island. In Proceedings of the International Symposium on The Ecology and Limnology of the Malili Lakes, Bogor, Indonesia, 20–23 March 2006; pp. 29–38.
48. Chapman, P. The ecology of caves in the Gunung Mulu National Park, Sarawak. *Trans. British Cave Research Assoc.* **1982**, *9*, 142–162.
49. Brancelj, A.; Boonyanusith, C.; Watiroyram, S.; Sanoamuang, L.O. The groundwater-dwelling fauna of Southeast Asia. *J. Limnol.* **2013**, *72*, e16. [CrossRef]
50. Kawakatsu, M.; Mitchell, R.W. Two new freshwater cavernicole planarians (Turbellaria, Tricladida, Paludicola) from Sulawesi (Celebes), Indonesia. *Spec. Bull. Jpn. Soc. Coleopterol.* **1995**, *4*, 81–104.

51. WoRMS Editorial Board. World Register of Marine Species. Available online: https://www.marinespecies.orgatVLIZ (accessed on 15 June 2021).
52. Harvey, M.S. Whip Spiders of the World, Version 1.0. Western Australian Museum, Perth. 2013. Available online: http://www.museum.wa.gov.au/catalogues/whip-spiders (accessed on 13 June 2021).
53. World Spider Catalog. Version 22.5. Natural History Museum Bern. Available online: http://wsc.nmbe.ch (accessed on 5 July 2021).
54. Bu, Y.; Souza, M.F.V.R.; Mayoral, J. New and interesting palpigrades (Arachnida, Palpigradi) of the genera *Koeneniodes* Silvestri, 1913 and *Prokoenenia* Börner, 1901 from Asia. *Zootaxa* **2021**, *4990*, 45–64. [CrossRef]
55. Bellinger, P.F.; Christiansen, K.A.; Janssens, F. Checklist of the Collembola of the World 1996–2021. Available online: http://www.collembola.org (accessed on 10 June 2021).
56. Sendra, A.; Jiménez-Valverde, A.; Rochat, J.; Legros, V.; Gasnier, S.; Cazanove, G. A new and remarkable troglobitic *Lepidocampa* Oudemans, 1890 species from La Réunion Island, with a discussion on troglobiomorphic adaptations in campodeids (Diplura). *Zoologischer Anzeiger* **2017**, *266*, 95–104. [CrossRef]
57. Faille, A. Les Coléoptères troglobies de l'île de Sulawesi (Indonésie); description du mâle du Paussidae cavernicole *Eustra saripaensis* Deuve, 2002 (Coleoptera). *Bull. Soc. Entomol. Fr.* **2010**, *115*, 375–380.
58. Deuve, T. Sur une population différenciée de *Mateuellus troglobioticus* (Deuve, 1990) dans le sud de Sulawesi (Col., Caraboidea, Harpalidae, Pterostichinae, Abacetini). *Bull. Soc. Entom. Fr.* **2010**, *115*, 310.
59. Cigliano, M.M.; Braun, H.; Eades, D.C.; Otte, D. Orthoptera Species File. Version 5.0/5.0. Available online: http://Orthoptera.SpeciesFile.org (accessed on 5 July 2021).
60. Integrated Taxonomic Information System (ITIS). Available online: www.itis.gov (accessed on 4 July 2021).
61. Botosaneanu, L. New stygobiontic isopods (Isopoda: Cirolanidae, Anthuridae) from caves in Sulawesi, Indonesia. *Bull. Inst. R. Sci. Nat. Belg.* **2003**, *73*, 91–105.
62. Wowor, D.; Rahmadi, C. Bab 8 Krustasea. In *Fauna Karst dan gua Maros, Sulawesi Selatan*; Suhardjono, Y.R., Ubaidillah, R., Eds.; LIPI Press: Jakarta, Indonesia, 2012; pp. 165–190.
63. Cai, Y.; Ng, P.K.L. *Marosina*, a New Genus of Troglobitic Shrimps (Decapoda, Atyidae) from Sulawesi, Indonesia, with Descriptions of Two New Species. *Crustaceana* **2005**, *78*, 129–139.
64. Cai, Y.; Ng, P.K.L. The freshwater shrimps of the genera Caridina and Parisia from karst caves of Sulawesi Selatan, Indonesia, with descriptions of three new species (Crustacea: Decapoda: Caridea: Atyidae. *J. Nat. Hist.* **2009**, *43*, 1093–1114. [CrossRef]
65. von Rintelen, K.; Glaubrecht, M.; Schubart, C.; Wessel, A.; von Rintelen, T. Adaptive radiation and ecological diversification of Sulawesi's ancient lake shrimps. *Evolution* **2010**, *64*, 3287–3299. [CrossRef]
66. von Rintelen, K.; Page, T.J.; Cai, Y.; Roe, K.; Stelbrink, B.; Kuhajda, B.R.; Iliffe, T.M.; Hughes, J.; von Rintelen, T. Drawn to the dark side: A molecular phylogeny of freshwater shrimps (Crustacea: Decapoda: Caridea: Atyidae) reveals frequent cave invasions and challenges current taxonomic hypotheses. *Mol. Phylogenet. Evol.* **2012**, *63*, 82–96. [CrossRef]
67. Chia, O.K.S.; Ng, P.K.L. The freshwater crabs of Sulawesi, with descriptions of two new genera and four new species (Crustacea: Decapoda: Brachyura: Parathelphusidae). *Raffles Bull. Zool.* **2006**, *54*, 381–428.
68. Ng, P.K.L. *Cancrocaeca xenomorpha*, new genus and species, a blind troglomorphic freshwater hymenosomatid (Crustacea: Decapoda: Brachyura) from Sulawesi, Indonesia. *Raffles Bull. Zool.* **1991**, *39*, 59–73.
69. Hoese, D.F.; Kottelat, M. *Bostrychus microphthalmus*, a new microphthalmic cavefish from Sulawesi (Teleostei: Gobiidae). *Ichthyol. Explor. Freshwaters* **2005**, *16*, 183–191.
70. Hadiaty, R.K. Bab 5 Ikan. In *Fauna Karst dan gua Maros, Sulawesi Selatan*; Suhardjono, Y.R., Ubaidillah, R., Eds.; LIPI Press: Jakarta, Indonesia, 2012; pp. 89–113.
71. Proudlove, G. Subterranean Fishes of the World. Available online: https://cavefishes.org.uk (accessed on 14 June 2021).
72. Marwoto, R.M.; Isnaningsih, N.R. Bab 6 Molluska. In *Fauna Karst dan gua Maros, Sulawesi Selatan*; Suhardjono, Y.R., Ubaidillah, R., Eds.; LIPI Press: Jakarta, Indonesia, 2012; pp. 115–148.
73. Brescovit, A.D.; Zampaulo, R.D.A.; Cizauskas, I. The first two blind troglobitic spiders of the genus *Ochyrocera* from caves in Floresta Nacional de Carajás, state of Pará, Brazil (Araneae, Ochyroceratidae). *Zookeys* **2021**, *1031*, 143. [CrossRef] [PubMed]
74. Miller, J.; Rahmadi, C. A troglomorphic spider from Java (Araneae, Ctenidae, *Amauropelma*). *Zookeys* **2012**, *163*, 1–11. [CrossRef] [PubMed]
75. Deeleman-Reinhold, C.L. The Ochyroceratidae of the Indo-Pacific region (Araneae). *Raffles Bull. Zool.* **1995**, *43* (Suppl. 2), 1–103.
76. Huber, B.A. Revision of the genus *Spermophora* Hentz in Southeast Asia and on the Pacific Islands, with descriptions of three new genera (Araneae: Pholcidae). *Zool. Meded.* **2005**, *79*, 61–114.
77. Huber, B.A.; Caspar, K.R.; Eberle, J. New species reveal unexpected interspecific microhabitat diversity in the genus *Uthina* Simon, 1893 (Araneae: Pholcidae). *Invertebr. Syst.* **2019**, *33*, 181–207. [CrossRef]
78. Abrams, K.M.; Huey, J.A.; Hillyer, M.J.; Humphreys, W.F.; Didham, R.K.; Harvey, M.S. Too hot to handle: Cenozoic aridification drives multiple independent incursions of Schizomida (Hubbardiidae) into hypogean environments. *Mol. Phylogenet. Evol.* **2019**, *139*, 106532. [CrossRef]
79. Rahmadi, C. Bab 9 Arthropoda Gua. In *Fauna Karst dan gua Maros, Sulawesi Selatan*; Suhardjono, Y.R., Ubaidillah, R., Eds.; LIPI Press: Jakarta, Indonesia, 2012; pp. 191–214.
80. Kraepelin, K. *Das Tierreich 8—Scorpiones und Pedipalpi*; R. Friedländer und Sohn: Berlin, Germany, 1899; pp. 1–265.

81. Harvey, M.S.; West, P.L.J. New species of *Charon* (Amblypygi, Charontidae) from northern Australia and Christmas Island. *J. Arachnol.* **1998**, *26*, 273–284.
82. Rahmadi, C.; Harvey, M.S.; Kojima, J.I. The status of the whip spider subgenus *Neocharon* (Amblypygi: Charontidae) and the distribution of the genera *Charon* and *Stygophrynus*. *J. Arachnol.* **2011**, *39*, 223–229. [CrossRef]
83. Golovatch, S.I.; Geoffroy, J.J.; Mauriès, J.P.; van den Spiegel, D. Review of the millipede genus *Eutrichodesmus* Silvestri, 1910 (Diplopoda, Polydesmida, Haplodesmidae), with descriptions of new species. *Zookeys* **2009**, *12*, 1–46. [CrossRef]
84. Dalens, H. 12. Données préliminaires sur les Isopodes terrestres récoltés dans les grottes de Sulawesi et des Moluques. In *Expédition Thaï-Maros 86, Rapport Spéléologique et Scientifique*; APS: Toulouse, France, 1987; pp. 129–132.
85. Suhardjono, Y.R. Bab 11 Ekorpegas. In *Fauna Karst dan gua Maros, Sulawesi Selatan*; Suhardjono, Y.R., Ubaidillah, R., Eds.; LIPI Press: Jakarta, Indonesia, 2012; pp. 227–246.
86. Deharveng, L.; Suhardjono, Y.R. *Pseudosinella maros* sp. n., a troglobitic Entomobryidae (Collembola) from Sulawesi Selatan, Indonesia. *Rev. Suisse Zool.* **2004**, *111*, 979–984. [CrossRef]
87. Cipola, N.G.; Oliveira, J.V.L.C.; Bellini, B.C.; Ferreira, A.S.; Lima, E.C.A.; Brito, R.A.; Stievano, L.C.; Souza, P.G.C.; Zeppelini, D. Review of Eyeless *Pseudosinella* Schäffer (Collembola, Entomobryidae, Lepidocyrtinae) from Brazilian Caves. *Insects* **2020**, *11*, 194. [CrossRef]
88. Deharveng, L. 4. La faune souterraine de Batu Lubang. In *Expédition Batukarst 88, Rapport Spéléologique et Scientifique*; APS: Toulouse, France, 1989; pp. 37–46.
89. Deharveng, L.; Whitten, T.; Leclerc, P. 5.13. Caves of Papua. In *The Ecology of Papua*; Marshall, A., Beehler, B., Eds.; Periplus: Budapest, Hungary, 2007; pp. 1064–1082.
90. Condé, B. Campodéides des grottes des Célèbes (Insectes, Diploures). *Mém. Biospéol.* **1992**, *19*, 155–158.
91. Allegrucci, G.; Trewick, S.A.; Fortunato, A.; Carchini, G.; Sbordoni, V. Cave crickets and cave weta (Orthoptera, Rhaphidophoridae) from the southern end of the World: A molecular phylogeny test of biogeographical hypotheses. *J. Orthoptera Res.* **2010**, *19*, 121–130. [CrossRef]
92. Rheindt, F.E.; Norman, J.A.; Christidis, L. Extensive diversification across islands in the echolocating *Aerodramus* swiftlets. *Raffles Bull. Zool.* **2014**, *62*, 89–99.
93. Suyanto, A.; Wiantoro, S. Bab 3 Kelelawar. In *Fauna Karst dan gua Maros, Sulawesi Selatan*; Suhardjono, Y.R., Ubaidillah, R., Eds.; LIPI Press: Jakarta, Indonesia, 2012; pp. 53–76.
94. Chapman, P. Species diversity in a tropical cave ecosystem. *Proc. Univ. Bristol Spelaeol. Soc.* **1983**, *16*, 201–213.
95. Coiffait, H. Les Coléoptères du sol. *Vie et Milieu* **1958**, *9* (Suppl. 7), 1–204.
96. Beron, P. Comparative study of the invertebrate cave faunas of Southeast Asia and New Guinea. *Hist. Nat. Bulg.* **2015**, *21*, 169–210.
97. Souza Silva, M.; Ferreira, R.L. The first two hotspots of subterranean biodiversity in South America. *Subterr. Biol.* **2016**, *19*, 1–21. [CrossRef]
98. Trajano, E.; Gallão, J.E.; Bichuette, M.E. Spots of high diversity of troglobites in Brazil: The challenge of measuring subterranean diversity. *Biodivers. Conserv.* **2016**, *25*, 1805–1828. [CrossRef]
99. Jeannel, R.; Racovitza, E.G. Biospeologica XXXIX. Enumération des grottes visitées 1913–1917 (sixième série). *Arch. Zool. Exp. Gen.* **1918**, *57*, 203–470.
100. Deharveng, L.; Lips, J.; Rahmadi, C. Focus on guano. In *The Natural History of Santo: Caves and Soils*; Bouchet, P., Le Guyader, H., Pascal, O., Eds.; MNHN; IRD; PNI: Paris, France, 2011; pp. 300–305.
101. Hanitsch, R. On some cave-dwelling Blattids from Celebes. *Tijdschr. Entomol.* **1932**, *75*, 264–265.

diversity

Article

The Coume Ouarnède System, a Hotspot of Subterranean Biodiversity in Pyrenees (France)

Arnaud Faille [1,*] and Louis Deharveng [2]

1   Department of Entomology, State Museum of Natural History, 70191 Stuttgart, Germany
2   Institut de Systématique, Évolution, Biodiversité (ISYEB), UMR7205, CNRS, Muséum National d'Histoire Naturelle, Sorbonne Université, EPHE, 75005 Paris, France; dehar.louis@wanadoo.fr
*   Correspondence: arnaud.faille@smns-bw.de

Abstract: Located in Northern Pyrenees, in the Arbas massif, France, the system of the Coume Ouarnède, also known as Réseau Félix Trombe—Henne Morte, is the longest and the most complex cave system of France. The system, developed in massive Mesozoic limestone, has two distinct resurgences. Despite relatively limited sampling, its subterranean fauna is rich, composed of a number of local endemics, terrestrial as well as aquatic, including two remarkable relictual species, *Arbasus caecus* (Simon, 1911) and *Tritomurus falcifer* Cassagnau, 1958. With 38 stygobiotic and troglobiotic species recorded so far, the Coume Ouarnède system is the second richest subterranean hotspot in France and the first one in Pyrenees. This species richness is, however, expected to increase because several taxonomic groups, like Ostracoda, as well as important subterranean habitats, like MSS ("Milieu Souterrain Superficiel"), have not been considered so far in inventories. Similar levels of subterranean biodiversity are expected to occur in less-sampled karsts of central and western Pyrenees.

Keywords: troglobionts; stygobionts; cave fauna

Citation: Faille, A.; Deharveng, L. The Coume Ouarnède System, a Hotspot of Subterranean Biodiversity in Pyrenees (France). *Diversity* **2021**, *13*, 419. https://doi.org/10.3390/d13090419

Academic Editors: Michael Wink and Spyros Sfenthourakis

Received: 7 August 2021
Accepted: 27 August 2021
Published: 31 August 2021

1. Introduction

Stretching at the border between France and Spain, the Pyrenees are known as one of the subterranean hotspots of the world [1]. This remarkable diversity is unevenly distributed along the Pyrenean range, reaching its highest value on the northern slope of central and western Pyrenees.

The Arbas massif is located on the northern slope of central Pyrenees, about 70 km south of Toulouse, at the limit of the departments of Haute-Garonne in the west and Ariège in the east (Figure 1). Extending the Lestelas massif to the west, it develops north of the Bouigane valley, east of the Ger valley, and south of rolling hills of Comminges. It ranges from 500 m to 1608 m in altitude.

The massif is formed by relatively complex series of Mesozoic limestones, generally overlaid by massive Urgonian limestones [2]. Under high rainfall exceeding 2000 mm a year, the massif is mostly covered by beech forest, with small stands of fir locally. Water transfers are very rapid due to a well-organized drainage and high surface karstification [2]. Not under threat and not under formal protection measures, the biological richness of the massif is nevertheless remarkable, and has been labeled as Zone Naturelle d'Intérêt Écologique, Faunistique et Floristique (ZNIEFF) *Massifs d'Arbas, Paloumère et Cornudère* (national ID: 730011048).

The main cave system of the massif is the Réseau Félix Trombe—Henne Morte, from the name of the French engineer and caver Félix Trombe, also known for his pioneer works on solar energy. This large system is commonly called Coume Ouarnède or Coumo d'Hyouernedo by cavers. It was explored for years by the famous speleologist Norbert Casteret who often referred to these explorations in his writings (see [3] for a list of references) and remains a well-known cave system for speleologists worldwide.

229

Figure 1. Location of the Arbas massif, in central French Pyrenees.

Various taxon-centered studies since more than one century ago progressively brought to light that the Coume Ouarnède system hosted a high diversity of cave animals. In the only global synthesis of its fauna, published in 1982, 29 cave-restricted species were listed from the system, including 18 stygobionts and 11 troglobionts [4]. Today, that is, 40 years later, there are 38 cave-restricted species recorded, including 21 stygobionts and 17 troglobionts, making of the Coume Ouarnède the richest cave fauna of the Pyrenees, which is itself a major European hotspot. The richness of the Pyrenean range can be explained by its biogeographical history [5]. The richness of the Coume Ouarnède can be explained by the great extent of the cave system and the diversity of its subterranean habitats. It may also result from the relatively good knowledge we have of its fauna, as the system is close to the renowned Subterranean Laboratory of Moulis where a number of biospeologists from all over the world have worked for decades.

In the framework of the special issue of the journal "Diversity" that deals with world hotspots of subterranean biodiversity, the Coume Ouarnède clearly deserves some attention. We shall provide in this paper an updated and comprehensive checklist of its subterranean fauna, put in its ecological and biogeographical context.

2. The Coume Ouarnède System

The Coume Ouarnède system is the longest subterranean system of France, and one of the most famous sites regarding French speleology. Today, it has a development of more than 112 km for a depth of 1020 m, with 57 inter-connected caves, while the massif of Arbas has about 500 caves in total (Figures 2 and 3) ([6]; S. Clément pers. comm. 05.2021). The network offers a great variety of geomorphological features, with countless fossil and active galleries and shafts, some of large dimensions (Pont de Gerbaut, Pène Blanque), and subterranean rivers [7].

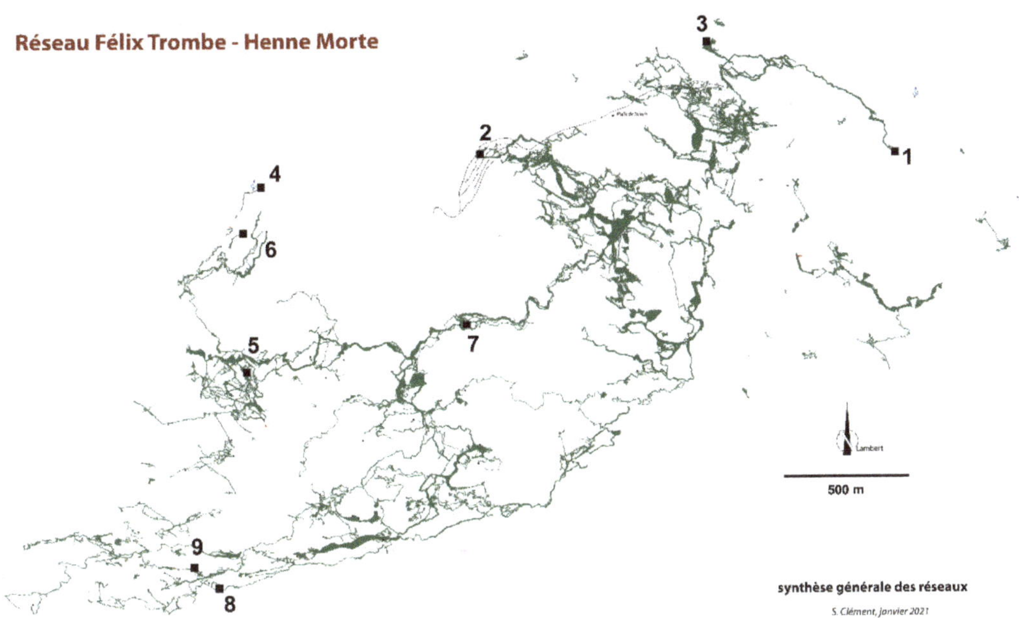

Figure 2. Coume Ouarnède. Synthesis of the networks (modified from [7] and S. Clément pers. comm.). 1–9: Localities with three or more listed species, by decreasing species richness. 1: Goueil di Her; 2: Grotte de Pène Blanque; 3: Poudac Gran; 4: Hount deras Hechos; 5: Henne Morte; 6: Puits du Mistral; 7: Gouffre du Pont de Gerbaut; 8: Trou Mile; 9: Gouffre Raymonde.

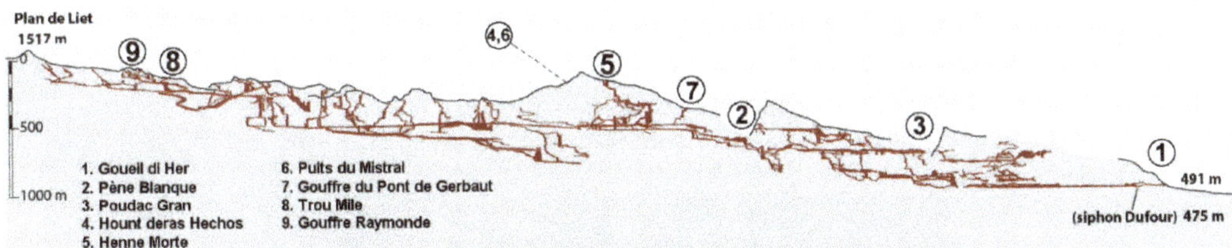

Figure 3. A transverse view of the system, with the location of the nine caves from which more than three taxa are listed (modified from [7]). Cave numbers as in Figure 2.

The Coume Ouarnède system has two distinct resurgences: the Goueil di Her at Arbas, and the Hount deras Hechos at Herran, 2.5 km to the west (*"Hount-des-Heretchos"* sensu [8,9], that is, «the source of the ash trees» after [10]). The Goueil di Her is the resurgence of the main hydrological system, the Réseau Félix Trombe. The Hount deras Hechos is the resurgence of the smaller Henne Morte system, regarded as a secondary derivation of the main system [2,7]. The two systems are interconnected.

3. Methods

Data synthetized here are drawn from the literature, that is, species lists [4,9,11–14] and taxonomic literature. Most specimens were collected in nine caves or cave complexes of the system: Gouffre de la Henne Morte (and spring nearby), Gouffre du Pont de Gerbaut, Poudac Gran, Gouffre Raymonde, Grotte de Pène Blanque, Goueil di Her, Hount deras Hechos (cave and spring), Puits du Mistral, and Trou Mile (Figures 2 and 3). These caves are now interconnected, except Poudac Gran. Most of the collections in the Goueil di Her, the richest of these caves, were made between the entrance and the first sump.

Species names and species validity have been checked from [15] and public databases [16,17]. Species ecological status has been inferred from the taxonomic literature, from [18], and from [19] for spiders.

Abbreviations and terms defining species ecology that are used in the text are defined below:

Endogean or euedaphic species: living or assumed to live only in deep soil, often common at cave entrances.

Eutroglophile: a species with permanent populations inside and outside caves.

Hyporheic species: living or assumed to live only in interstitia of sediment beneath and alongside streams.

MSS: Milieu Souterrain Superficiel (often translated as "Superficial Underground Compartment" in the literature).

Stygobiont: a species living or assumed to live only in groundwater.

Subtroglophile: a species spending a part of its life cycle in caves.

Troglobiont or troglobiotic species: living or assumed to live only in caves or in the MSS.

Troglophile or troglophilic species: living inside as well as outside caves.

Only troglobionts, stygobionts and the most important troglophiles are considered in this paper. Guano and shallow subterranean habitats have not been sampled, though these habitats are present and promising in the massif d'Arbas [20].

4. The History of Biological Explorations

The Goueil di Her cave, the eye of Hell ("*uèlh d'in hèrn*") in the Gascon dialect, is the main resurgence of the Coume Ouarnède system. The first explorations of the cave are related in [21]. In 1908, 30 years later, Jeannel visited the cave, and did the first biological collections, during a speleological expedition conducted by E.A. Martel which aimed at studying the hydrological characteristics of Pyrenees [22,23]. He introduced the cave in a dramatic way [23]: «*Près d'Arbas se trouve une étrange caverne, le Goueil di Her, redouté des habitants du pays, parce qu'après les pluies une puissante rivière souterraine jaillit sous pression hors de la grotte, produisant une détonation qui s'entend à plusieurs kilomètres*». "Near Arbas is a strange cave, the Goueil di Her, feared by the inhabitants of the country, because after the rains a powerful subterranean river gushes under pressure out of the cave, producing a detonation which can be heard several kilometers away". During subsequent trips, Jeannel, Racovitza and several biologists continued to regularly sample this cave (Figure 4).

Figure 4. René Jeannel at the entrance of the Goueil di Her cave, modified from [11].

A summary of their collections was published in the "Enumérations des Grottes Visitées" [22,24,25]. In 1908, Jeannel visited several caves of the Arbas massif, and sampled their fauna (Gouffre du Pont de Gerbaut, Grotte de Pène Blanque, Poudac Gran, Hount deras Hechos, Grotte de Gourgue) [22]. In particular, he returned repeatedly to the Goueil di Her (1908, 1910, 1912). An important integrative work was done by F. Trombe and his collaborators (1945–1947), based on a compilation of data in the fields of speleology, karstology, climatology, hydrology, ecology, and biospeology of the Comminges area, with a strong focus on the Arbas massif karst [12,26] Some years later, the Pène Blanque cave fauna was the subject of another contribution [13]. The last and most important investigations for aquatic subterranean fauna were done by Lescher-Moutoué, Gourbault and Rouch, who studied the composition, distribution and ecology of the stygobiotic fauna and characterized abiotic parameters of the Goueil di Her habitats in great detail [9,14]. A synthesis of our knowledge on the hydrology and biospeology of the massif was published by Bou [4,27]. Aside from these fundamental works, data available in the literature are few, based on punctual samples in a small number of caves.

5. The Fauna

The Goueil di Her was ranked among the world hotspots of subterranean biodiversity, with 26 cave-restricted species taxa (14 aquatics, 12 terrestrial species) in [28,29], then with 29 species [30]. This last number is actually an underestimation of the biological richness of the cave, and of the whole system. Our knowledge of the Coume Ouarnède terrestrial biodiversity relies on disparate sampling surveys in a few caves and pits, that often focused on peculiar groups only, mostly beetles; conversely, aquatic fauna has been the object of an intensive taxonomic and ecological study in karst and hyporheic habitats, but limited to the first part of the Goueil di Her cave itself and to the hyporheic zone of the stream that emerges from this cave [9].

Although the system counts more than 50 entrances, faunistic records come from a few of them: nine caves have three cave species or more, but among them, only two were investigated thoroughly, and provided more than seven cave-restricted species: the Pène Blanque cave (eight troglobionts, no stygobionts) and the Goueil di Her, the main resurgence of the system and by far the richest (17 troglobionts and 13 stygobionts). Fourteen stygobionts were collected in the hyporheic of streamlet under Goueil di Her. Faunistic data are detailed in Tables 1 and 2.

5.1. Stygobiotic Taxa

The works of Lescher-Moutoué, Gourbault, and Rouch are the main sources of information about stygobionts of the system [9,14]. They mainly focused on crustacea of Goueil di Her cave and of the hyporheic of the stream down to 2 km from the resurgence, that is, a limited number of habitats compared to those of the whole system. In most cases, Copepoda were the dominant group in abundance and diversity, rarely surpassed by Ostracoda for abundance. The taxonomic coverage of the collected fauna is globally good, but Ostracoda remain unidentified and only two species of Gastropoda were mentioned. Below, we browse the most interesting stygobiotic species of the Coume Ouarnède system.

5.1.1. Tricladida

Plagnolia vandeli, the only representative of the genus *Plagnolia*, is a blind white flatworm endemic of a few karsts in central northern Pyrenees. Metabolism of the species has been shown to be strongly reduced compared to epigean species and is associated with a considerable lengthening of all biological processes, such as regeneration and life expectancy [31].

5.1.2. Gastropoda

Both listed species, *Moitessieria simoniana* and *Islamia moquiniana*, are wide-range endemics of subterranean aquifers, the former in eastern Pyrenees and Montagne Noire, the later in eastern Pyrenees [32].

Table 1. List of stygobionts and troglobionts of the Coume Ouarnède system (CO), Grotte de Lestelas (L), Grotte de Gourgue (G). *, ectoparasite; hab, habitats; c, cave; h, hyporheic; s, spring; x, species present in CO, L or G.

Group	Species	hab	CO	L	G
Stygobionts					
Suctorida: Choanophryidae	*Echinophrya stenaselli* Matjašič, 1963 *	c		x	x
Tricladida: Planariidae	*Plagnolia vandeli* de Beauchamp & Gourbault, 1964	c	x		
Clitellata: Haplotaxidae	*Delaya leruthi* (Hrabě, 1958)	c	x	x	
Gastropoda: Hydrobiidae	*Islamia moquiniana* (Dupuy, 1851)	h, c	x		
Gastropoda: Moitessieriidae	*Moitessieria simoniana* (Saint-Simon, 1848)	c	x		
Amphipoda: Niphargidae	*Niphargus foreli* Humbert, 1877	s	x		
Amphipoda: Niphargidae	*Niphargus robustus* Chevreux, 1901	c	x		
Amphipoda: Niphargidae	*Niphargus pachypus* Schellenberg, 1933	h	x		
Amphipoda: Salentinellidae	*Parasalentinella rouchi* Bou, 1971	h, c	x		
Amphipoda: Salentinellidae	*Salentinella* sp.	h	x		
Syncarida: Bathynellidae	*Bathynella* sp.	h, c	x		
Isopoda: Asellidae	*Proasellus racovitzai* Henry & Magniez, 1972	h, c	x		x
Isopoda: Stenasellidae	*Stenasellus virei hussoni* Magniez, 1968	c	x	x	x
Copepoda: Ameiridae	*Nitocrella gracilis* Chappuis, 1955	c	x		
Copepoda: Ameiridae	*Parapseudoleptomesochra subterranea* (Chappuis, 1928)	h, c	x		
Copepoda: Canthocamptidae	*Ceuthonectes gallicus* Chappuis, 1928	h, c	x		
Copepoda: Canthocamptidae	*Elaphoidella infernalis* Rouch, 1970	h, c	x		
Copepoda: Cyclopidae	*Diacyclops languidoides* (Lilljeborg, 1901)	c	x		
Copepoda: Cyclopidae	*Graeteriella* (*Paragraeteriella*) sp.	h	x		
Copepoda: Cyclopidae	*Speocyclops anomalus* Chappuis & Kiefer, 1952	c	x	x	
Copepoda: Cyclopidae	*Speocyclops racovitzai* (Chappuis, 1923)	c	x		
Copepoda: Parastenocarididae	*Parastenocaris dianae* Chappuis, 1955	h	x		
Troglobionts					
Acari: Rhagidiidae	*Troglocheles vandeli* Zacharda, 1987	c	x		
Araneae: Linyphiidae	*Birgerius microps* (Simon, 1911)	c		x	
Araneae: Leptonetidae	*Leptoneta microphthalma* Simon, 1873	c	x	x	x
Pseudoscorpiones: Neobisiidae	*Neobisium* (*Blothrus*) *abeillei* Simon, 1872	c	x		
Pseudoscorpiones: Neobisiidae	*Neobisium* (*Blothrus*) sp.	c			x
Pseudoscorpiones: Neobisiidae	*Neobisium* (*Neobisium*) *cavernarum* (L. Koch, 1873)	c		x	
Opiliones: Cladonychiidae	*Arbasus caecus* (Simon, 1911)	c	x	x	x
Opiliones: Ischyropsalididae	*Ischyropsalis pyrenaea* Simon, 1873	c		x	
Chilopoda: Lithobiidae	*Lithobius cavernicola* Fanzago, 1877	c		x	
Diplopoda: Blaniulidae	*Blaniulus lorifer* (Brölemann, 1921)	c	x	x	
Diplopoda: Blaniulidae	*Blaniulus troglobius* (Brölemann, 1921)	c	x	x	x
Diplopoda: Glomeridae	*Spelaeoglomeris jeanneli* Brölemann, 1913	c	x		x
Isopoda: Trichoniscidae	*Scotoniscus macromelos macromelos* Racovitza, 1908	c	x	x	x
Collembola: Entomobryidae	*Pseudosinella theodoridesi* Gisin & Gama, 1969	c	x	x	
Collembola: Oncopoduridae	*Oncopodura tricuspidata* Cassagnau, 1964	c	x	x	
Collembola: Tomoceridae	*Tomocerus problematicus* Cassagnau, 1964	c	x		
Collembola: Tomoceridae	*Tritomurus falcifer* Cassagnau, 1958	c	x		
Coleoptera: Carabidae	*Aphaenops bucephalus* (Dieck, 1869)	c	x	x	
Coleoptera: Carabidae	*Aphaenops cerberus bruneti* Jeannel, 1926	c	x	x	
Coleoptera: Carabidae	*Aphaenops crypticola* (Linder, 1859)	c	x		
Coleoptera: Carabidae	*Aphaenops ehlersi* (Abeille de Perrin, 1872)	c	x	x	x
Coleoptera: Carabidae	*Aphaenops tiresias tiresias* (Piochard de la Brûlerie, 1872)	c	x	x	
Coleoptera: Carabidae	*Geotrechus trophonius trophonius* (Abeille de Perrin, 1872)	c	x		
Coleoptera: Leiodidae	*Speonomus* (*Machaeroscelis*) *infernus arbasanus* Jeannel, 1924	c	x		x
Coleoptera: Leiodidae	*Speonomus* (*Machaeroscelis*) *infernus infernus* (Dieck, 1869)	c		x	
Coleoptera: Leiodidae	*Speonomus* (*Speonomus*) *ehlersi* (Abeille de Perrin, 1872)	c		x	
Fungi: Laboulbeniaceae	*Rhachomyces aphaenopsis* Thaxter, 1905 *	c	x	x	

Table 2. List of troglophiles of the Coume Ouarnède system (CO), Grotte de Lestelas (L), Grotte de Gourgue (G). ecol, ecology; endo, endogean; TPeu, eutroglophiles; TPsub, subtroglophiles; x, species present in CO, L or G.

Group	Species	ecol	CO	L	G
Araneae: Agelenidae	*Eratigena inermis* (Simon, 1870)	TPeu	x		x
Araneae: Leptonetidae	*Leptoneta infuscata* Simon, 1873	TPeu	x		x
Araneae: Linyphiidae	*Troglohyphantes marqueti* (Simon, 1884)	TPeu	x		
Araneae: Tetragnathidae	*Meta menardi* (Latreille, 1804)	TPeu	x		x
Araneae: Tetragnathidae	*Metellina merianae* (Scopoli, 1763)	TPeu	x		x
Opiliones: Ischyropsalididae	*Ischyropsalis luteipes* Simon, 1873	TPeu	x		
Chilopoda: Lithobiidae	*Lithobius piceus piceus* L. Koch, 1862	TPeu	x		
Chilopoda: Lithobiidae	*Lithobius pilicornis pilicornis* Newport, 1844	TPeu	x		
Chilopoda: Lithobiidae	*Lithobius troglodytes* Latzel, 1886	TPeu	x	x	x
Chilopoda: Lithobiidae	*Lithobius microps* Meinert, 1868	TPeu			x
Chilopoda: Lithobiidae	*Lithobius macilentus* L. Koch, 1862	TPeu	x	x	x
Isopoda: Trichoniscidae	*Phymatoniscus tuberculatus* ssp.	endo		x	
Isopoda: Trichoniscidae	*Phymatoniscus tuberculatus arbassanus* Vandel, 1948	endo			x
Isopoda: Trichoniscidae	*Oritoniscus trajani trajani* Vandel, 1933	endo			x
Collembola: Hypogastruridae	*Schaefferia decemoculata* (cf.) (Stach, 1939)	TPeu	x		
Collembola: Onychiuridae	*Onychiuroides pseudogranulosus* (Gisin, 1951)	TPeu	x	x	
Collembola: Arrhopalitidae	*Pygmarrhopalites pygmaeus* (Wankel, 1860)	TPeu		x	
Coleoptera: Carabidae	*Geotrechus orpheus consorranus* (Dieck, 1870)	endo	x	x	x
Coleoptera: Carabidae	*Laemostenus oblongus oblongus* (Dejean, 1828)	TPeu	x		
Coleoptera: Curculionidae	*Raymondionymus perrisi* (Grenier, 1864)	endo		x	
Coleoptera: Leiodidae	*Bathysciola lapidicola simplex* Coiffait, 1959	endo	x		
Coleoptera: Leiodidae	*Bathysciola lapidicola lapidicola* (Saulcy, 1872)	endo		x	x
Coleoptera: Leiodidae	*Bathysciola schivedlei schioedtei* (Kiesenwetter, 1850)	endo	x	x	x
Coleoptera: Leiodidae	*Bathysciola asperula* (Fairmaire, 1858)	endo		x	
Coleoptera: Leiodidae	*Bathysciola ovata* (Kiesenwetter, 1850)	endo			x
Coleoptera: Leiodidae	*Choleva angustata* (Fabricius, 1781)	TPsub	x		
Coleoptera: Staphylinidae	*Linderia cristata* (Saulcy, 1872)	endo		x	x
Coleoptera: Staphylinidae	*Linderia bidentata* (Dodero, 1919)	endo			x
Coleoptera: Staphylinidae	*Leptotyphlus (Leptotyphlus) anchorifer* Coiffait, 1957	endo			x
Coleoptera: Carabidae	*Hypotyphlus pandellei* (Saulcy, 1867)	endo		x	x
Coleoptera: Staphylinidae	*Octavius capdeviellei* Orousset, 1979	endo			x
Diptera: Mycetophilidae	*Speolepta leptogaster* (Winnertz, 1863)	TPsub	x		
Lepidoptera: Geometridae	*Triphosa tauteli* Leraut, 2009	TPsub	x		
Trichoptera: Limnephilidae	*Stenophylax permistus* McLachlan, 1895	TPsub	x		

5.1.3. Amphipoda

Five species of Amphipoda in three different genera are present in the system. *Parasalentinella rouchi*, from the hyporheic zone of the Arbas and Escalette streams, is a small species known from the hyporheic of a few stations in Ariège and eastern Haute-Garonne [33]; it is present in this habitat downstream of Goueil di Her resurgence, where it sometimes occurs together with *Salentinella petiti*.

Niphargus are present but less common in Pyrenees than in most other French regions [34]. Three species of *Niphargus* have been collected in the system, in three different habitats: *N. pachypus* in the hyporheic downstream of the resurgence [4], a species of the group *longicaudatus* Costa, 1851, probably *N. robustus*, inside the Goueil di Her [4,34] and *N. foreli* at a spring in the upper part of the system [35]. Further sampling as well as taxonomic work would be necessary to confirm these findings.

5.1.4. Isopoda

Proasellus racovitzai was described from the Goueil di Her where it coexists with *Stenasellus virei hussoni*. It is considered endemic of this cave [36]. *Stenasellus virei hussoni* (Figure 5), a "carnivore facultative omnivore" after [37], is widespread in caves of the northern part of central Pyrenees. This typical stygobiont has a life span of 15–20 years [38].

Figure 5. *Stenasellus virei hussoni*, a stygobiotic species common in hypogean waters of the Coume Ouarnède system (specimen from the Tute de Jovis cave, in the Sourroque massif, east of the Arbas massif; body 10 mm long; reproduced with permission from S. Huang).

5.1.5. Copepoda

The Coume Ouarnède is especially rich in Copepoda, with nine stygobiotic species recorded so far, of which six occur at the Goueil di Her. A large number of epigean species are associated to these strictly hypogean species, for example 11 troglophilic or trogloxenic species for four troglobiotic ones among harpacticoids are cited from Goueil di Her [14]. Copepoda are by far dominant in number and diversity in all hypogean compartments.

5.2. Terrestrial Taxa

5.2.1. Acari

Rhagidiidae, a family of tiny predatory mites, comprises the largest number of troglobionts among Acari of temperate caves. *Troglocheles vandeli* is a troglomorphic mite only known so far from its type locality, the Trou Mile in Arbas [39]. The remarkable blind, slender, and transparent mite mentioned in [40] in the Goueil di Her might be this species. Given the rarity of cave Rhagidiidae, the presence of a second troglobiotic species of this family, *Rhagidia (Deharvengiella) troglomorphica* Zacharda, 1987, in the Grotte de la Buhadère, a cave of the Source Bleue system adjacent to the southwestern part of the Arbas massif, may indicate that these rare mites may be more diversified than expected.

5.2.2. Araneae

The Coume Ouarnède system, as well as the Arbas and surrounding massifs, are riche in troglophilic spiders, but very poor in troglobiotic ones. A single troglobiotic spider, *Leptoneta microphthalma*, is recorded from system, and present in several caves of the massif d'Arbas and surrounding karsts. It has six reduced eyes. *L. infuscata*, a troglophilic but mostly cave dwelling species, is also frequent in the caves of the region [41].

5.2.3. Opiliones

Arbasus caecus is a relictual species of harvestmen, the only species in the genus *Arbasus* (Figure 6). The superfamily Travunioidea, to which *Arbasus* belongs, is a north temperate lineage known from North America, Europe and East Asia (Japan, Korea) and counts four families and 24 genera, some of them troglobiotic [42]. The peculiar morphology of *Arbasus* makes difficult its attribution to a family, and it is regarded as a representative of either Travuniidae [43] or Cladonychiidae [42]. Genetic data are still lacking to confirm this attribution.

Figure 6. *Arbasus caecus*, a monospecific genus of harvestmen Travuniidae described from Pène Blanque cave (specimen from Artigouli cave, Estadens; body 2 mm long; reproduced with permission from C. Vanderbergh).

First described in the genus *Phalangodes* by Simon [44], *Arbasus caecus* was subsequently included in a new genus, *Arbasus*, by Roewer in his revision of European Opiliones [45]. Contrary to what was mentioned by Bou [4], the species was not described from Gourgue cave, where it is present, but on a single specimen collected in the Pène Blanque cave, probably by R. Jeannel [22,44]. Since then, the species was found in Goueil di Her, as well as a few other caves in karsts surrounding the Arbas massif: Riusec cave of the Source Bleue system [12], Cap de Payssas cave at Juzet-d'Izaut [46], the Lestelas cave [47] of the Caussanous system [2], and more recently, the Artigouli cave at Estadens (C. Vanderbergh pers. comm.), a cave of the small Peyrein system northwest of Arbas [48].

5.2.4. Chilopoda

One of the three troglophilic *Lithobius*, *L. troglodytes*, has a strong affinity for cave habitats as more than 70% of its 73 French records listed by Iorio in 2014 were collected in caves [49]. The two other species are more abundant outside, rather than inside caves.

5.2.5. Diplopoda

All three species listed in Table 1 are detritivores. Four Blaniulidae (two species and four subspecies) have been reported in various caves of the Coume Ouarnède system in the literature. A taxonomic re-examination of this material would be necessary to validate these forms. We provisionally listed only the two species, *Blaniulus lorifer* and *B. troglobius*, regardless of their subspecies. The third species, *Spelaeoglomeris jeanneli*, is a small pill-millipede only known from the Arbas massif (Goueil di Her, Grotte de Gourgue, Grotte de Paloumère and Poudac Gran). This species is the only troglobiont endemic of the Arbas massif [4,11,50].

5.2.6. Isopoda Oniscida

The species *Scotoniscus macromelos*, endemic of central Pyrenees, is split into nine parapatric subspecies, all strictly troglobiotic. *S. macromelos macromelos* is the easternmost form, endemic of a cluster of karstic massifs including Arbas and Lestelas [51–53].

5.2.7. Collembola

Collembola are usually the dominant arthropod species in caves. They are detritivore, but troglobiotic species mostly ingest clay, like many deep soil species, perhaps feeding on the micro-organisms it contains, while many litter species ingest fungi mycelium.

Three cave-restricted species are known from the system. At least two additional ones from two genera are expected to occur, as they are known in caves of other systems of the region: a *Micronychiurus* and a blind *Pseudosinella*. No species from the guano has been cited from the system, but they probably exist, and a guanobiotic endemic Hypogastruridae, like those known from the Grotte de Mont de Chac and Grotte de Payssa nearby can also be expected.

Pseudosinella theodoridesi. It is a common species in the caves of Ariège and Haute-Garonne. Its eyes and pigmentation are reduced, but present. Its appendages are, however, clearly elongated. Several forms differing by eye reduction have been recognized in the caves of the region [54], but their taxonomic status is unclear.

Oncopodura tricuspidata. Described from the grotte de Pène Blanque, this troglomorphic species has populations in several caves of central Pyrenees, which are likely to be undescribed species [55].

Tritomurus falcifer. This highly troglomorphic species is blind, depigmented, and has long antennae and elongate claw, in contrast to all other western European Tomoceridae. It is geographically isolated from the two other species of its genus which are located in caves of the Dinarides [56] and can be qualified as relictual. The species is the only troglobiont of the Coume Ouarnède system to live in hygropetric habitats [57], where it moves relatively slowly, often associated to *Aphaenops ehlersi*; though equipped with a long furca, "the large Collembola, special to" the Goueil di Her "jump with difficulty and prefer to run away: they walk by sweeping the ground with their very long antennae which strike the roughness of the ground while folding back." [40]. It is known from the Goueil di Her and Trou Mile in the system, but also from a few caves in the surroundings of the Arbas massif, that is, Grotte du Béguet, Grotte de la Buhadère and Grotte de Riusec.

5.2.8. Coleoptera

The system is rich in 11 species of Coleoptera, among which six are troglobitic (genera *Aphaenops* and *Speonomus*, both endemic of the Pyrenees), three are endogean and two are troglophilic. Though narrow endemics, all Coume Ouarnède beetle species in their current acceptation have distribution areas exceeding the Coume Ouarnède and even the Arbas massif. None of the species are endemic of the Arbas massif itself, but all the troglobiotic ones have a narrow distribution in Arbas and surrounding massifs, like the Leiodidae *Speonomus (Machaeroscelis) infernus arbasanus* (Figure 7) [58].

Figure 7. *Speonomus (Machaeroscelis) infernus arbasanus*, a hypogean Leiodidae present in the Arbas massif (specimen from Lespugue cave, Saleich; body 2.5 mm long; reproduced with permission from S. Huang).

The genus *Aphaenops*, endemic of the Pyrenees, is composed of many troglobiotic and a few endogean species. Five species are quoted from the system, all strictly troglobiotic. These *Aphaenops* are easily recognizable morphologically and were, until recently, distributed in different subgenera. Three of them are common (*A. ehlersi, A. t. tiresias, A. cerberus bruneti*) in the Goueil di Her cave, while the two other ones are rare.

Aphaenops (formerly *Hydraphaenops*) *ehlersi* (Figure 8) has a particularly interesting ecology, very different from the other species of the genus present in the Goueil di Her. During a visit of the cave just after a storm and a flood in July 1914, Jeannel observed numerous exemplars of "*Hydraphaenops*" (actually *A. ehlersi*) walking on wet clay around the sump [59], that were usually not present or rare in this habitat. He was convinced that he found here the real way of life of the *Hydraphaenops*-like species, which are always extremely rare in most of caves of Pyrenees. He hypothesized the existence of a terrestrial phreatic habitat restricted to the deep parts of the karsts, to which some species are subservient, coming out to the riverbanks of subterranean rivers only after flooding [60,61].

Aphaenops (formerly *Arachnaphaenops*) *tiresias* is a wide-range endemic that is present in several karsts in the Pyrenees of Ariège and Haute-Garonne.

Aphaenops (formerly *Cerbaphaenops*) *cerberus* is widespread in Ariège and Haute-Garonne, where it is differentiated in many genetically divergent populations, some taxonomically recognized as subspecies, the splits between the populations reflecting largely the fragmentation of the karst itself [62]. The *Aphaenops cerberus bruneti* populations of Coume Ouarnède and of Lestelas, *locus typicus* of the subspecies are slightly different morphologically. Because of morphological similarities, it was suggested that the Coume Ouarnède population, the westernmost of *A. cerberus*, might actually belong to the species *A. jauzioni* Faille, Déliot, and Quéinnec, 2007 which was described from Grotte d'Artigouli, a cave in a small isolated limestone outcrop 5 km northwest from Goueil di Her [63,64].

Figure 8. *Aphaenops ehlersi*, specimen from Goueil di Her (body 4.2 mm long; reproduced with permission from C. Vanderbergh).

Aphaenops (formerly *Cephalaphaenops) bucephalus* is known from many caves of Ariège and Haute-Garonne, where it is always rare [65] (Figure 9).

Figure 9. *Aphaenops bucephalus*, a species endemic from the Ariège and Haute-Garonne departments, always rare in its distribution area, and known by only a few records in Goueil di Her cave (specimen from Mount cave, Juzet-d'Izaut; body 5.6 mm long; reproduced with permission from S. Huang).

Aphaenops crypticola is quoted from a single record under the name *A. parallelus* Coiffait, 1954, and its identification needs confirmation [66].

One species of the Coume Ouarnède system initially described as distinct subspecies (*A. tiresias* ssp. *proserpina* Jeannel, 1909) was recently synonymized with its nominal subspecies. Another one, *A. ehlersi* ssp. *longiceps* Jeannel, 1926, is no more recognized as a valid subspecies. As a result, none of the recognized *Aphaenops* of the Coume Ouarnède is currently restricted to the system, but all remain relatively narrow regional endemics.

Among the three endogean species found in the system, *Geotrechus orpheus consorranus* belongs to a genus very close to *Aphaenops*, but with as many endogean forms as troglobiotic ones. All *Geotrechus* are blind and depigmented, but none exhibit strong appendage elongation. The species found in the Coume Ouarnède, often endogean, is also present in Grotte de Lestelas. However, this last cave also has another species of the same genus, *G. trophonius trophonius*, which is rare.

5.2.9. Laboulbeniomycetes

Pyrenean Trechinae frequently carry on their integument a species of fungi of the Laboulbeniaceae family. Laboulbeniomycetes are a group of ascomycete fungi that utilizes arthropods for nutrition and/or dispersal. The species *Rhachomyces aphaenopsis* is known on a large number of Pyrenean cave Trechinae (see [67] for a complete list of the hosts) and was found on three species of Trechinae occurring in the Coume Ouarnède system: *Aphaenops cerberus bruneti*, *A. tiresias tiresias* and *A. ehlersi* [67,68].

6. Species Richness in a Regional Context

The Coume Ouarnède system is the second hotspot of subterranean biodiversity of France; it is second to Cent Fonts spring in total species richness and second to Grotte de Lestelas in number of troglobionts (Table 3). Aside from its 21 stygobionts and 17 troglobionts, the system is also home to 20 troglophiles, including three species of endogean beetles (Table 2).

Table 3. Number of obligate subterranean species of the richest caves and aquifers of France, and of three rich caves close to the Coume Ouarnède system (Grotte de la Buhadère, Grotte de Lestelas and Grotte de Gourgue). Tb, number of troglobionts; Sb, number of stygobionts; *, caves close to Coume Ouarnède.

Site	Region	Tb	Sb	Source
Cent Fonts spring	Massif Central	0	43	[69]
Triadou aquifer	Massif Central	0	34	[30]
Baget aquifer and karst	Pyrenees	9	24	[30]
Grotte de Lestelas *	Pyrenees	19	3	Table 1
Coume Ouarnède hyporheic and caves	Pyrenees	17	21	Table 1
Pierre-Saint-Martin system	Pyrenees	14	4	[70]
Grotte de la Buhadère *	Pyrenees	14	2	[71]
Grotte de Gourgue *	Pyrenees	8	2	Table 1

The Coume Ouarnède system shares most of its species with three karstic systems geographically close to it [2], that is, the Grotte de Gourgue, 200 m from Goueil di Her, but not related to the Coume Ouarnède system; the Lestelas cave of the Cassaunous system; and the Buhadère cave of the Source Bleue system. The differences in faunistic richness are primarily due to undersampling of these three caves, in particular for aquatics. None of the three cited caves have been sampled for microcrustaceans, which account for more than half of the total number of species in the Coume Ouarnède system. More generally, inventories of the hyporheic fauna would be required to precise the distribution limits and the degree of endemicity of the microfauna of subterranean aquifers in most parts of the Pyrenees. The few stygobionts encountered in the three caves are present in the Coume Ouarnède, except *Echinophrya stenaselli*, ectoparasite on *Stenasellus* pleopods [72].

The troglobiont dataset is less uneven. In the Grotte de Gourgue, very close to the Goueil di Her, all troglobionts, but one, are shared with the Coume Ouarnède system. The missing one is a rare Pseudoscorpion, *Neobisium* (*Blothrus*) sp. The Grotte de Lestelas, more thoroughly sampled, shares 11 species with the Coume Ouarnède, but has 8 species not found in this system. Among them, two species have vicariants in the Coume Ouarnède.

Tomocerus problematicus is replaced by *Tritomurus falcifer* (though very different morphologically, they are mutually exclusive in the region). *Speonomus (M.) infernus infernus* is represented by a different subspecies, *S. (M.) infernus arbasanus* in the Coume Ouarnède. The other six species, that live in habitats undersampled in the Coume Ouarnède system, could be potentially found in the system, like Pseudoscorpiones or *Ischyropsalis pyrenaea*, a Pyrenean endemic, which occurs in several caves close to the system: Lestelas, Lespugue among others [46]. Comparisons with Grotte de la Buhadère would require deeper taxonomic investigations, but this cave has at least two rare troglobiotic micro-arthropods that could be potentially present in the Coume Ouarnède system: a Rhagidiidae and a new Oncopodura springtail of a species group different from *O. tricuspidata*. Discoveries of additional troglobionts in the Coume Ouarnède system is therefore foreseeable.

Conversely, several of the troglobionts (*Spelaeoglomeris jeanneli* for instance) present in the Coume Ouarnède system and lacking in Lestelas, are susceptible to be found in this last cave. Given the lack of all-taxa inventories in most caves of Pyrenees, it is likely that caves of richness similar to that of the Coume Ouarnède exist in the central and western Pyrenees. In the same line, some groups which are widespread in subterranean ecosystems of the Pyrenean range have not been reported from the Coume Ouarnède system nor from the three caves of its surroundings mentioned above, such as Diplura, frequent troglobionts in Pyrenean caves [73], or Ostracoda among stygobionts. This last group is widespread and diversified in groundwaters of central Pyrenees, but its species have been rarely identified.

At the scale of France, between-site comparisons are more biased. The highest species richness known so far, that of Cent-Fonts spring, and the third and fourth richest ones, those of Triadou and of the Baget, do not compare with the Coume Ouarnède dataset as they are focused on interstitial aquatic fauna based on exceptional sampling efforts during years. The current pattern at country level is that hotspots of subterranean diversity are located in two regions: central and western Pyrenees on one hand for both stygobionts and troglobionts, and southeastern Massif Central on the other hand for stygobionts. Within these regions, the richest caves are clearly also the most heavily sampled.

7. Conclusions and Perspectives

The faunistic inventory presented in this paper has to be put in perspective. Its limitation is due to several ecological, spatial, taxonomical and methodological gaps. Guano and shallow subterranean habitats have been only marginally sampled or not sampled at all. A single site, the Goueil di Her, has been the object of thorough all-taxa sampling; however, its cave habitats, though highly interesting, are quite unusual among the caves of the region due to frequent flooding of all the sampled passages. Taxonomic coverage has been insufficient for snails and springtails, while Ostracoda and Diplura have not been identified. At least, mostly basic sampling techniques have been used for terrestrial and aquatic fauna except at Goueil di Her. Overall, this may explain why several cave species encountered in the surroundings of the Arbas massif have not yet been recorded from the Coume Ouarnède system. Filling these gaps would significantly increase the subterranean biodiversity of the system. On the other hand, sampling surrounding karsts, finetuning distribution limits, and testing genetic differentiation of populations of stygobionts would obviously help in understanding the observed geographical patterns. In this respect, investigations are currently being carried out, that focus on the major biodiversity issues raised above, that is, sampling a few targeted caves located in the upper Coume Ouarnède system and shallow subterranean habitats of its basin, and caves in the three surrounding systems identified by the BRGM [2]. The basic sampling methods were completed by bait-based techniques in oligotrophic and aquatic habitats.

In spite of these limitations, this work confirms the richness of the Coume Ouarnède system and of its most interesting cave, the Goueil di Her. Springtails and beetles are dominant, like in all central and western Pyrenean karsts (Table 1) [18], and these two groups clearly illustrate its difference with non-Pyrenean regions of France in terms of global subterranean species richness. The presence of two relictual troglomorphic species in the Coume Ouarnède fauna is a second illustration of its exceptional biodiversity. This raises fascinating questions about the origin and age of this fauna, as much as similarly isolated relicts do not exist in the karsts of central Pyrenees surrounding the Arbas and its satellite massifs.

Author Contributions: Both authors A.F. and L.D. participated in all phases of the realization and writing of the paper. Both authors have read and agreed to the published version of the manuscript.

Funding: This research received no external funding.

Institutional Review Board Statement: Not applicable.

Data Availability Statement: Not applicable.

Acknowledgments: We thank Christian Vanderbergh and Sunbin Huang for providing pictures of some hypogean taxa. We are particularly grateful to Sylvestre Clément (Speleo-Club du Comminges, Arbas) for sharing with us his deep knowledge of the Coume Ouarnède system and for providing an up-to-date general topography of the system.

Conflicts of Interest: The authors declare no conflict of interest.

References

1. Culver, D.C.; Deharveng, L.; Bedos, A.; Lewis, J.J.; Madden, M.; Reddell, J.R.; Sket, B.; Trontelj, P.; White, D. The mid-latitude biodiversity ridge in terrestrial cave fauna. *Ecography* **2006**, *29*, 120–128. [CrossRef]
2. BRGM. Couledoux Arbas—Estelas. In *Atlas des Potentialités Aquifères des Formations Pyrénéennes Projet Potapyr—BRGM/RP-66912-FR, (Technical Document)*; BRGM: Orléans, France, 2017; pp. 1–27.
3. Chabert, C. Bibliographie. In *La Coumo d'Hyouernedo—Réseau Félix Trombe—Henne Morte—Massif d'Arbas*; Duchêne, M., Drillat, P.A., Eds.; Groupe Spéléologique des Pyrénées: Toulouse, France, 1982; pp. 335–340.
4. Bou, C. Biospéologie. In *La Coumo d'Hyouernedo—Réseau Félix Trombe—Henne Morte—Massif d'Arbas*; Duchêne, M., Drillat, P.A., Eds.; Groupe Spéléologique des Pyrénées: Toulouse, France, 1982; pp. 195–204.
5. Jeannel, R. Le peuplement des Pyrénées. *Rev. Fr. Entomol.* **1947**, *14*, 53–104.
6. Clément, S.; Vennarecci, P. *Réseau Félix Trombe—Henne Morte. Massif d'Arbas—Pyrénées Centrales. Synthèse Topo explo du Karst d'Arbas*; CDS Haute-Garonne: Toulouse, France, 2003; pp. 1–351.
7. Clément, S. La Coume Ouarnède, le plus grand réseau souterrain de France. In *Grottes et karsts de France*; Audra, P., Ed.; Karstologia Mémoires: Lyon, France, 2010; Volume 19, pp. 278–279.
8. Trombe, F. L'exploration du Gouffre de la Henne Morte, commune d'Arbas (Haute-Garonne). *Ann. Spéléol.* **1948**, *3*, 25–48.
9. Lescher-Moutoué, F.; Gourbault, N. Recherches sur les eaux souterraines. Etude écologique du peuplement des eaux souterraines de la zone de circulation permanente d'un massif karstique. *Ann. Spéléol.* **1970**, *25*, 765–848.
10. Martel, E.A. Rapport sur l'exploration souterraine hydrologique des Pyrénées en 1908. In *Ann. Minist. Agriculture Paris*; Imprimerie Nationale: Paris, France, 1910; Volume 38, pp. 1–96.
11. Jeannel, R. *Faune cavernicole de la France*; Lechevalier: Paris, France, 1926; pp. 1–334.
12. Dresco, E.; Nègre, J. Recherches souterraines dans les Pyrénées Centrales Années 1945 à 1947. Chapitre IV. Résultats Biospéléologiques. *Ann. Spéléol.* **1947**, *2*, 149–164.
13. Derouet, L.; Dresco, E. Etudes sur la grotte de Pèneblanque. I. Faune et Climats. *Notes Biospéol.* **1955**, *10*, 123–131.
14. Rouch, R. Recherches sur les eaux souterraines. 14—Peuplement par les Harpacticides d'un drain situé dans la zone de circulation permanente. *Ann. Spéléol.* **1971**, *26*, 107–133.
15. INPN. Muséum National d'Histoire Naturelle. Available online: https://inpn.mnhn.fr (accessed on 5 July 2021).
16. World Register of Marine Species, WoRMS Editorial Board. Available online: https://www.marinespecies.org (accessed on 15 June 2021).
17. World Spider Catalog. Version 22.5. Natural History Museum Bern. Available online: http://wsc.nmbe.ch (accessed on 5 July 2021).
18. Juberthie, C. France. In *Encyclopaedia Biospeologica, Tome 1*; Juberthie, C., Decu, V., Eds.; Société de Biospéologie: Moulis, France, 1994; pp. 665–692.
19. Mammola, S.; Cardoso, P.; Ribera, C.; Pavlek, M.; Isaia, M. A synthesis on cave-dwelling spiders in Europe. *J. Zoolog. Syst. Evol. Res.* **2018**, *56*, 301–316. [CrossRef]
20. Culver, D.C.; Pipan, T.; Gottstein, S. Hypotelminorheic—A unique freshwater habitat. *Subterr. Biol.* **2006**, *4*, 1–7.
21. Filhol, C.; Jeanbernat, E.; Timbal-Lagrave, E. *Exploration scientifique du Massif d'Arbas*; Imprimerie de Louis & Jean Matthieu Douladoure: Toulouse, France, 1874; Volume 2, pp. 367–477.
22. Jeannel, R.; Racovitza, E.G. Biospeologica XVI. Enumération des grottes visitées, 1908–1909 (Troisième série). *Arch. Zool. Exp. Gen.* **1910**, *5ème série, tome V*, 489–536.
23. Jeannel, R. Quarante années d'Explorations souterraines. *Notes Biospéologiques* **1950**, *6*, 1–94.

24. Jeannel, R.; Racovitza, E.G. Biospeologica XXIV. Enumération des grottes visitées, 1909–1911 (Quatrième série). *Arch. Zool. Exp. Gen.* **1912**, *5ème série, tome IX*, 501–667.

25. Jeannel, R.; Racovitza, E.G. Biospeologica LIV. Enumération des grottes visitées, 1918–1927 (Septième série). *Arch. Zool. Exp. Gen.* **1929**, *68*, 293–608.

26. Trombe, F. Recherches souterraines dans les Pyrénées centrales. *Ann. Spéléol.* **1947**, *2*, 67–164.

27. Bou, C. Hydrogéologie. In *La Coumo d'Hyouernedo—Réseau Félix Trombe—Henne Morte—Massif d'Arbas*; Duchêne, M., Drillat, P.A., Eds.; Groupe Spéléologique des Pyrénées: Toulouse, France, 1982; pp. 205–214.

28. Culver, D.C.; Sket, B. Hotspots of subterranean biodiversity in caves and wells. *J. Caves Karst Stud.* **2000**, *62*, 11–17.

29. Deharveng, L. Diversity in the Tropics. In *Encyclopedia of Caves*; Culver, D.C., White, W.D., Eds.; Elsevier: London, UK, 2005; pp. 166–170.

30. Deharveng, L.; Bedos, A. Chapter 18—Diversity Patterns in the Tropics. In *Encyclopedia of Caves*; White, W.B., Culver, D.C., Pipan, T., Eds.; Academic Press—Elsevier: Amsterdam, The Netherlands, 2019; pp. 146–162.

31. Gourbault, N. Recherches sur les Triclades Paludicoles hypogés. In *Mémoires du Muséum National d'Histoire Naturelle*; Éditions du Muséum: Paris, France, 1972; Volume 73, pp. 1–249.

32. Bertrand, A. Atlas préliminaire de répartition géographique des mollusques stygobies de la faune de France (Mollusca: Rissoidea: Caenogastropoda). *Doc. Malacol.* **2004**, *h.s. 2*, 1–81.

33. Bou, C. Recherches sur les eaux souterraines 16. Parasalentinella rouchi n. g., n. sp., des eaux souterraines des Pyrénées françaises (Amphipoda, Gammaridae). *Ann. Spéléol.* **1971**, *26*, 481–494.

34. Ginet, R. Stations de *Niphargus* pyrénéens. *Notes Biospéol.* **1956**, *11*, 17–22.

35. Balazuc, J. Les Amphipodes troglobies et phréatobies de la faune gallo-rhénane. *Arch. Zool. Exp. Gen.* **1954**, *91*, 153–193.

36. Henry, J.P.; Magniez, G. Observations sur un Aselle obscuricole de France: *Proasellus racovitzai* n. sp. (Crustacea Isopoda Asellota). *Int. J. Speleol.* **1972**, *4*, 171–188. [CrossRef]

37. Racovitza, E.G. Biospeologica LXX. Asellides (Première série). *Arch. Zool. Exp. Gen.* **1950**, *87*, 1–94.

38. Henry, J.P.; Magniez, G. Introduction pratique à la systématique des organismes des eaux continentales françaises. 4: Crustacés Isopodes (principalement Asellotes). *Bull. Soc. Linn. Lyon* **1983**, *52*, 319–357. [CrossRef]

39. Zacharda, M. New taxa of Rhagidiidae (Acari: Prostigmata) from Pyrenean caves. *Can. J. Zool.* **1987**, *65*, 2051–2056. [CrossRef]

40. Jeannel, R.; Racovitza, E.G. Biospeologica XXXIII. Enumération des grottes visitées 1911–1913 (cinquième série). *Arch. Zool. Exp. Gen.* **1914**, *53*, 325–558.

41. Dresco, E. Etude des *Leptoneta*. *Leptoneta convexa* Sim. et *microphtalma* Sim. (Araneae, Leptonetidae). *Bull. Soc. Hist. Nat. Toulouse* **1980**, *116*, 146–149.

42. Derkarabetian, S.; Starrett, J.; Tsurusaki, N.; Ubick, D.; Castillo, S.; Hedin, M. A stable phylogenomic classification of Travunioidea (Arachnida, Opiliones, Laniatores) based on sequence capture of ultraconserved elements. *Zookeys* **2018**, *760*, 1–36. [CrossRef]

43. Kury, A.B.; Mendes, A.C. Taxonomic status of the European genera of Travuniidae (Arachnida: Opiliones: Laniatores). *Munis Entomol. Zool.* **2007**, *2*, 1–14.

44. Simon, E. Biospeologica XXIII. Araneae et Opiliones (3ème série). *Arch. Zool. Exp. Gen.* **1911**, *49*, 177–206.

45. Roewer, C.F.R. Biospeologica LXII. Opiliones (fünfte série)—Zugleich eine Revision aller bisher bekannten europäischen Laniatores. *Arch. Zool. Exp. Gen.* **1935**, *78*, 1–96.

46. Juberthie, C.; Jauzion, I.; Gers, C. Faune de la grotte du Cap de Payssas. *Mém. Biospéol.* **1988**, *15*, 215.

47. Delfosse, E. Catalogue préliminaire des Opilions de France métropolitaine (Arachnida Opiliones). *Bull. Phyllie* **2004**, *20*, 34–52.

48. BRGM. Sauveterre-de-Comminges Gar-Cagire. In *Atlas des Potentialités Aquifères des Formations Pyrénéennes Projet Potapyr—BRGM/RP-66912-FR, (Technical Document)*; BRGM: Orléans, France, 2017; pp. 1–24.

49. Iorio, E. Catalogue biogéographique et taxonomique des chilopodes (Chilopoda) de France métropolitaine [Biogeographic and taxonomic catalogue of the centipedes (Chilopoda) of metropolitan France. *Mém. Soc. Linn. Bordx.* **2014**, *15*, 1–372.

50. Manfredi, P. Contributo alla conscenza dei Miriapodi cavernicoli della Francia (Diplopodi). In *Premier Congrès International de Spéléologie, Paris, France, 1953, Tome III*; CNRS: Paris, France, 1956; pp. 283–288.

51. Vandel, A. *Faune de France 64—Isopodes Terrestres (Première Partie)*; Lechevalier: Paris, France, 1960; pp. 1–416.

52. Deharveng, L.; Dalens, H.; Bedos, A.; Souqual, M.C. Les Isopodes Terrestres Endémiques de l'Europe de l'Ouest. 2012. Available online: http://endemica.mnhn.fr (accessed on 10 July 2021).

53. Séchet, E.; Noël, F. Catalogue commenté des Crustacés Isopodes terrestres de France métropolitaine (Crustacea, Isopoda, Oniscidea). *Mém. Soc. Linn. Bordeaux* **2015**, *16*, 1–156.

54. Christiansen, K.A.; Bouillon, M. An Evolutionary and Ecological Analysis of the Terrestrial Arthropods of Caves in the Central Pyrenees, Part One: Ecological Analysis with Special Reference to Collembola. *NSS Bull.* **1978**, *40*, 103–113.

55. Deharveng, L. Collemboles cavernicoles—VIII—Contribution à l'étude des Oncopoduridae. *Bull. Soc. Entomol. Fr.* **1988**, *92*, 133–147.

56. Yu, D.; Deharveng, L.; Lukic, M.; Wei, Y.; Hu, F.; Liu, M. Molecular phylogeny and trait evolution in an ancient terrestrial arthropod lineage: Systematic revision and implications for ecological divergence (Collembola, Tomocerinae). *Mol. Phylogenet. Evol.* **2021**, *154*, 106995. [CrossRef]

57. Lukic, M.; Houssin, C.; Deharveng, L. A new relictual and highly troglomorphic species of Tomoceridae (Collembola) from a deep Croatian cave. *ZooKeys* **2010**, *69*, 1–16.

58. Coiffait, H. Enumération des grottes visitées 1950–1957 (9ème série). Biospeologica LXXVII. *Arch. Zool. Exp. Gen.* **1959**, *97*, 209–465.
59. Jeannel, R.; Racovitza, E.G. Biospeologica XXXIX. Enumération des grottes visitées 1913–1917 (sixième série). *Arch. Zool. Exp. Gen.* **1918**, *57*, 203–470.
60. Jeannel, R. *Les Fossiles Vivants des Cavernes*; Gallimard: Paris, France, 1943; p. 321.
61. Jeannel, R. Le sous-genre *Hydraphaenops* Jeannel (Coleoptera Trechidae). *Notes Biospéol.* **1948**, *3*, 17–27.
62. Faille, A.; Tänzler, R.; Toussaint, E.F.A. On the way to speciation: Shedding light on the karstic phylogeography of the micro-endemic cave beetle *Aphaenops cerberus* in the Pyrenees. *J. Hered.* **2015**, *106*, 692–699. [PubMed]
63. Faille, A. Endémisme et Adaptation à la Vie Cavernicole Chez les Trechinae Pyrénéens (Coleoptera: Caraboidea). Approches Moléculaire et Morphométrique. Ph.D. Thesis, Muséum National d'Histoire Naturelle, Paris, France, 2006; pp. 1–318.
64. Faille, A.; Déliot, P.; Queinnec, E. A new cryptic species of *Aphaenops* (Coleoptera, Trechinae) from french Pyrenean cave: Congruence between morphometrical and geographical data confirm species isolation. *Ann. Soc. Entomol. Fr.* **2007**, *43*, 363–370. [CrossRef]
65. Deliot, P.; Delay, B. Variabilité biométrique et morphologique d'*Aphaenops bucephalus* (Coléoptère Trechinae). *Mém. Biospéol.* **1983**, *10*, 285–294.
66. Coiffait, H. Coléoptères troglobies pyrénéens nouveaux ou peu connus. *Ann. Spéléol.* **1969**, *24*, 557–561.
67. Santamaria, S.; Faille, A. *Rhachomyces* (Ascomycota, Laboulbeniales) parasites on cave-inhabiting Carabidae beetles from Pyrenees. *N. Hedwig.* **2007**, *85*, 159–186. [CrossRef]
68. Lepesme, P. Catalogue des Laboulbéniales de la collection François Picard. *Bull. Mus. Natl. Hist. Nat.* **1941**, *13*, 481–488.
69. Olivier, M.J.; Martin, D.; Bou, C.; Prié, V. Interprétation du suivi hydrobiologique de la faune stygobie réalisé sur le système karstique des Cent Fonts lors du pompage d'essai. In *Système karstique des Cent Fonts Simulation de Scénarios d'Exploitation et de Gestion de la Ressource, Rapport Final*; BRGM: Montpellier, France, 2006; pp. 235–265.
70. Faille, A.; Bourdeau, C.; Deharveng, L. Weak impact of tourism activities on biodiversity in a subterranean hotspot of endemism and its implications for the conservation of cave fauna. *Insect Conserv. Divers.* **2015**, *8*, 205–215. [CrossRef]
71. Deharveng, L.; Bedos, A. The cave fauna of Southeast Asia. Origin, evolution and ecology. In *Ecosystems of the World 30. Subterranean Ecosystems*; Wilkens, H., Culver, D.C., Humphreys, W.F., Eds.; Elsevier: Amsterdam, The Netherlands, 2000; pp. 603–632.
72. Matjasic, J. Une nouvelle *Choanophrya* (Ciliata Suctoria) sur *Stenasellus virei*. *Ann. Spéléol.* **1963**, *18*, 267–270.
73. Sendra, A.; Antić, D.; Barranco, P.; Borko, Š.; Christian, E.; Delić, T.; Fadrique, F.; Faille, A.; Galli, L.; Gasparo, F.; et al. Flourishing in subterranean ecosystems: Euro-Mediterranean Plusiocampinae and Tachycampolds (Diplura, Campodeidae). *Eur. J. Taxon.* **2020**, *591*, 1–138. [CrossRef]

Communication

Biodiversity of the Huautla Cave System, Oaxaca, Mexico

Oscar F. Francke, Rodrigo Monjaraz-Ruedas † and Jesús A. Cruz-López *,‡

Colección Nacional De Arácnidos, Departamento de Zoología, Instituto de Biología, Universidad Nacional Autónoma de México, Ciudad Universitaria, Coyoacán, Mexico City C. P. 04510, Mexico; offb@ib.unam.mx (O.F.F.); roy_monrue@hotmail.com (R.M.-R.)

* Correspondence: thelyphonidito@gmail.com
† Current address: San Diego State University, San Diego, CA 92182, USA.
‡ Current address: Instituto Nacional de Investigaciones Agrícolas y Pecuarias del Valle de Oaxaca, Santo Domingo Barrio Bajo, Etla C. P. 68200, Mexico.

Abstract: Sistema Huautla is the deepest cave system in the Americas at 1560 m and the fifth longest in Mexico at 89,000 m, and it is a mostly vertical network of interconnected passages. The surface landscape is rugged, ranging from 3500 to 2500 masl, intersected by streams and deep gorges. There are numerous dolinas, from hundreds to tens of meters in width and depth. The weather is basically temperate subhumid with summer rains. The average yearly rainfall is approximately 2500 mm, with a monthly average of 35 mm for the driest times of the year and up to 500 mm for the wettest month. All these conditions play an important role for achieving the highest terrestrial troglobite diversity in Mexico, containing a total of 35 species, of which 27 are possible troglobites (16 described), including numerous arachnids, millipedes, springtails, silverfish, and a single described species of beetles. With those numbers, Sistema Huautla is one of the richest cave systems in the world.

Keywords: troglobitics; arachnids; insects; millipedes

Citation: Francke, O.F.; Monjaraz-Ruedas, R.; Cruz-López, J.A. Biodiversity of the Huautla Cave System, Oaxaca, Mexico. *Diversity* **2021**, *13*, 429. https://doi.org/10.3390/d13090429

Academic Editors: Tanja Pipan, David C. Culver and Louis Deharveng

Received: 29 June 2021
Accepted: 25 August 2021
Published: 6 September 2021

Publisher's Note: MDPI stays neutral with regard to jurisdictional claims in published maps and institutional affiliations.

1. Introduction

Caves are some of the most adverse environments on earth, as the restricted access to food and the extreme conditions of darkness and humidity make these habitats very challenging for living organisms [1]. Despite this, many taxa have colonized these subterranean environments, including arthropods and vertebrates [2,3]. Some of these animals have become fully established in cave systems; therefore, they have evolved specific morphological adaptations (troglomorphisms) such as cuticular depigmentation, elongation of appendages, and reduction or loss of eyes [4]. Troglobitic animals (those who exhibit troglomorphisms) are excellent models for studies of the evolution, e.g., cave adaptations such as morphological convergences among distant lineages [5–8].

Caves are extremely important in terms of diversity, due to the number of endemic species inhabiting these environments [2]. For example, in the class Arachnida, troglobites are known for 9 of the 11 extant orders; only Thelyphonida (vinegaroons) and Solifugae (camel spiders or wind scorpions) lack cave representatives [9], and the Huautla System has a fair representation of all other Arachnida groups.

Cave explorations in the Systema Huautla have taken place since the mid-1960s, and multiple discoveries in terms of speleology but also in biological diversity have been reported [1–8]. However, this exploration is yet to be completed as new expeditions recently have concluded in extending the length of the system (see http://www.mexicancaves.org/maps/0104 (accessed on 27 August 2021)) and have also increased the number of new species inhabiting the system [1,2,8,9].

As part of the Special Issue "Hotspots of Subterranean Biodiversity" in the Diversity journal, we have decided to put together a list of the numerous taxa inhabiting one of the most important cave systems in Mexico revisiting all of the available literature. This list exemplifies the large biological diversity of the system, and furthermore, this resource can be used as reference for future works not only on diversity but evolution, conservation, and inspiration for cave exploration in the several unexplored cave systems in Mexico.

2. Sistema Huautla

This is the deepest cave system in the Americas at 1560 m, and the fifth longest in Mexico at 89,000 m, and it is a mostly vertical network of interconnected passages, a few of which can be traversed by humans, but there are undoubtedly many narrow cracks and crevices inaccessible to cavers. Located at Huautla de Jiménez, on the Sierra Mazateca in the northwestern part of the State of Oaxaca, the caves have been explored by skilled speleologists since 1966, with international and local support. The topography of the cave is formed by multiple pits, stretch passages, and crevices (See http://www.mexicancaves. org/maps/0104 (accessed on 27 August 2021).

The Sierra Mazateca is basically a massive Cretaceous karst formation, up to 5000 m thick in some areas, overlain by older intrusive volcanic rocks. The landscape is rugged, ranging from 3500 to 2500 masl, intersected by streams and deep gorges. There are numerous dolinas, from hundreds to tens of meters in width and depth. Rural land in Mexico is mostly owned by the community living there (called "ejidos"), and permission is required from the local authorities before any exploration.

The weather is basically temperate subhumid, with a marked dry season from March to May. The average yearly rainfall is approximately 2500 mm, with a monthly average of 35 mm for the driest times of the years and up to 500 mm for the wettest month. The average monthly temperature is 23.6 °C, and the minimum average monthly temperature is 9.4 °C for the coldest month of the year and 13 °C for the warmest month, whereas the monthly daily average is of 14 °C for the coldest month and 29 °C for the warmest month [10].

The major caves are each associated with one of the larger dolinas, and the walls of the dolinas also have side entrances and pits at varying depths. There are at least 28 entrances and 6 major caves. The system continues to be pushed to new depths and lengths on a yearly basis. The six major caves in the system are: Sótano de San Agustín, Sótano del Río Iglesia, La Grieta, Sótano de Agua de Carrizo, Li Nita, and Nita Nanta (Figure 1). The Huautla system contains 50 species, with possibly 27 or more troglobites total (16 described), including arachnids, millipedes, springtails, silverfish, and a beetle [11–14]. In Li Nita, a new troglobitic scorpion of the genus *Typhlochactas* Mitchel, 1971 was collected in 2014. A small colony of vampire bats roosts not far inside one of the three entrances to Sótano del Río Iglesia and another unidentified bat colony roost just inside Sótano de San Agustín. In its deepest parts (~1000 m), Huautla system contains long, deep sumps, which makes the exploration and collecting even harder, as diving experience is needed. Three caves were impacted by garbage dumping, including medical waste with syringes, but the dumping has ceased, and the PESH cavers (Proyecto Espeleológico Sistema Huautla) have begun a clean-up effort [11–14].

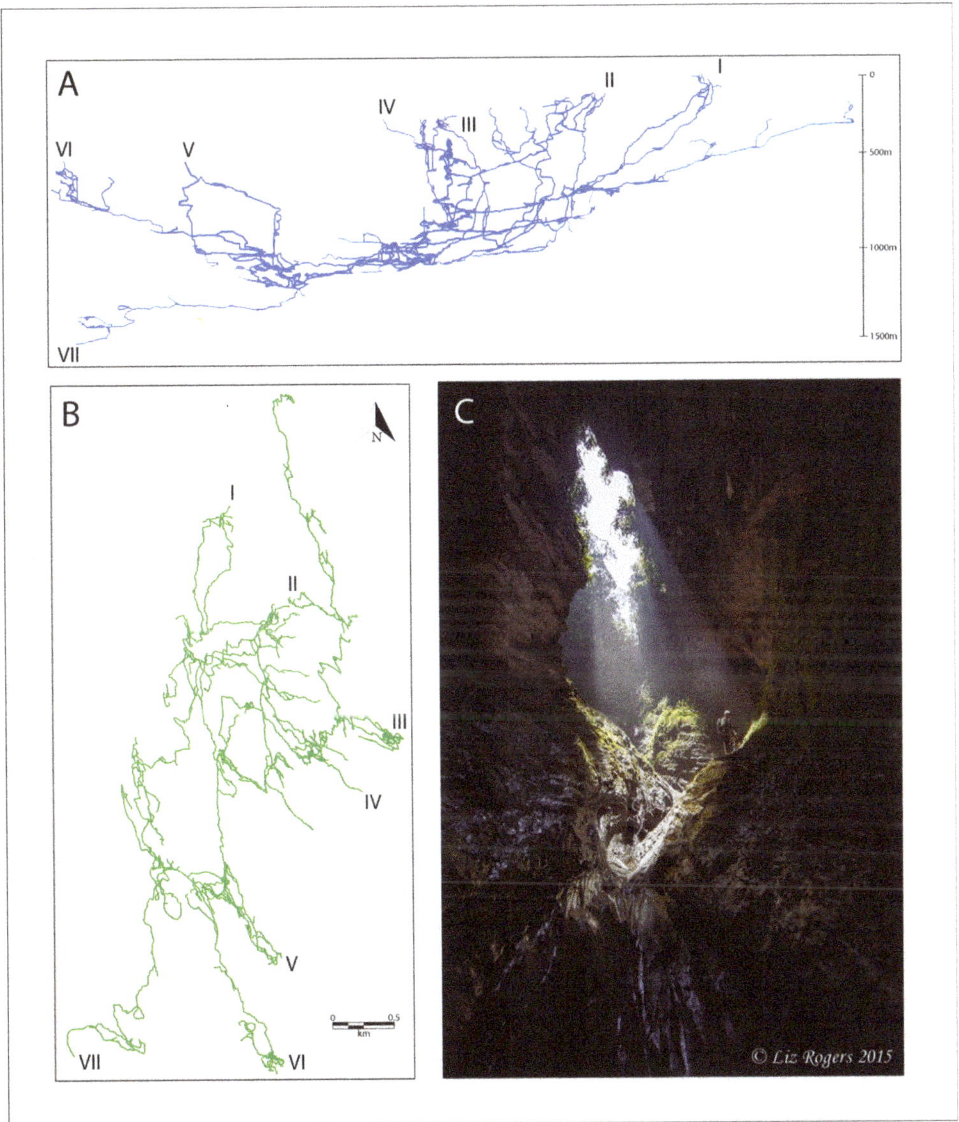

Figure 1. Sistema Huautla maps. (**A**) profile view, (**B**) plan view, (**C**) Sótano de San Agustín entrance. I, Nita Nanta entrance. II, Li Nita entrance. III, La Grieta entrance. IV, Sótano de Carrizo entrance. V, Sótano de San Agustín entrance. VI, Río Iglesia entrance. VII, Sump 9. Cave maps provided by William Steele. Photo on Figure C taken by Liz Rogers, 2015.

3. Biological Diversity

Although troglobionts have been collected in the Huautla Cave System since the beginning explorations, it has only been in the past eight years that real collecting efforts have been conducted by a group of arachnologists from the National Arachnid Collection at Universidad Nacional Autónoma de México (CNAN, UNAM). We were kindly invited by the organizers of the PESH expeditions on a yearly basis, exploring over 40 caves in the area. The PESH Expeditions are organized in the months of April–May of each year, during the dry season to minimize the risk of being trapped underground by flooding;

the entire expedition lasts one month, with about 100 cavers from around the world participating. We usually participate during the third week of the expedition, when the speleologists have already rigged several of the deeper caves; therefore, we can concentrate our efforts strictly on collecting with minimal time spent securing ropes. Usually, three to four members go into a cave with the support of several speleologists, and two or three remain collecting on the surface so that we can understand better the origin and evolutions of the troglobionts inside a given cave. Some species, such as the highly troglomorphic scorpion *Alacran tartarus* Francke, 1982, have been collected as shallow as −60 m and as deep as −920 m in several different caves; therefore, presumably, it moves into the deeper, common passage region of the system and can then disperse upwards through the numerous shafts. However, others, such as the troglobitic tarantula *Hemirrhagus grieta* (Gertsch, 1982), are confined to the middle depths (from −300 to −600 m) of a single cave (Cueva de La Grieta).

It is remarkable that no stygobionts have been collected in the system, even though there have been numerous diving expeditions to explore the deepest sumps, from both ends, i.e., from inside the cave and from the resurgence 7 km away. There are several reports, photographs and at least one video showing *Alacran* walking/running underwater; and the only other living animals reported at those depths are tadpoles (immature anurans), undoubtedly carried down during the flash floods of the rainy season. The tips of the pedipalp fingers in *Alacran* have sharp elongated hooks, which presumably can be an adaptation to catch those slippery prey items underwater, although nobody has seen a scorpion feeding on a tadpole underwater (or anywhere else).

The Arachnid collection in Mexico City has 19 species of the 27 known troglobionts in the Huautla Cave System, earning it one of the top ranks as the most diverse explored underground biotas in the world [15,16]. Sistema Huautla is the third-most diverse cave in Mexico, after Cueva de la Mina, Tamaulipas (24 troglobionts out of 60 species), and Sistema Purificación, also in Tamaulipas (19 troglobionts out of 103 species), followed in fourth place by Sistema de los Sabinos, San Luis Potosi, with 14 troglobionts out of 127 species (William Elliott, pers. comm.). A detailed list of the troglobionts in Sistema Huautla is given in Table 1. Underground collecting during the rainy season may increase the number of recorded troglobionts for all cave systems, as many terrestrial arthropods are known to overwinter in inactive stages during the drier and colder winter months, such as mites and ticks (Acari); however, due to the risk of flooding, this task is basically impossible to achieve.

It is important to emphasize that in the Huautla Cave System, seven different species of troglomorphic tarantulas of the genus *Hemirrhagus* Simon, 1903 have been collected thus far (several still undescribed): two of them from the same cave (La Grieta), though they are found at different depths and have not been collected together. The other five species are from separate entrances into the system and do not appear to be closely related, presumably representing independent invasions from epigean ancestor(s).

At least two species of bats (Mammalia: Chiroptera) have been photographed from caves in the Sistema Huautla; one is a leaf-nosed bat belonging to the family Phyllostom-atidae. We have not been able to collect in the guano piles associated with those roosting areas, and it is likely there might be some additional troglobionts there.

4. Class Arachnida

4.1. Order Araneae

Spiders are among the most common arachnids into caves. According to available estimates, around 1000 troglomorphic spiders worldwide have been classified as troglo-bionts [9,17,18]. They are located from the entrance to the deepest galleries of the caves, although not all the species found in the caves are exclusive to the underground environ-ments. Around 25–30% are accidental (trogloxenes) and appear in the entrance area; about 50% are regularly found in caves but also in the epigean environment (troglophiles); and between 20–25% are strictly cave inhabitants (troglobitic) [19,20]. In the Huautla Cave

System seven families, eleven genera and thirteen species of spiders have been recorded. The most diverse spider families are Pholcidae and Theraphosidae. In Mexico, pholcids spiders are the most common spider family found in caves, mainly in the entrances and median-depth zone of caves and grottoes. Huber (2018) [21] reported about 86 species of troglomorphic pholcid species worldwide, including 21 eyeless species and 21 species with strongly reduced eyes. Most troglomorphic pholcids spiders are representatives of only two genera: *Anopsicus* Chamberlin and Ivie, 1938 and *Metagonia* Simon, 1893 [21] with *Metagonia* represented in the Huautla Cave System (Table 1). Mexico is by far the richest country in terms of troglomorphic pholcids. This apparent dominance may partly be due to collectors' and taxonomists' biases [21], mainly towards the northern cave systems (e.g., Purification and Cuetzalan Cave Systems). Mygalomorph spiders are generally poorly represented in the cave faunas of the world. Mexico holds the second highest diversity of species of tarantulas worldwide behind Brazil, and Mexico is the richest country in tarantula cave species. The genus *Hemirrhagus* is endemic to Mexico and has 27 described species, with epigean, troglophile, and troglobitic species [22]. In the Huautla Cave System, *Hemirrhagus grieta* Gertsch, 1982 and *H. billsteelei* Mendoza and Francke, 2018 have been described; however, an additional five probably new species have been recently collected (Table 1; Figures 2A,D and 3D). Although important contributions about the spider fauna of the cave system have been published in the last five years, the Huautla Cave System is still poorly known and explored with respect to their spider fauna diversity.

4.2. Order Opiliones

In Mexico, a total of 265 species of harvestmen (Opiliones) has been recorded, representing the four extant suborders ([23] and Cruz-López, pers. obsv.). The Mexican fauna of Opiliones is remarkable because about 20% are troglobites or troglophiles, and estimates indicate that there is still a high percentage of undescribed species for the country, especially those that inhabit different cave systems [23].

Until 2003, only two species of Stygnopsidae were reported for the Huautla system; *Hoplobunus mexicanus* (Roewer, 1915) and *Karos gratiosus* Goodnight and Goodnight, 1971 [24]. However, recent taxonomic revisions of several genera of Stygnopsidae have revealed that the original taxonomic determinations made by Goodnight and Goodnight during 1953–1973 are erroneous [25,26]. Currently, *H. mexicanus* is in fact an undescribed species of *Stygnopsis* Sørensen, 1932 (Figure 2C), and all previous records published of *K. gratiosus* now correspond to *Huasteca kardia* Cruz-López and Francke, 2019. Additionally, Cruz-López et al. (2019) [27] described the endemic genus *Minisge* Cruz-López, Monjaraz-Ruedas and Francke, 2019, which includes two highly troglomorphic species: the shallower *Minisge sagai* Cruz-López, Monjaraz-Ruedas and Francke, 2019 (Figure 3F) and the deeper *Minisge kanoni* Cruz-López, Monjaraz-Ruedas and Francke, 2019. It is hypothesized [27] that both species of *Minisge* colonized the Huautla System independently, with *M. kanoni* being the oldest one, colonizing the caves about 3.3 Mya ago, whereas the colonization by *M. sagai* occurred very recently. It is remarkable that this species inhabiting the shallow regions of the cave system does not present a large genetic diversification or structure, among the populations in the many caves it has been found.

Figure 2. Troglobiont fauna of Sistema Huautla. (**A**) *Hemirrhagus* sp., (**B**) *Paraphrynus grubbsi* Cokendolpher and Sissom, 2001, (**C**) *Stygnopsis* sp., (**D**) *Hemirrhagus billsteelei* Mendoza and Francke, 2018, (**E**) *Rhachodesmus digitatus* Causey, 1971. Photos by Jean Krejca, Figure (**D**) taken and modified from Mendoza and Francke, 2018.

Figure 3. Troglobiont fauna of Sistema Huautla. (**A**) *Typhlochactas* sp., (**B**) *Sphaeriodesmus iglesia* Shear, 1986, (**C**) *Typhlochactas* sp. collecting method, (**D**) *Hemirrhagus grieta* Gertsch, 1982, (**E**) *Alacran tartarus* Francke, 1982, (**F**) *Minisge sagai* Cruz-López et al. 2019. Photos by Jean Krejca and CNAN-IBUNAM historic record.

Another remarkable species inhabiting the Huautla System is an undescribed species of an undescribed genus of the subfamily Gagrellinae (Eupnoi: Sclerosomatidae), which seems to be closely related to *Parageaya* Mello-Leitão, 1933, but currently, only considered as a member of Gagrellinae with troglomorphic features such as body pale color. Other record of troglomorphic species in the system is an undescribed *Stygnomma* Roewer, 1912. However, since the revision of Stygnommatidae by Pérez-González (2009) [28], *Stygnomma* was recognized as a polyphyletic assemblage, with the fauna of southern Mexico belonging in fact to members to the family Biantidae, but, unfortunately, a taxonomic act has not yet been formalized for this group.

Finally, near of the entrances of several caves, we have found specimens of an undescribed species of *Isaeus* Sørensen, 1932, but this species does not exhibit any troglomorphic features.

4.3. Order Amblypygi

Whip spiders are conspicuous arachnids that are very common in caves and most of the times very easy to spot by people visiting caves. The case of the Sistema Huautla is not different, and there are at least two different species reported in several caves of the system belonging to the genus *Paraphrynus* Moreno, 1940. The only species described as *Paraphrynus grubbsi* Cokendolpher and Sissom, 2001 (Figure 2B) was reported for several caves including Nita Lajao, Sótano de San Agustín, Cueva del Escorpión, etc., suggesting that this species is widely distributed in the system and can probably move from cave to cave. Additionally, Cokendolpher and Sissom (2001) [29] reported an undescribed species of *Paraphrynus* which was considerably smaller and morphologically distinct from *P. grubbsi* which has well-developed eyes, suggesting that it could be an epigean species not adapted to caves which lives in the surroundings of Sistema Huautla and can be found occasionally in the caves [29].

4.4. Order Pseudoscorpiones

A few small pseudoscorpions have been collected in the Huautla Cave System [13], but they have not been fully identified yet.

4.5. Order Schizomida

Short-tailed whip scorpions are represented in the cave system by a single species *Baalrog magico* (Monjaraz-Ruedas and Francke, 2018), which was originally described under the genus *Stenochrus* Chamberlin, 1922 and then transferred to the newly described genus *Baalrog* Monjaraz-Ruedas, Prendini and Francke, 2019, which harbors species restricted to the cave environment. Other members of genus *Baalrog* are distributed in cave systems in Valle Nacional, Oaxaca, and Atoyac, Veracruz [30]. Although there are other undescribed species of schizomids in the Sierra Mazateca outside the caves, those belong to different genera and are apparently not related to the species distributed inside Sistema Huautla [31].

4.6. Order Scorpiones

The Huautla Cave System is unusual among Mexican caves as it is the only one that harbors two species of troglobiont scorpions, both belonging to the family Typhlochactidae. *Typhlochactas* n. sp. is a small species about 2 cm long, totally eyeless and unpigmented, known from a single specimen collected about −100 m depth in one of the major caves (Figure 3A,C). The genus *Typhlochactas* has eight described troglomorphic species and a couple of new ones waiting to be described. This is the first record for the Huautla System [32].

The second one is *Alacran tartarus* (Figure 3E), which is about 8 cm long and is known from more than a dozen specimens collected in several caves in the system. It is a tan brown in color and is also completely eyeless. Two other species are known in the genus: one from the state of Puebla from a cave system that was presumably at one time interconnected to Systema Huautla and the other one from Oaxaca about 50 km away, both known from single collection events, as some caves in Mexico are not visited very frequently [33].

5. Class Diplopoda

Five endemic troglobitic species of millipedes belonging to four genera, four families, and four separate orders have been collected inside the Huautla Cave System. Perhaps, the most spectacular is a large yellow and bluish-green rhachodesmid which is quite abundant in the aptly named Millipede Cave (Figures 2E and 3B).

6. Class Insecta

There are five orders of insects reported from the Huautla Cave System, none of which are very abundant. There are two species of ground-dwelling beetles, two species of springtails, and one of each cave cricket, silverfish and dipluran (Table 1). Videos taken inside several of the caves in the system show flying insects glowing from the lights of the cavers, but as far as we know, they have not been collected and identified.

Table 1. A detailed list of the troglobionts in Sistema Huautla is given. Abbreviations: TB, troglobiont; SB, stygobiont; TP, troglophile.

Class	Order	Family	TB or SB	TP	Genus	Species	Authors	Source	Endemic
Arachnida	Amblypygi	Phrynidae	1		*Paraphrynus*	*grubbsi*	Cokendolpher and Sissom, 2001	[29]	yes
	Amblypygi	Phrynidae		1	*Paraphrynus*	sp. nov.		[29]	yes
	Araneae	Ctenidae		?	*Ctenus*	sp.		[15]	yes?
	Araneae	Dipluridae		?	Undet.	undet.		[18]	yes?
	Araneae	Euctenizidae		?	*Aptostychus*	*sabinae*	Valdez-Mondragón and Cortes-Roldan, 2016	[34]	yes?
	Araneae	Mysmenidae		?	*Maymena*	sp.		[15]	yes?
	Araneae	Nesticidae		1	*Eidmannella*	*pallida*	(Emerton, 1875)	[15]	?
	Araneae	Nesticidae		1	*Gaucelmus*	*calidus*	(Gertsch, 1971)	[15]	yes?
	Araneae	Pholcidae		?	*Ixchela*	*panchovillai*	Valdez-Mondragón, 2020	[20]	yes
	Araneae	Pholcidae		?	*Modisimus*	sp. nov.		[15]	yes?
	Araneae	Pholcidae		?	*Metagonia*	sp. nov		[15]	yes?
	Araneae	Pholcidae		?	*Pholcophora*	sp. nov.		[15]	yes?
	Araneae	Theraphosidae	1		*Hemirrhagus*	*grieta*	Gertsch, 1982	[22]	yes
	Araneae	Theraphosidae	1		*Hemirrhagus*	*billsteelei*	Mendoza and Francke, 2018	[22]	yes
	Araneae	Theraphosidae	1		*Hemirrhagus*	sp. nov.	(Basketball Cave)	[22]	yes
	Araneae	Theraphosidae	1		*Hemirrhagus*	sp. nov.	(Church Cave)	[22]	yes
	Araneae	Theraphosidae	1		*Hemirrhagus*	sp. nov.	(Thirty Skeleton Cave)	[22]	yes
	Araneae	Theraphosidae	1		*Hemirrhagus*	sp. nov.	(Li Nita Cave)	[22]	yes
	Araneae	Theraphosidae	1		*Hemirrhagus*	sp. nov.	(Nita Nanta Cave)	[22]	yes
	Opiliones	Sclerosomatidae/ Gagrellinae	1		Gen. nov.	sp. nov.	(aff. *Parageaya*)	[26]	yes
	Opiliones	Stygnommatidae	1		*Stygnomma*	sp. nov.	new	[26]	yes
	Opiliones	Stygnopsidae	1		*Huasteca*	*kardia*	Cruz-Lopez and Francke, 2019	[26]	yes
	Opiliones	Stygnopsidae	1		*Minisge*	*kanoni*	Cruz-Lopez et al., 2019	[27]	yes
	Opiliones	Stygnopsidae	1		*Minisge*	*sagai*	Cruz-Lopez et al., 2019	[27]	yes
	Opiliones	Stygnopsidae	1		*Stygnopsis*	sp. nov.		[26]	yes
	Pseudoscorpiones	Chernetidae?	1		Undet.	undet.		[13]	yes
	Schizomida	Hubbardiidae	1		*Baalrog*	*magico*	(Monjaraz-Ruedas and Francke, 2018)	[31]	yes
	Scorpiones	Typhlochactidae	1		*Alacran*	*tartarus*	Francke, 1982	[33]	yes
	Scorpiones	Typhlochactidae	1		*Typhlochactas*	sp. nov	(Li Nita Cave)	[13]	yes
Diplopoda	Chordoneumatida	Cleidogonidae	1		*Cleidogona*	*baroqua*	Shear, 1982	[35]	yes
	Polydesmida	Sphaeriodesmidae	1		*Sphaeriodesmus*	*grubbsi*	Shear, 1986	[35]	yes
	Polydesmida	Sphaeriodesmidae	1		*Sphaeriodesmus*	*iglesia*	Shear, 1986	[35]	yes
	Rhachodesmidae	Rhachodesmidae	1		*Rhachodesmus*	*digitatus*	Causey, 1971	[35]	yes
	Spirostrepsida	Cambalidae	1		*Mexicambala*	*fishi*	Causey, 1971	[36]	yes
Insecta	Coleoptera	Carabidae		1	*Mexisphodrus*	*urquijoi*	(Hendrichs et al., 1978)	[36]	yes?
	Coleoptera	Staphylinidae		1	*Belonuchus*	sp.		[36]	yes?
	Collembola	Entomobryidae	1		*Pseudosinella*	*bonita*	Christiansen, 1973	[36]	yes
	Collembola	Entomobryidae	1		*Pseudosinella*	*huautla*	Christiansen, 1982	[37]	yes

Table 1. *Cont.*

Class	Order	Family	TB or SB	TP	Genus	Species	Authors	Source	Endemic
	Diplura	Campodeidae		1	Undet.	sp.		[11]	yes?
	Zygentoma	Nicoletiidae	1		*Anelpistina*	*specusprofundi*	Espinasa and Fisher, 2006	[38]	yes
	Orthoptera	Rhaphidophoridae		1	Undet.	undet.		video PESH 2019	yes?
Crustacea	Isopoda	Trichoniscidae		1	Undet.	undet.		Reddell, MexBio Files	yes
	Isopoda	Armadillidae		1	Undet.	undet.		Bill Steele, per. Com.	yes
			27	8					
			35 total						

: ? = there is no enough data about the biological ecology of the species. Yes? = there is no enough data about geographical distribution.

7. Class Crustacea

As mentioned earlier, no stygobionts have been observed in the Huautla Cave System, and there are two unidentified species of rolly-pollies (Isopoda) found in the cave, neither one with troglomorphic features.

8. Conservation

As we have shown, caves are important harbors of biodiversity and should be considered as habitats of high priority for conservation due the high number of endemic species inhabiting them. Their importance for evolutionary studies makes them natural laboratories for the study of evolution. Caves are important water reservoirs, which help maintain stability on the entire environments, as they store and provide water during the dry seasons. In recent years, we have noticed an increase in cave conservation activities in three areas. First, several cave entrances were used to dump garbage by the local residents; however, that practice has been actively discouraged by members of the PESH Expeditions, and in many cases, it has ceased altogether. Second, these same expedition participants have played a significant role in cleaning up some of those cave entrances that had been severely affected. Finally, some of the cave entrances are now protected by locked fences, and access is only allowed with the proper authorization from the local authorities. None of the cave entrances are commercially exploited, and very few speleologists visit the caves, as they are technically challenging and remote. Cave explorations as well as research should emphasize the importance of conservation, combating the myths behind caves as they are highly perpetuated in the communities as places of evil, which only results in the destruction of these important environments. Finally, an extensive multidisciplinary effort is necessary to incorporate Huautla System into conservation agendas, in order to propose the system as a protected area given the biological relevance mentioned above. In addition, one of the most important points for the conservation of cave species is the wide gap of knowledge that exists on the species inhabiting these systems, since generally only very few observations during explorations were reported from Huautla System.

Author Contributions: The three authors participated equally during the field work and speleological expeditions, curation of the material, and in writing the manuscript. All authors have read and agreed to the published version of the manuscript.

Funding: This research received no external funding.

Institutional Review Board Statement: Not applicable.

Informed Consent Statement: Not applicable.

Data Availability Statement: Not applicable.

Acknowledgments: We are thankful to W. Steele, T. Shifflet, M. Minton, I. Romms, and S. Davlantes for assistance in entering the caves: to J. Mendoza, G. Contreras. D. Barrales, and A. Guzman for collecting in the caves; J. Krejka provided some photos and collected with us; A. Valdez assisted with the identification of spiders; and W. Elliott, G. Roewer, J. Reddell, and C. Jewel shared with us information on the cave inhabitants. Finally, we express our gratitude to T. Pipan for inviting us to contribute to this Special Issue on subterranean biodiversity.

Conflicts of Interest: Authors declare no conflict of interest.

References

1. Hüppop, K. Adaptation to Low Food. In *Encyclopedia of Caves*, 2nd ed.; White, W.B., Culver, D.C., Eds.; Academic Press: London, UK, 2012; pp. 1–9.
2. Howarth, F.G.; Hoch, H. Adaptative shifts. In *Encyclopedia of Caves*, 2nd ed.; White, W.B., Culver, D.C., Eds.; Academic Press: London, UK, 2012; pp. 9–17.
3. Culver, D.C. Karst environment. *Zeits. Geomorph* **2016**, *60*, 103–117. [CrossRef]
4. Trajano, E.; Cobolli, M. Evolution of lineages. In *Encyclopedia of Caves*, 2nd ed.; White, W.B., Culver, D.C., Eds.; Academic Press: London, UK, 2012; pp. 295–304.
5. Culver, D.C.; Kane, T.C.; Fong, D.W. *Adaptations and Natural Selection in Caves: The Evolution of Gammarus Minus*, 1st ed.; Harvard University Press: Cambridge, MA, USA, 1995; p. 223.
6. Sbordoni, V.; Allegrucci, C.; Cesaroni, D. Population genetic structure, speciation and evolutionary rates in cave-dwelling organisms. In *Ecosystems of the World*, 1st ed.; Culver, D.C., Humphreys, W.F., Eds.; Elsevier: San Diego, CA, USA, 2000; pp. 450–483.
7. Jeffery, W.R. Adaptive evolution of eye degeneration in the Mexican blind cavefish. *J. Heredity* **2005**, *96*, 185–196. [CrossRef]
8. Cruz-López, J.A.; Proud, D.; Pérez-González, A. When troglomorphism dupes taxonomists: Morphology and molecules reveal the first pyramidopid harvestman (Arachnida, Opiliones, Pyramidopidae) from the New World. *Zool. J. Linn. Soc.* **2016**, *177*, 602–620. [CrossRef]
9. Ribera, C. Arachnida: Araneae (Spiders). In *Encyclopaedia of Caves and Karst Sciences*, 1st ed.; Gunn, J., Ed.; Taylor and Francis: London, UK, 2004; pp. 71–73.
10. CONAGUA. *Estadísticas del Agua en México*; Comisión Nacional del Agua: Mexico City, Mexico, 2016.
11. Palacios-Vargas, J.G.; Reddell, J.R. Actualización del inventario cavernícola (estigobiontes, estigófilos y troglobios) de México. *Mundos Subterr.* **2013**, *24*, 33–95.
12. Mendoza-Marroquín, J.I. Taxonomic revision of Hemirrhagus Simon, 1903 (Araneae: Theraphosidae, Theraphosinae), with description of five new species from Mexico. *Zool. J. Linn. Soc.* **2014**, *170*, 634–689. [CrossRef]
13. Krejca, J. Biological Exploration during the 2015 Proyecto Espeleológico Sistema Huautla Expedition. *Nat. Speleol. Soc. News.* **2015**, *73*, 4–28.
14. Steele, C.W.; Shifflett, T.E. Huautla cave system (Sistema Huautla), Mexico. In *Encyclopedia of Caves*, 2nd ed.; White, W.B., Culver, D.C., Eds.; Academic Press: London, UK, 2012; pp. 527–536.
15. Deharveng, L.; Bedos, A. Biodiversity in the tropics. In *Encyclopedia of Caves*, 3rd ed.; White, W.B., Culver, D.C., Pipan, T., Eds.; Academic Press: London, UK, 2019; pp. 146–162.
16. Ozimec, R.; Lučić, I. The Vjetrenica cave (Bosnia & Herzegovina)—One of the World's most prominent biodiversity hotspots for cave-dwelling fauna. *Subterr. Biol.* **2010**, *7*, 17–23.
17. Reddell, J.R. Spiders and related groups. In *Encyclopedia of Caves*, 2nd ed.; Culver, D.C., White, W.B., Eds.; Elsevier: Amsterdam, The Netherlands, 2005; pp. 786–797.
18. Mammola, S.; Isaia, M. Spiders in caves. *Proc. Roy. Soc. Lond. B* **2017**, *284*, 20170193. [CrossRef]
19. Gunn, J. *Encyclopedia of Caves and Karst Science*, 1st ed.; Taylor & Francis: New York, NY, USA, 2004; p. 1940.
20. Valdez-Mondragón, A. COI mtDNA barcoding and morphology for species delimitation in the spider genus Ixchela Huber (Araneae, Pholcidae), with the description of two new species from Mexico. *Zootaxa* **2020**, *4747*, 54–76. [CrossRef]
21. Huber, B.A. Cave-dwelling pholcid spiders (Araneae, Pholcidae): A review. *Subterr. Biol.* **2018**, *26*, 1–18. [CrossRef]
22. Mendoza-Marroquin, J.; Francke, O.F. Five new cave-dwelling species of Hemirrhagus Simon 1903 (Araneae: Theraphosidae, Theraphosinae), with notes on the generic biogeography and novel morphological features. *Zootaxa* **2018**, *4407*, 451–482. [CrossRef]
23. Kury, A.B.; Cokendolpher, J.C. Opiliones. In *Biodiversidad, Taxonomía y Biogeografía de Artrópodos de México: Hacia una Síntesis de su Conocimiento*; Llorente-Bousquets, J., González-Soriano, E., Papavero, N., Eds.; CONABIO: Mexico City, Mexico, 2000; Volume II, pp. 137–157.
24. Kury, A.B. Annotated catalogue of the Laniatores of the New World (Arachnida, Opiliones). *Rev. Iber. Aracno.* **2003**, *7*, 337.
25. Cruz-López, J.A.; Francke, O.F. Total evidence phylogeny of the North American harvestman family Stygnopsidae (Opiliones: Laniatores: Grassatores) reveals hidden diversity. *Invert. Syst.* **2017**, *31*, 317–360. [CrossRef]
26. Cruz-López, J.A.; Francke, O.F. New species of the cave-dwelling genus Huasteca (Opiliones: Stygnopsidae: Karosinae), from Northern Oaxaca, Mexico, with a SEM survey of the sexually dimorphic areas on legs and structures related to chelicera in the genus. *J. Nat. Hist.* **2019**, *53*, 1451–1464. [CrossRef]

27. Cruz-López, J.A.; Monjaraz-Ruedas, R.; Francke, O.F. Turning to the dark side: Evolutionary history and molecular species delimitation of a troglomorphic lineage of armoured harvestman (Opiliones: Stygnopsidae). *Arthro. Syst. Phyl.* **2019**, *77*, 285–302.
28. Pérez-González, A. Revisão sistemática e análise filogenética de Stygnommmatidae (Arachnida: Opiliones: Laniatores). Ph.D. Thesis, Universidade Federal do Rio de Janeiro, Museu Nacional, Rio de Janeiro, Brazil, 2006; p. 308.
29. Cokendolpher, J.C.; Sissom, W.D. A new troglobitis Paraphrynus from Oaxaca, Mexico (Amblypygi, Phrynidae). *Texas Mem. Mus. Speleo. Monog.* **2001**, *5*, 17–23.
30. Monjaraz-Ruedas, R.; Prendini, L.; Francke, O.F. Systematics of the short-tailed whipscorpion genus Stenochrus Chamberlin, 1922 (Schizomida: Hubbardiidae), with descriptions of six new genera and five new species. *Bull. Am. Mus. Nat. Hist.* **2019**, *435*, 1–91. [CrossRef]
31. Monjaraz-Ruedas, R.; Francke, O.F. Five new species of Stenochrus (Schizomida: Hubbardiidae) from Oaxaca, Mexico. *Zootaxa* **2018**, *4374*, 189–214. [CrossRef] [PubMed]
32. Vignoli, V.; Prendini, L. Systematic revision of the troglomorphic North American scorpion family Typhlochactidae (Scorpiones: Chactoidea). *Bull. Am. Mus. Nat. Hist.* **2009**, *326*, 1–94. [CrossRef]
33. Santibáñez-López, C.E.; Francke, O.F.; Prendini, L. Shining a light into the World's deepest caves: Phylogenetic systematics of the troglobiotic scorpion genus Alacran Francke, 1982 (Typhlochactidae: Alacraninae). *Invert. Syst.* **2014**, *28*, 643–664. [CrossRef]
34. Valdez-Mondragón, A.; Cortez-Roldán, M.R. On the trapdoor spiders of Mexico: Description of the first new species of the spider genus Aptostichus from Mexico and the description of the female of Eucteniza zapatista (Araneae, Mygalomorphae, Euctenizidae). *ZooKeys* **2016**, *641*, 81–102. [CrossRef]
35. Bueno-Villegas, J.; Sierwald, P.; Bond, J.E. Diplopoda. In *Biodiversidad, Taxonomía y Biogeografía de Artrópodos de México: Hacia una Síntesis de su Conocimiento*; Llorente-Bousquets, J., Morrone, J.J., Ordóñes, O.Y., Fernández, I.V., Eds.; CONABIO: Mexico City, Mexico, 2004; Volume IV, pp. 137–157.
36. Reddell, J.R. A review of the cavernicole fauna of Mexico, Guatemala and Belize. *Mus. Speleol. Mon.* **1981**, *27*, 69–257.
37. Palacios-Vargas, J.G. Biodiversidad de Collembola (Hexapoda: Entognatha) en México. *Rev. Mex. Biod.* **2014**, *85*, 220–231. [CrossRef]
38. Espinasa, L.; Fisher, A. A cavernicole species of the genus Anelpistina (Insecta: Zygentoma: Nicoletiidae) from San Sebastian cave, Oaxaca, Mexico. *Proc. Entomol. S Was.* **2006**, *108*, 655–660.

diversity

MDPI

Article

A Hotspot of Arid Zone Subterranean Biodiversity: The Robe Valley in Western Australia

Huon L. Clark [1], Bruno A. Buzatto [1,2,3] and Stuart A. Halse [1,*]

[1] Bennelongia Environmental Consultants, 5 Bishop Street, Jolimont, WA 6014, Australia; huon.clark@bennelongia.com.au (H.L.C.); bruno.buzatto@bennelongia.com.au (B.A.B.)
[2] Department of Biological Sciences, Macquarie University, 6 Wally's Walk, Macquarie Park, NSW 2109, Australia
[3] School of Biological Sciences, University of Western Australia, Crawley, WA 6009, Australia
* Correspondence: stuart.halse@bennelongia.com.au

Abstract: Knowledge of subterranean fauna has mostly been derived from caves and streambeds, which are relatively easily accessed. In contrast, subterranean fauna inhabiting regional groundwater aquifers or the vadose zone (between surface soil layers and the watertable) is difficult to sample. Here we provide species lists for a globally significant subterranean fauna hotspot in the Robe Valley of the Pilbara region, Western Australia. This fauna was collected from up to 50 m below ground level using mining exploration drill holes and monitoring wells. Altogether, 123 subterranean species were collected over a distance of 17 km, comprising 65 troglofauna and 58 stygofauna species. Of these, 61 species were troglobionts and 48 stygobionts. The troglofauna occurs in small voids and fissures in mesas comprised mostly of an iron ore formation, while the stygofauna occurs in the alluvium of a river floodplain. The richness of the Robe Valley is not a localized aberration, but rather reflects the richness of the arid Pilbara region. While legislation in Western Australia has recognized the importance of subterranean fauna, mining is occurring in the Robe Valley hotspot with conditions of environmental approval that are designed to ensure species persistence.

Keywords: stygobiont; troglobiont; conservation; groundwater; arid zone; mining

Citation: Clark, H.L.; Buzatto, B.A.; Halse, S.A. A Hotspot of Arid Zone Subterranean Biodiversity: The Robe Valley in Western Australia. *Diversity* **2021**, *13*, 482. https://doi.org/10.3390/d13100482

Academic Editors: Tanja Pipan, David C. Culver and Louis Deharveng

Received: 9 September 2021
Accepted: 29 September 2021
Published: 30 September 2021

Publisher's Note: MDPI stays neutral with regard to jurisdictional claims in published maps and institutional affiliations.

1. Introduction

Subterranean habitats have little to no light, scarce organic matter, constantly high humidity, and a much more stable temperature than the surface [1]. Unsurprisingly, the vast majority of studies of troglofauna and stygofauna have been in caves that people can access or in streambeds (the hyporheic zone). Both types of habitats allow researchers relatively easy access to explore and collect animals [2,3]. It is often overlooked, however, that much more extensive habitats likely to support subterranean fauna occur worldwide. For stygofauna, these habitats occur in unconfined regional groundwater aquifers. For troglofauna, they occur in the vadose zone that extends from c. 2 m below the ground surface to the watertable. Abundant small spaces for subterranean animals are present in geologies ranging from detritals (e.g., colluvium, alluvium, scree, etc.) to sedimentary rocks, often at tens of meters depth [4–6]. Sampling troglofauna and stygofauna from these environments, which have no connection at the human scale to the surface, is obviously much more difficult than exploring large cave systems or digging in streambeds and the documentation of their biodiversity has lagged behind the general study of subterranean fauna.

Recognition that Western Australia supports rich stygofauna communities in arid zone aquifers began in the 1990s, with sampling at Cape Range on the mid-west coast and in the Pilbara and Yilgarn regions [7,8] (Figure 1). The richness of stygofauna across most Pilbara landscapes has been confirmed by a large regional survey [9,10] and many surveys associated with the environmental impact assessment of mining projects [11,12]. Recognition that equally rich troglofauna communities are present did not begin until

259

stygofauna surveys in the Robe Valley during 2004 and 2005 also collected troglofauna, mostly schizomids [13,14]. Knowledge of both troglofauna and stygofauna in the Pilbara has continued to grow in the past 15 years, and it was recently estimated that 1511 troglofauna and 1329 stygofauna species occur in the region [5], with the caveat that the precision of these estimates is limited by a weak taxonomic framework for certain fauna groups.

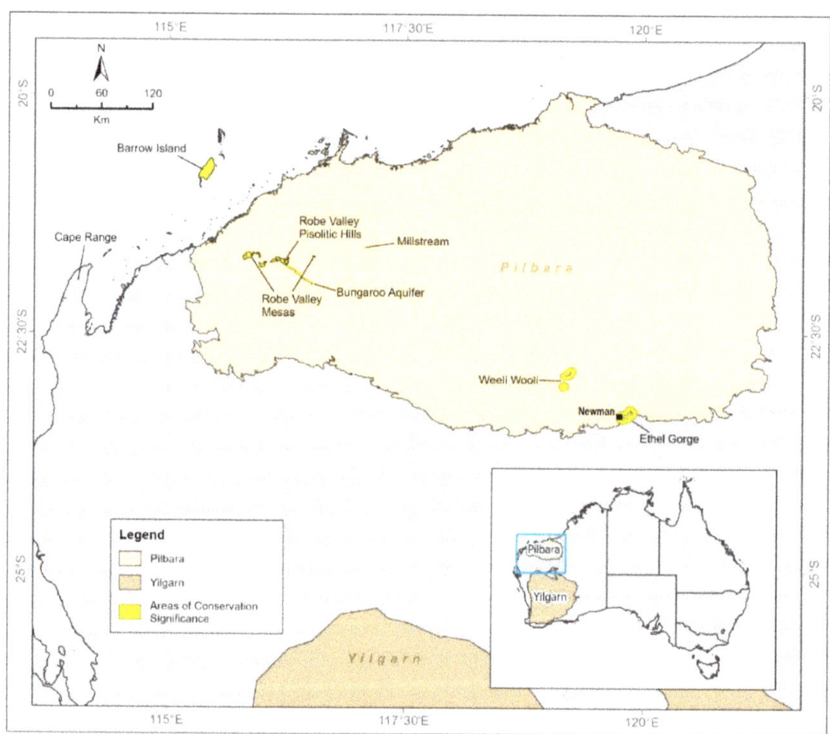

Figure 1. The Pilbara region of Western Australia, showing subterranean fauna communities recognized as significant for conservation. The Robe Valley hotspot is at the western end of the Robe Valley Pisolitic Hills.

The richness of subterranean fauna in the Pilbara has been explained as partly a result of many aquatic and mesic-adapted terrestrial invertebrate species escaping the harshness of surface habitats during the Miocene aridification by moving underground, sometimes with dramatic subsequent radiation [15,16]. Schizomids are a well-documented case; these typically forest litter-dwelling animals moved into subterranean habitats as forest habitat was being lost, and then radiated in these habitats [17,18]. Equally impressive radiations have occurred, however, in stygofauna groups such as candonid ostracods [19], syncarids [20] and amphipods [21,22], which are essentially subterranean groups and would not have colonized from the surface. Information from a regional survey suggests that, as a result of these radiations, there are about 30% more stygofauna than surface aquatic invertebrate species in the Pilbara [23]. As far as we are aware, this greater richness of subterranean fauna is unique, but comparative data for other regions are scarce because of the rarity of published regional inventories.

While the Pilbara can be regarded as an area of globally significant radiation of subterranean fauna, here we focus on the richness of troglofauna and stygofauna in the western part of the Robe River catchment (Figure 2). We call this the Robe Valley, although it is a mostly flat landscape. The area has previously been identified as a hotspot for troglofauna [24] and stygofauna [25]. Cape Range, to the south-west of the Pilbara, the offshore Barrow Island, and the inland Ethel Gorge have also been identified as globally significant hotspots in, or near, the Pilbara ([24], Figure 1).

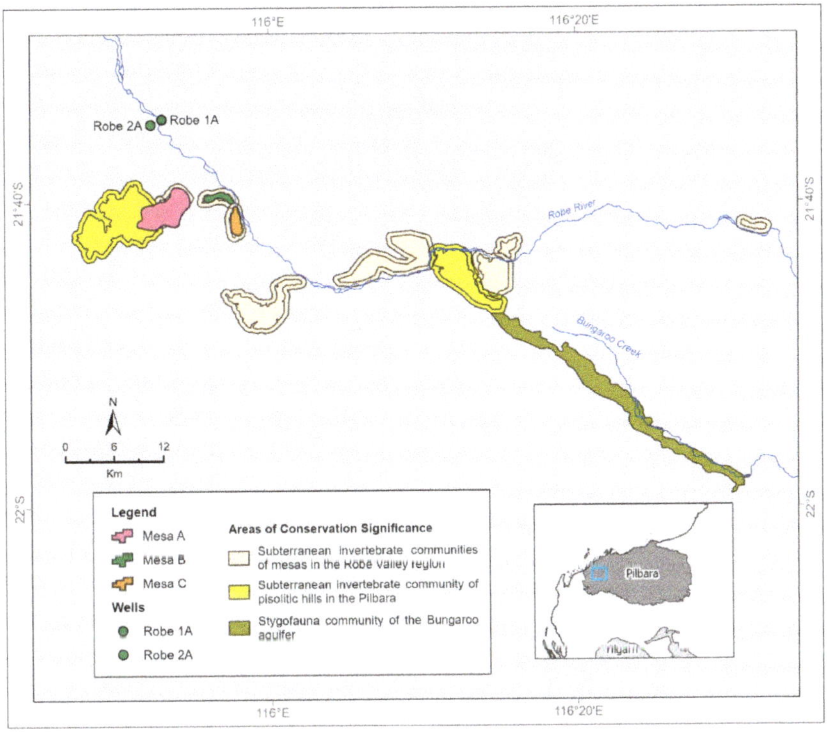

Figure 2. Components of the Robe Valley subterranean fauna hotspot.

In addition to providing species lists for sites in the Robe Valley, we have three goals in this manuscript. One is to highlight the importance of vadose zone and groundwater aquifers for troglofauna and stygofauna. The second is to illustrate that, in many parts of the world, huge numbers of subterranean species remain to be discovered and described. The third is to point out that water extraction and mining, particularly for iron ore but also gold, base metals, lithium, potash and rare earth metals, may sometimes threaten the conservation of subterranean communities. At the same time, we recognize that it is the wells and exploration drill holes of these developments that provide the portal that has enabled the discovery of the otherwise inaccessible and hidden subterranean fauna of the Pilbara region.

2. Materials and Methods

We confirm previous identification of the Robe Valley as a subterranean fauna hotspot by compiling lists of species for discrete areas or sites within this hotspot. For troglofauna, we provide lists for three small mesas (Mesas A, B and C) on the west flank of the Robe River, approximately 47 km from the river mouth (Figure 2). The mesas represent topographically distinct features in the landscape (Figure 3), and their outlines as defined here can be seen in Figure 2. For stygofauna, we provide lists of species collected from two wells (Robe 2A and 1A) near the Robe River as it crosses the coastal plain, approximately 8 km north of the mesas. The wells are portals to a stygofauna community for which the geographical boundary is not obvious at the surface.

Figure 3. Upper panel, Mesa A, showing hardcap on top of the mesa; inset A, a scrape net; inset B, a troglofauna trap. Lower panel, Robe 2A well showing steel collar set in concrete; inset A, a stygofauna net. Photos by Claire Stephenson, Mike Scanlon and Melanie Fulcher.

Troglofauna were sampled at the mesas for the purposes of environmental impact assessments and monitoring compliance with the conditions for mining. Most of the sampling was undertaken by Biota Environmental Sciences, although in some years other consultants (including Bennelongia) or mining company staff in conjunction with consultants, undertook the work. The predominant sampling method was trapping, whereby between one and four cylinders of PCV containing moist leaf material were suspended at varying depths in a series of vertical exploration drill holes. The drill holes (diameter 130 mm) had 2–3 m of casing at the surface (called a collar) to prevent the collapse of sand and loose rock into the hole, but the remainder of the hole was in direct contact with the surrounding substrate. Traps were left in place for 6–8 weeks. A smaller number of samples were collected by a technique called scraping, whereby a net was lowered to the bottom of the exploration hole and scraped along the substrate surface while being retrieved [26]. Approximately 1100 trap or scrape samples were collected from Mesa A (21.66072 °S, 115.89547 °E; hole depth range: 8–88 m) in 11 years of sampling between 2004 and 2017. This equates to about 450 sampling events (i.e., site/date combinations) because multiple traps were set in most holes, and from 2014 onwards holes were also scraped. Sampling effort was lower at the other two mesas, with approximately 400 samples and 180 sampling events during seven years between 2005 and 2016 at Mesa B (21.65894°S, 115.94547°E; depth range: 15–70 m) and 300 samples and 180 sampling events in three years of sampling between 2005 and 2017 at Mesa C (21.68928°S, 115.96206°E; depth range: 15–125 m). Troglofauna traps were set in an iron ore formation called channel iron deposit (CID) at depths between 5 m and 70 m below ground level, with the median depth being 20 m. Animals were recovered from depths up to 50 m.

Stygofauna were collected mainly by net hauling at Robe 2A (well G70730102, 21.58123S°S, 115.87047°E) and Robe 1A (well G70730101, 21.57542°S, 115.88266°E). These wells are 14 m and 23 m deep, with standing water level fluctuating between 4–9 m and 6–10 m below ground level, respectively. This level of groundwater fluctuation is characteristic of Pilbara floodplains. The wells had a steel collar extending approximately 1 m above and below the ground surface and a 100 mm diameter PVC casing inside the collar that extended to the end of the hole. The casing was slotted continuously below the watertable. A small, weighted plankton net, with a diameter only slightly smaller than the cased hole, was lowered to the base of the hole, agitated vigorously to stir 'benthic' fauna into the water column, and then retrieved. Two sizes of mesh (50 μm and 150 μm) were used in different hauls. Pump samples were also collected on two dates, whereby three times the bore volume was pumped through a 50 μm-mesh net [9]. Twelve sampling events occurred at Robe 2A in five years of sampling between 2002 and 2007, while eight sampling events occurred in the three years between 2002 and 2004 at Robe 1A.

Determining Species Numbers and Subterranean Affinity

The taxonomic framework for the identification of most groups of subterranean fauna in the Pilbara is poor. As a result, even when there was competent species-level discrimination by different survey teams, the compilation of species lists for the hotspots was difficult. Species-level discrimination was mostly achieved using morphology, often coupled with molecular sequencing, but also by sequencing alone. Very few species have been described to date, so most species were identified by codes linked to voucher specimens or molecular sequences. In a few cases, when Western Australian Museum staff work on a group and verified identifications, a museum voucher name was used. More often codes were generated by individual consultancies. Representative specimens of most putative species are lodged in the Western Australian Museum.

Another part of the process of compiling the species lists was determining the subterranean affinity of specimens. This is more important for troglofauna, because aquatic species collected from groundwater beneath a completely dry terrestrial surface rely entirely on subterranean habitat for persistence in that setting. On the other hand, scrapes and traps usually collect an order of magnitude more surface animals than troglofauna. While most surface drop-ins can be easily identified as such, some surface soil species lack pigmentation and eyes as a result of their specialization to life in topsoil and leaf litter, and it can difficult to distinguish between them and troglofauna. We considered species to be troglofauna if they are known only from scrapes or traps and possess some troglomorphic traits in addition to lack of pigmentation and eyes. Pauropods and symphylans are the groups with greatest potential for surface species to be treated as troglofauna but soil and litter sampling in the Pilbara has collected very few species of either group [26; Bennelongia, unpublished data], suggesting few species in these groups are drop-ins. Mites and collembolans were omitted from the lists, reflecting advice from the Western Australian Museum to the environmental impact assessment process many years ago that few species in these groups are subterranean. We recognise that this is incorrect for some species.

3. Environment and Geology

The Pilbara Craton is one of the oldest and most stable geological regions in the world, with basement rocks dating as far back as 3.5 billion years [27]. Uniquely, this basement rock in the form of granite and greenstone terrains remains largely exposed [28], particularly in the north of the craton [5]. Throughout the south, and to a lesser extent in the north, the basement rock is overlain by depositional sedimentary rocks from the Archaean to Proterozoic time periods [29]. The major ranges of hills in the Pilbara (i.e., the Hamersley and Chichester Ranges) consist largely of iron-rich sedimentary rocks, mostly banded iron formations. Less resistant rock formations (such as the dolomitic Wittenoom formation) have been highly weathered, with geologies such as CID being deposited in the resultant drainage lines [29].

Despite geological stability, the Pilbara climate has changed substantially over the past 20 million years from cool and wet to hot and arid as Australia separated from Antarctica and began moving northwards [16]. The Pilbara now has very hot summers, moderate winters, and low but highly variable annual rainfall. Mean daily maximum temperatures exceed 35 °C from October to March, and the hottest recorded Pilbara day is 50.5 °C (at Mardie near the mouth of the Robe River in February 1998). Mean annual rainfall at Mardie is 275 mm, with 54% of this rain falling between January and March, mostly from cyclones or monsoonal rain. A further 33% falls between April and June, mostly from late cyclones or large cold fronts that extend to the Pilbara from the south at the start of winter. The main source of groundwater recharge is cyclones [30].

The prospective habitats for troglofauna in the Robe Valley, and the Pilbara generally, mostly comprise centimeter to millimeter-scale fissures and vugs. In some ways, these Pilbara habitats are similar to the small microcaverns radiating out from limestone caves, although limestone microcaverns often become blocked with detritus and are unsuitable for troglofauna [31]. The Pilbara habitats are perhaps more directly comparable with canga and other ferruginous formations in South America, although caves are more common in South American formations and are the focus of most studies [32].

The major troglofauna habitat in the Robe Valley lies within the flat-topped mesas that typically stand 30–50 m above the surrounding landscape. The flat top consists of lateralised hardcap, with a thickness of up to 10 m (Figure 3). This cap has a near vertical face at the edges of the mesas, below which is a sloping face of detritals. The core geology of the mesas is CID. Details of the genesis of this CID in the palaeochannel of the Robe River are contentious, but it is agreed CID began to be laid down after sediment in-filled the Robe River palaeovalley during the Palaeogene, then larger areas of CID formed in the Miocene [33,34]. Subsequent surface flow events eroded the in-fill and left the mesas observed today [33]. The typical stratigraphy of the mesas consists of a weathered hardcap (loosely analogous to Brazilian canga) above layers of CID. The CID is predominately made up of iron-rich pisoids, which are 2–10 mm rounded bodies that are bonded together by a geothitised secondary matrix. Over time, this sedimentary rock has become weathered, principally through groundwater leaching, leaving vugs and other voids in the matrix. Some of the voids have gradually combined to form larger spaces, resulting in limited connectivity of spaces throughout the matrix. Most of these subterranean spaces now sit above the watertable and have been colonised by troglofauna. Mesa A has an approximate area of 5.7 km^2, Mesa B 1.7 km^2 and Mesa C 2.3 km^2. Relative humidity in the subterranean habitats of Robe River mesas is 100% [5].

Near the surface, the troglofauna habitat in sections of the Robe River mesas, and in the Pilbara more generally, may be considered to have similarities with Milieu Souterrain Superficial (MSS) habitat, especially on valley slopes [35]. However, this is not the case at depth, where the troglofauna sampling was undertaken. The mesas, and Pilbara as a whole, mostly consist of consolidated rock habitat, including bedrock, rather than scree and loose rock.

The habitat for stygofauna in the Robe Valley is provided mainly by aquifers in CID or alluvium. Robe 2A and 1A lie in an alluvial aquifer associated with the active channel of the Robe River as it crosses the Pilbara coastal plain; CID occurs in the vicinity if not at the wells themselves. The surface provides little indication of the rich stygofauna community below (Figure 3). Depth to the watertable varies from 4–10 m, depending on season and year, with salinity in the top metre of groundwater being 2400–5500 µS/cm and temperature ranging from 30.5–32.4 °C during two years. The Robe River flows, on average, less than once a year and most of the riverbed is dry the remainder of the time. Flood volumes can be large (mean annual flow is 108 GL [36]).

4. Results

Altogether, 65 species of troglofauna have been collected from Mesas A, B and C, with 35 species in Mesa A, 30 species in Mesa B and 17 species in Mesa C (Table 1, Figure 4). Two species, the millipede *Lophoturus madecassus* and the pseudoscorpion *Tyrannochthonius aridus*, are also known from surface habitats and are troglophiles [37,38]. Two other species, the hemipteran *Cixiidae* sp. B02 and the dipteran *Allopnyxia* sp. B01, are treated as troglophiles. It has been shown genetically that they are widespread across the Pilbara. Adult *Allopnyxia* sp. B01 has a vestigial eye, consisting of a dark crescent with a few variable-sized ommatidia and, therefore, may have some surface dispersal. *Cixiidae* sp. B02 has obligate subterranean nymphs with eyes that become larger as they approach the adult stage. Adults are capable of surface dispersal (Bennelongia, unpublished data). All other species are troglobionts known only from Mesas A, B and C, and usually from only one of these mesas. They are extreme short-range endemics.

Table 1. Troglofauna species collected from Mesas A, B and C. Records with an asterisk represent identifications above species level, but they are usually aligned with a related species to reduce table size. Species in bold have special conservation protection. Comments show species delimited using CO1 results, with divergence values from other species indicated where available. # indicates species are troglophiles, whereas all other species are likely troglobionts.

Taxonomic Group	Family	Species	A	B	C	Comment
Chilopoda	Cryptopidae	Cryptopidae 'Helix-CHI023'	✓*	✓		CO1
		Cryptopidae 'Helix-CHI026'			✓	CO1
	Scolopendridae	*Cormocephalus* sp.		✓*		
	Geophilidae	*Geophilidae* sp.	✓*			
Diplopoda	Haplodesmidae	Haplodesmidae 'Helix-DIHA001'	✓*	✓	✓	CO1
		Haplodesmidae 'Helix-DIHA005'		✓		CO1
	Lophoproctidae	*Lophoturus madecassus* (Marquet & Conde, 1950)#	✓	✓		
		Scolopendrellidae 'Helix-SYM026'	✓	✓*		
Pseudoscorpiones	Chthoniidae	**Lagynochthonius asema Edward & Harvey, 2008**	✓			
		Tyrannochthonius aridus Edward & Harvey, 2008#	✓	✓		
		Tyrannochthonius basme Edward & Harvey, 2008		✓		
		Tyrannochthonius 'MesaA'	✓			
		Chthoniidae 'Helix_PCH047'		✓		CO1, >20%
		Chthoniidae 'Helix_PCH049'		✓		CO1, 10%
		Chthoniidae 'Helix_PCH050'		✓		CO1
		Chthoniidae 'Helix_PCH058'			✓	CO1
	Olpiidae	Olpiidae 'POL013'	✓			
	Hyidae	Hyidae 'Helix-PH001'		✓		CO1, 5%
		Hyidae 'Helix-PH008'		✓		CO1, 5%
	Syarinidae	*Ideoblothrus linnaei* Harvey & Leng, 2008	✓			

Diversity 2021, 13, 482

Table 1. *Cont.*

Taxonomic Group	Family	Species	A	B	C	Comment
		Ideoblothrus pisolitus Harvey & Edward, 2007		✓		CO1, 6%
		***Ideoblothrus* 'MesaA'**	✓			
		Ideoblothrus 'MesaA2'	✓			CO1,>9%
		Ideoblothrus 'MesaA3'	✓			CO1, >9%
		Ideoblothrus 'MesaA4'	✓			CO1, >18%
Araneae	Gnaphosidae	Gnaphosidae 'Helix-AG001'			✓	
	Oonopidae	*Prethopalpus scanloni* Baehr et al., 2012	✓			
		Prethopalpus 'ARA051'		✓		WAM
	Pholcidae	*Trichocyclus* sp. 'MesaA'	✓			
	Theridiidae	Theridiidae 'Helix-AT001'	✓		✓	CO1
Schizomida		***Draculoides anachoretus* (Harvey et al., 2008)**	✓			
		***Draculoides bythius* (Harvey et al., 2008)**		✓	✓	
		Hubbardiidae gen. nov. 'Helix-SCH052'		✓	✓	CO1, WAM
Isopoda	Armadillidae	*Troglarmadillo* 'Helix-ISA005'		✓		CO1
		Troglarmadillo 'Helix-ISA008'		✓		CO1
		Armadillidae 'Helix—ISA054'			✓	CO1
		Armadillidae 'Helix—ISA056'			✓	CO1
		Armadillidae 'MesaAOES19'	✓			
	Oniscidae	*Hanoniscus* 'MesaAOES22'	✓			
		Philosciidae 'Helix-ISP052'	✓			CO1
		Philosciidae 'Helix-ISP053'		✓		CO1
Diplura	Campodeidae	Campodeidae 'Helix-DCA001'		✓	✓*	CO1, 6%
		Campodeidae 'Helix-DCA002'		✓		CO1
		Campodeidae sp. B6	✓			
	Japygidae	*Japygidae* sp.	✓*	✓*	✓*	
	Parajapygidae	Parajapygidae 'Helix-DPA003'		✓	✓	
		Parajapygidae 'Helix-DPA004'			✓	CO1, 12–19%
		Parajapygidae 'Helix-DPA006'			✓	CO1
		Parajapygidae 'Helix-DPA008'			✓	CO1
	Projapygidae	*Projapygidae* sp.	✓*			
Zygentoma	Nicoletiidae	Atelurinae 'Helix-TA009'		✓		CO1
		Hemitrinemura sp. B4	✓			
		Trinemura 'MesaA1'	✓			
		Trinemura 'MesaA2'	✓			
		Nicoletiinae 'Helix-TN010'		✓		CO1
		Nicoletiinae 'Helix-TN012			✓	CO1
Dictyoptera	Nocticolidae	*Nocticola* 'OES11'	✓			
Hemiptera	*Cixiidae* sp. B02	*Cixiidae* sp. B02#	✓	✓		
Coleoptera	Carabidae	*Gracilanillus hirsutus* Giachino et al., 2021	✓	✓		
		Angustanillus striatipennis Giachino et al., 2021	✓			
		Angustanillus armatus Giachino et al., 2021		✓		
	Ptiliidae	*Ptinella* sp. B01	✓			
	Staphylinidae	?Staphylinidae 'MesaKOES2'	✓			
	Curculionidae	Cryptorhynchinae 'CCU004'	✓	✓	✓	WAM
Diptera	Sciaridae	*Allopnyxia* sp. B01#	✓			

Only 12 of the 65 troglofauna species at the mesas are described, but assignments to at least 34 other species have been supported by sequencing, and another three have been through the process of receiving museum codes. Many additional species, such as the cockroach *Nocticola* 'OES11', the spider *Trichocyclus* sp. 'MesaA', various beetles, the diplurans *Japygidae* sp. and *Projapygidae* sp., and the centipedes *Cormocephalus* sp. and *Geophilidae* sp., are the single occurrence of their families in a mesa (Table 1), which ensures it is valid to recognize each of these taxa as representing at least one unit in the species list.

Figure 4. Troglofauna at the mesas. (**A**) *Trinemura* 'MesaA1'; (**B**) *Ptinella* sp. B01; (**C**) *Japygidae* sp.; (**D**) *Troglarmadillo* 'Helix-ISA005'; (**E**) *Allopnyxia* sp. B01; (**F**) *Tyrannochthonius aridus*; (**G**) *Draculoides* sp.; (**H**) *Prethopalpus scanloni*. Photos by Jane McRae.

The most difficult issue when compiling species lists was reconciling the multiple species names used in different surveys. Some problems were caused by different levels of identification (earlier surveys tended to have less taxonomic information available and hence less rigorous identifications). Others resulted from lack of clear diagnoses associated with each voucher name, leading to conflicting views about which species animals belonged to. For example, delimitation of the five species of *Ideoblothrus* at Mesa A in an area <6 km^2 has been confused, particularly in relation to *Ideoblothrus* 'MesaA', which was described but not formally named [39]. The 'type' animal could not be sequenced, but some other animals identified by the Western Australian Museum as *Ideoblothrus* 'MesaA' appear genetically to be *Ideoblothrus linnaei*, which was described in the same paper. Whether some museum identifications were wrong, or whether *Ideoblothrus* 'MesaA' should be treated as *Ideoblothrus linnaei*, remains unclear. Delimitation is further complicated by co-occurrence with *Ideoblothrus* 'MesaA2', 'MesaA3' and 'MesaA4'. *Ideoblothrus* 'MesaA2' and 'MesaA3' are separated from other species by a borderline 9% sequence divergence in CO1 (Table 1).

Five troglofauna species at Mesa A and one at Mesa B are listed in Western Australian species conservation legislation or associated mechanisms. These are two schizomids (*Draculoides anachoretus* and *D. bythius*) that are listed as Vulnerable under the Western Australian *Biodiversity Conservation Act 2016* and three pseudoscorpions (*Lagynochthonius asema*, *Tyrannochthonius* 'MesaA' and *Ideoblothrus* 'MesaA') that are informally listed by the government conservation agency as Priority 1 species (defined as poorly known from few locations and potentially under threat). In addition, the troglofauna communities and their habitats at Mesas A, B and C, as well as nearby mesas in the Robe Valley, are listed informally by the conservation agency as a priority ecological community, namely the Priority 1 *Subterranean invertebrate communities of mesas in the Robe valley region*. Another Priority 1 troglofauna community, the *Subterranean invertebrate community of pisolitic hills in the Pilbara*, occurs immediately west of the mesa community (Figure 2). Priority 1 communities are poorly known but considered to have restricted occurrence and to be potentially threatened.

At least 58 species of stygofauna have been collected from the Robe 2A and 1A wells, with 46 species at Robe 2A and 31 species at Robe 1A (Table 2, Figure 5). All species collected are reliant on groundwater for their persistence at the site, although the cyclopoid copepod *Apocyclops dengizicus* is cosmopolitan and would usually be classed as a stygophile [40]. Twenty-nine of the species are described, largely through taxonomy funded in association with a large biodiversity survey of the Pilbara [41]. An additional species, the thermosbaenid *Halosbaena* sp. PL, is undescribed, but has been well studied genetically [42]. A lack of diagnostic information for the multiple names used (in different surveys) for syncarid taxa resulted in them being reduced to two species, Bathynellidae sp. and *Atopobathynella* 'A'. This probably underestimates the number of syncarid species present. There is also taxonomic and ecological uncertainty associated with oligochaete records. The *Enchytraeus*, Phreodrilidae and Tubificidae species names in Table 2 represent morphological clades to which immature animals are assigned, and each clade potentially contains multiple species. The groundwater dependence of enchytraeids is uncertain, and they perhaps should be regarded as amphibious species [43], although collected commonly in stygofauna sampling in Western Australia and apparently subterranean. Similarly, some tubificids are stygophiles, and the groundwater dependence of rotifers, nematodes and the leech is unknown. In total, 48 species are recognised as stygobionts.

Table 2. Stygofauna species collected from wells Robe 2A and Robe 1A. Records with an asterisk represent identifications above species level. # indicates species are stygophiles, ‡ indicates uncertain status; all other species are likely stygobionts.

Taxonomic group	Family	Species	2A	1A
Rotifera	-	*Bdelloidea* sp. ‡		✓*
	Philodinidae	*Dissotrocha* sp. ‡	✓*	
Nematoda	-	*Nematoda* sp. 02 (PSS) ‡		✓
		Nematoda sp. 03 (PSS) ‡	✓	
		Nematoda sp. 11 (PSS) ‡	✓	
Annelida	-	*Hirudinea* sp. ‡	✓*	
	Enchytraeidae	*Enchytraeus* sp. AP PSS1 ‡	✓	
		Enchytraeus sp. AP PSS2 ‡	✓	
	Phreodrilidae	*Phreodrilidae* sp. AP DVC	✓	✓
		Phreodrilidae sp. AP SVC	✓	
	Tubificidae	Tubificidae 'stygo type 2A' #	✓	
	Nereididae	*Namanereis pilbarensis* Glasby et al., 2014	✓	
Acarina	Arrenuridae	*Arrenurus* sp. nov. 2 (PSS)	✓	
	Mideopsidae	*Guineaxonopsis* sp. S01 group	✓	✓
	Pezidae	*Peza* sp.	✓*	
	Unionicolidae	*Unionicolidae* sp. B02	✓	
Ostracoda	Limnocytheridae	*Gomphodella hirsuta* Karanovic, 2006	✓	✓
	Candonidae	*Deminutiocandona aenigma* Karanovic, 2007	✓	
		Humphreyscandona fovea Karanovic & Marmonier, 2003	✓	✓
		Humphreyscandona imperfecta Karanovic, 2005	✓	✓
		Humphreyscandona pilbarae Karanovic & Marmonier, 2003		✓

Table 2. *Cont.*

Taxonomic group	Family	Species	2A	1A
		Humphreyscandona woutersi Karanovic & Marmonier, 2003	✓	✓
		Pierrecandona posteriorrecta Karanovic, 2007	✓	
		Pilbaracandona rosa Karanovic, 2007	✓	✓
		Areacandona astrepte Karanovic, 2007	✓	✓
		Areacandona cylindrata Karanovic, 2007	✓	
		Areacandona '4' (PSS)		✓
		Areacandona fortescueiensis Karanovic, 2007		✓
		Kencandona verrucosa Karanovic, 2007		✓
		Candonidae '2' (PSS)		✓
		Candonidae gen. 4 '1'		✓
Copepoda	Ridgewayiidae	*Stygoridgewayia trispinosa* Tang et al., 2008	✓	✓
	Cyclopidae	*Apocyclops dengizicus* (Lepeshkin, 1900) #		✓
		Halicyclops rochai Karanovic, 2006	✓	✓
		Orbuscyclops westaustraliensis Karanovic, 2006	✓	
		Diacyclops einslei De Laurentiis et al., 1999		✓
		Diacyclops h. humphreysi x unispinosus (see Karanovic, 2006)	✓	
		Diacyclops humphreysi Pesce & De Laurentiis, 1996	✓	
		Diacyclops h. unispinosus Karanovic, 2006	✓	
		Diacyclops sobeprolatus Karanovic, 2006	✓	
	Ectinosomatidae	*Pseudectinosoma galassiae* Karanovic, 2006	✓	✓
	Diosaccidae	*Schizopera robcriverensis* Karanovic, 2006	✓	✓
	Ameiridae	*Megastygonitocrella unispinosa* Karanovic, 2006	✓	✓
		Megastygonitocrella trispinosa Karanovic, 2006	✓	
	Parastenocarididae	*Parastenocaris jane* Karanovic, 2006	✓	
Syncarida	Bathynellidae	*Bathynellidae* sp.	✓*	
	Parabathynellidae	*Atopobathynella* 'A'	✓	✓
Thermosbaenacea	Halosbaenidae	*Halosbaena* sp. PL (see Page et al., 2016)	✓	✓
Amphipoda	Bogidiellidae	*Bogidiellidae* sp.	✓*	✓
	Eriopisidae	*Nedsia* 'mcraei' King & Cooper, in press	✓	✓
		Nedsia 'robensis' King & Cooper, in press	✓	✓
		Eriopisidae gen. nov. 'Helix-AMM006'		✓
	Neoniphargidae	*Wesniphargus* 'Helix-AMN004'		✓
	Paramelitidae	*Pilbarus* sp.	✓*	✓*
Isopoda	Cirolanidae	*Haptolana yarraloola* Bruce, 2008	✓	
		Kagalana tonde Bruce, 2008	✓	
	Olibrinidae	*Adoniscus* sp.	✓*	
	Microcerberidae	*Microcerberidae* sp.	✓*	

Unlike the mesas, which have unique assemblages of species associated with easily delineated topographic features, the spatial extent of the stygofauna hotspot is unclear. Of the four wells within 2 km of Robe 2A, only Robe 1A was sampled by the team that sampled Robe 2A. The other wells yielded few species, but Robe 2A and 1A appear to have the same rich stygofauna community, with 61% of the species collected at Robe 1A also found at the better sampled Robe 2A. In addition, five described ostracod and copepod species, three undescribed ostracods, two undescribed amphipod species, a rotifer and a nematode not found at Robe 2A were collected from Robe 1A. The amphipods include a new and distinctive lineage of Eriopisidae that is widespread in coastal parts of the Pilbara and the low-lying Fortescue Valley, as well as a species of *Wesniphargus,* representing a family (Neoniphargidae) that is relatively uncommon in the Pilbara (Table 2, Figure 5).

Figure 5. Stygofauna at Robe 2A and 1A. (**A**) *Humphreyscandona fovea;* (**B**) *Wesniphargus* 'Helix-AMN004'; (**C**) *Namanereis pilbarensis;* (**D**) *Kagalana tonde;* (**E**) *Microcerberidae* sp.; (**F**) Eriopisidae gen. nov. 'Helix AMM006'; (**G**) *Nedsia* 'mcraei' King & Cooper in press; (**H**) *Halosbaena* sp. PL.; (**I**) *Peza* sp. Photos by Jane McRae.

Some described stygobiont species have relatively large linear ranges, e.g., the polychaete *Namanereis pilbarensis* (c. 500 km), the calanoid copepod *Stygoridgewayia trispinosa* (>300 km), the cyclopid copepods *Halicyclops rochai* (>300 km), *Diacyclops* species (400–600 km) and *Orbuscyclops westaustraliensis* (>800 km), as well as the harpacticoid copepods *Pseudectinosoma galassiae* (>300 km), *Schizopera roberiverensis* (c. 300 km), *Megastygonitocrella* species (600–700 km) and *Parastenocaris jane* (600 km, Bennelongia unpublished data). Ostracods tend to have smaller ranges, e.g., *Deminutiocandona aenigma* (c. 200 km), *Humphreyscandona* species (80–190 km), *Pierrecandona posteriorrecta* (100 km), *Pilbaracandona rosa* (120 km) and *Areacandona* species (50–60 km). Most other species can be considered short-range endemics [9].

The relatively small amount of comprehensive genetic work done on stygofauna (rather than comparisons of individual specimens a few kilometres apart to determine whether they are conspecific) suggests it is likely that some, if not all, of the stygofauna species currently regarded as widespread are, in fact, either cryptic species complexes or have very high intraspecific variability (or both). For example, what was previously regarded as *Halosbaena tulki* is now regarded as comprising up to five different species, including *Halosbaena* sp. PL [42].

No stygofauna species from Robe 2A and 1A are listed for protection under conservation legislation, and neither is the community. However, the community of an area to the southeast of the mesas, namely the Priority 1 *Stygofauna community of the Bungaroo aquifer*, has been listed (Figure 2). It contains at least 17 (30%) of the species present in the Robe 2A and 1A wells.

5. Discussion

The voids in groundwater aquifers and the vadose zone are amongst the least accessible continental habitats for fauna. Hence, it is unsurprising they harbour a subterranean fauna that is not as well characterized as that of streambeds and caves. Sampling yields are low, especially for troglofauna, where the combination of scraping and trapping often yields less than one animal per site and one species for every four sites sampled [26]. Net hauling for stygofauna usually yields about 50 animals and three species per site in areas where stygofauna is abundant [10]. A major contributing factor to these low yields is the disconnection between the sampler on the surface and the animals many metres below in a hidden matrix. In this situation, acquiring fine scale habitat and ecological information to improve sampling yields is difficult. Use of packers and optical viewers [44] can provide information on habitat preferences in small areas but is too expensive for large-scale deployment. Despite these difficulties, sampling of wells and drill holes has enabled documentation of a globally significant number of subterranean fauna species in the Robe Valley. Altogether, 123 species have been recorded over a distance of 17 km. Troglofauna sampling was concentrated in an area of approximately 16 km^2. The spatial extent of the stygofauna community documented at Robe 2A and 1A is unknown but probably <4 km^2. Of the 123 subterranean species collected, 61 species are considered troglobionts and at least 48 species are stygobionts.

When comparing species richness of the Robe Valley with other subterranean fauna hotspots, some of our assignments of species as troglobionts may seem controversial to readers who think of subterranean animals outside caves as deep soil fauna. However, all troglofauna sampling occurred in CID, which is consolidated rock. Unlike caves, where animals can position themselves along a subterranean gradient according to their ecological requirements, nearly all species collected in the Robe Valley mesas spend their full life cycles in environmental conditions typical of the deep sections of caves [5]. This is also seen in stygofauna, for which there is relatively little overlap in the assemblages of the hyporheic zone and regional aquifers [45]. We treat the stygofauna species that occur in both habitats as stygophiles.

Recognition of the Robe Valley as a hotspot for subterranean fauna is not new. The troglofauna community at Mesa A has previously been identified as a hotspot with 24 troglofauna species [24]. Additional data for Mesa A and the inclusion of Mesas B and C species explains the much larger number of species recognised in Table 1. The stygofauna community at Robe 2A has also previously been recognized as a hotspot with 32 stygobiont [25] or 54 stygofauna species [24], depending on how survey data were interpreted [10]. The stygobiont list [25] excluded rotifers, worms and mites, and the stygofauna list [24] included some taxa that are treated here as duplicate names. Table 2 includes data from well Robe 1A.

The landscape of the Pilbara region as a whole (approximately 250,000 km^2) is estimated to support 1511 troglofauna and 1329 stygofauna species [5], nearly all of which we classify as troglobionts or stygobionts. The subterranean fauna richness of the Robe Valley does not stand out in a major way from many other parts of the Pilbara, with a compilation of impact assessment surveys for troglofauna between 2007 and 2016 showing that the two major ranges of hills in the Pilbara to the east of the Robe Valley (Hamersley and Chichester) are also important for troglofauna [12]. In fact, survey data suggest the richest parts of the Hamersley Range contain a greater density of troglofauna species than the Robe Valley (Bennelongia, unpublished data). Broadscale stygofauna survey of the Pilbara between 2001 and 2005 identified nine areas, including the Robe Valley, as important for stygofauna [10]. Subsequent compilation of impact assessment surveys for stygofauna between 2007 and 2016 identified additional areas in the eastern Pilbara as important [12]. Overall, the Pilbara region appears to have a significance for subterranean fauna similar to that of the Dinaric karst in Europe, which has historically been regarded as the most speciose region globally [46,47]. The Pilbara has an estimated 1511 mostly troglobiont species [5] compared with 995 troglobionts in the Dinaric karst [46], and 1329 mostly stygobiont species [10] compared with 680 stygobionts in the Dinaric karst [46]. The Postojna-Planina cave system, the richest hotspot in the Dinaric karst, supports 71 stygobiont and 45 troglobiont species [48], which is similar to the numbers for the Robe Valley hotspot.

As is the case in most areas outside Western Europe and North America [49], especially when considering fauna outside caves, the majority of troglofauna and stygofauna species in the Pilbara are undescribed [10,26]. This is less pronounced for stygofauna because of extensive taxonomic work on ostracods [19] and copepods [50]. With the current state of taxonomy and the complexities of species delimitation for subterranean species, some described Pilbara species are known, or likely, to be species complexes in which several cryptically different species hide under a single name [42,51,52]. On the other hand, the disjunct occurrence of the well-studied cave eel *Ophisternon candidum* in the Pilbara and surrounds [53] highlights that care should be taken when assuming large-range taxa are species complexes.

The lack of formal species description holds back conservation in two ways. Firstly, accurately determining the number of species in an area is difficult without sound taxonomy. A confusion of names arises when relatively little time can be spent during inventory surveys delimiting near-cryptic species, and when various surveys independently apply their own codes for the same undescribed species. This confusion can create doubts about the validity of species counts, which in turn reduces the credibility of calls to conserve areas or communities. Molecular sequencing is very useful for assigning individual animals, especially if juvenile or physically damaged, to described or vouchered species for which sequence information is available. It can also assist with species delimitation, but it does not replace formal taxonomy or the use of morphology when recognising new species, especially among subterranean animals where intraspecific genetic distances are highly variable and often large [18,52,54]. In situations where a species is encountered only once in an area and is subsequently being compared to a single animal from another area, it can be difficult to determine genetically whether the animals represent different species, different evolutionarily significant units, or just a genetically variable species [55]. Only additional sampling between the two areas or across the possible range of the species can clarify the issue. Furthermore, different genes seem to work better for different taxa and over-splitting into different genetic species, rather than lumping of morphologically cryptic species, seems to be a risk if the 'wrong' gene is selected for sequencing [56]. In this context, we point out that COI sequencing without morphological examination of animals has produced more taxa than expected for pseudoscorpions at Mesa A (*Ideoblothrus*, 5 species) and Mesa B (Chthoniidae, 3 species). While this may have led to a slight overestimate of troglofauna species richness in the Robe Valley, other factors such as sampling adequacy are likely to have caused underestimation.

The second way in which lack of species description obstructs conservation is by hampering the accumulation of ecological knowledge. Species descriptions provide a framework for the collation of information about the ecology and occurrence of different subterranean groups. Synthesis of this knowledge enables more informed decision-making in the environmental impact assessment process and supports future management and conservation of subterranean fauna [57].

Recognition of conservation values of subterranean fauna has strongly influenced land management at Mesas A, B and C, where high conservation values coincide with high economic value of the subterranean habitat. A compromise between fauna conservation and economic development means that Mesa A is currently being mined for iron ore, while future mining of Mesas B and C has been approved by government [58]. However, at all three mesas an outer 'shell' approximately 80–200 m wide will remain un-mined the full way around the mesa in recognition of the importance of protecting troglofauna values. These shells should provide a sufficient volume of mesa to ensure persistence of the troglofauna species restricted to each mesa. There has been extensive monitoring at Mesa A since mining began (reported to the Environmental Protection Authority) to ensure that all troglofauna species known from that mesa are persisting in the un-mined shell. The documentation of stygofauna values has also been important, with early consideration of a borefield in the vicinity of Mesa 2a and 1A abandoned. These trade-offs are an example of a process that is likely to become increasingly important worldwide as awareness grows of the occurrence of subterranean fauna outside caves and, at the same time, there is increased demand for the products of mining.

Author Contributions: Conceptualization, H.L.C., S.A.H., B.A.B.; data curation, H.L.C.; writing—original draft preparation, S.A.H., B.A.B., H.L.C.; writing—review and editing, S.A.H., B.A.B.; visualization, H.L.C., S.A.H. All authors have read and agreed to the published version of the manuscript.

Funding: Funding for manuscript preparation was provided by Bennelongia Environmental Consultants.

Institutional Review Board Statement: Not applicable.

Data Availability Statement: Data are provided in Tables 1 and 2.

Acknowledgments: We thank the editors for inviting us to contribute to this special issue and accommodating our late contribution. We also thank the consultants who collected most of the data on which this paper is based. Vitor Marques assisted with mapping; Claire Stephenson, Mike Scanlon, Melanie Fulcher and Jane McRae provided photographs.

Conflicts of Interest: Bennelongia Environmental Consultants has sometimes undertaken monitoring of stygofauna and troglofauna in the Robe Valley for the Robe River Mining Company, including at Mesas A and B.

References

1. Moldovan, O.T.; Kovác, L.; Halse, S. *Cave Ecology*; Springer Nature: Cham, Switzerland, 2018; p. 545.
2. Culver, D.C.; Deharveng, L.; Bedos, A.; Lewis, J.J.; Madden, M.; Reddell, J.R.; Sket, B.; Trontelj, P.; White, D. The mid-latitude biodiversity ridge in terrestrial cave fauna. *Ecography* **2006**, *29*, 120–128. [CrossRef]
3. Schneider, K.; Culver, D.C. Estimating subterranean species richness using intensive sampling and rarefaction curves in a high density cave region in West Virginia. *J. Cave Karst Stud.* **2004**, *66*, 39–45.
4. Hahn, H.J.; Fuchs, A. Distribution patterns of groundwater communities across aquifer types in south-western Germany. *Freshwater Biol.* **2009**, *54*, 848–860. [CrossRef]
5. Halse, S.A. Subterranean fauna of the arid zone. In *On the Ecology of Australia's Arid Zone*; Lambers, H., Ed.; Springer Nature: Cham, Switzerland, 2018; p. 388.
6. Pipan, T.; López, H.; Oromí, P.; Polak, S.; Culver, D.C. Temperature variation and the presence of troglobionts in terrestrial shallow subterranean habitats. *J. Nat. Hist.* **2010**, *45*, 253–273. [CrossRef]
7. Pesce, G.L.; de Laurentiis, P.; Humphreys, W.F. Copepods from ground waters of Western Australia. II. The genus *Halicyclops* (Crustacea, Copepoda, Cyclopidae). *Rec. West. Aust. Mus.* **1996**, *18*, 77–85.
8. Humphreys, W.F. Relict stygofauna's living in sea salt, karst and calcrete habitats in arid northwestern Australia contain many ancient lineages. In *The Other 99%: The Conservation and Biodiversity of Invertebrates*; Ponder, W., Lunney, D., Eds.; Royal Zoological Society of New South Wales: Sydney, Australia, 1999; pp. 219–227.
9. Eberhard, S.M.; Halse, S.A.; Williams, M.R.; Scanlon, M.D.; Cocking, J.S.; Barron, H.J. Exploring the relationship between sampling efficiency and short range endemism for groundwater fauna in the Pilbara region, Western Australia. *Freshwater Biol.* **2009**, *54*, 885–901. [CrossRef]
10. Halse, S.A.; Scanlon, M.D.; Cocking, J.S.; Barron, H.J.; Richardson, J.B.; Eberhard, S.M. Pilbara stygofauna: Deep groundwater of an arid landscape contains globally significant radiation of biodiversity. *Rec. West. Aust. Mus. Suppl.* **2014**, *78*, 443–483. [CrossRef]

11. Humphreys, G.; Alexander, J.; Harvey, M.S.; Humphreys, W.F. The subterranean fauna of Barrow Island, north-western Australia: 10 years on. *Rec. West. Aust. Mus. Suppl.* **2013**, *83*, 145–158. [CrossRef]

12. Mokany, K.; Harwood, T.D.; Halse, S.A.; Ferrier, S. Riddles in the dark: Assessing diversity patterns for cryptic subterranean fauna of the Pilbara. *Divers. Distrib.* **2019**, *25*, 240–254. [CrossRef]

13. Biota. *Mesa A and Robe Valley Mesas Troglobitic Fauna Survey*; Biota Environmental Sciences: Leederville, Australia. 2006. Available online: https://www.epa.wa.gov.au/sites/default/files/PER_documentation/Vegetation%20and%20Flora%20and%20Fauna_Biota%202006f%20Subterranean%20Fauna%20Assessment.pdf (accessed on 9 September 2021).

14. Harvey, M.S.; Berry, O.; Edward, K.L.; Humphreys, G. Molecular and morphological systematics of hypogean schizomids (Schizomida:Hubbardiidae) in semiarid Australia. *Invertebr. Syst.* **2008**, *22*, 167–194. [CrossRef]

15. Guzik, M.T.; Austin, A.D.; Cooper, S.J.B.; Harvey, M.S.; Humphreys, W.F.; Bradford, T.; Eberhard, S.M.; King, R.A.; Leys, R.; Muirhead, K.A.; et al. Is the Australian subterranean fauna uniquely diverse? *Invertebr. Syst.* **2010**, *24*, 407–418. [CrossRef]

16. Byrne, M.; Yeates, D.K.; Joseph, L.; Kearney, M.; Bowler, J.; Williams, A.J.; Cooper, S.; Donnellan, S.C.; Keogh, S.; Leys, R.; et al. Birth of a biome: Insights into the assembly and maintenance of the Australian arid zone biota. *Mol. Ecol.* **2008**, *17*, 4398–4417. [CrossRef] [PubMed]

17. Abrams, K.M.; Huey, J.A.; Hillyer, M.J.; Humphreys, W.F.; Didham, R.K.; Harvey, M.S. Too hot to handle: Cenozoic aridification drives multiple independent incursions of Schizomida (Hubbardiidae) into hypogean environments. *Mol. Phylogenet. Evol.* **2019**, *139*, 106532. [CrossRef]

18. Harms, D.; Curran, M.K.; Klesser, R.; Finston, T.L.; Halse, S.A. Speciation patterns in complex subterranean environments: A case study using short-tailed whipscorpions (Schizomida: Hubbardiidae). *Biol. J. Linn. Soc.* **2018**, *125*, 355–367. [CrossRef]

19. Karanovic, I. Candoninae (Ostracoda) from the Pilbara region in Western Australia. *Crustaceana Monogr.* **2007**, *7*, 1–432.

20. Matthews, E.F.; Abrams, K.M.; Cooper, S.J.B.; Huey, J.A.; Hillyer, M.J.; Humphreys, W.F.; Austin, A.D.; Guzik, M.T. Scratching the surface of subterranean biodiversity: Molecular analysis reveals a diverse and previously unknown fauna of Parabathynellidae (Crustacea: Bathynellacea) from the Pilbara, Western Australia. *Mol. Phylogenet. Evol.* **2020**, *142*, 106643. [CrossRef]

21. Finston, T.L.; Johnson, M.S.; Humphreys, W.F.; Eberhard, S.M.; Halse, S.A. Cryptic speciation in two widespread subterranean amphipod genera reflects historical drainage patterns in an ancient landscape. *Mol. Ecol.* **2007**, *16*, 355–365. [CrossRef]

22. King, R.A.; Fagan-Jefferies, E.; Bradford, T.M.; Stringer, D.M.; Finston, T.L.; Halse, S.A.; Eberhard, S.M.; Humphreys, G.; Humphreys, W.F.; Austin, A.D.; et al. Cryptic diversity Down Under: Defining species in the subterranean amphipod genus *Nedsia* Barnard and Williams (Hadzioidea: Eriopisidae) from the Pilbara, Western Australia. *Invertebr. Syst.* **2021**, in press.

23. Pinder, A.M.; Halse, S.A.; Shiel, R.J.; McRae, J.M. An arid zone awash with diversity: Patterns in the distribution of aquatic invertebrates in the Pilbara region of Western Australia. *Rec. West. Aust. Mus. Suppl.* **2010**, *78*, 205–246. [CrossRef]

24. Eberhard, S.M.; Howarth, F.G. Undara lava cave fauna in tropical Queensland with an annotated list of Australian subterranean biodiversity hotspots. *Diversity* **2021**, *13*, 326. [CrossRef]

25. Deharveng, L.; Bedos, A. Biodiversity in the tropics. In *Encyclopedia of Caves*, 3rd ed.; White, W.B., Culver, D.C., Pipan, T., Eds.; Academic Press: New York, NY, USA, 2019; pp. 146–162.

26. Halse, S.A.; Pearson, G.B. Troglofauna in the vadose zone: Comparison of scraping and trapping results and sampling adequacy. *Subterr. Biol.* **2014**, *13*, 17–34. [CrossRef]

27. Buick, R.; Thornett, J.R.; McNaughton, N.J.; Smith, J.B.; Barley, M.E.; Savage, M. Record of emergent continental crust ~3.5 billion years ago in the Pilbara Craton of Australia. *Nature* **1995**, *375*, 574–577. [CrossRef]

28. Johnson, D.P. *The Geology of Australia*; Cambridge University Press: London, UK, 2004.

29. Johnson, S.L.; Wright, A.H. *Central Pilbara Groundwater Study*; Hydrogeological Record Series; Report HG 8; Water and Rivers Commission Resource Science Division: East Perth, WA, Australia, 2001; p. 124.

30. Skrzypek, G.; Dogramaci, S.; Page, G.F.M.; Rouillard, A.; Grierson, P.F. Unique stable isotope signatures of large cyclonic events as a tracer of soil moisture dynamics in the semiarid subtropics. *J. Hydrol.* **2019**, *578*, 124124. [CrossRef]

31. Howarth, F.G.; Moldovan, O.T. The ecological classifications of cave animals and their adaptations. In *Cave Ecology*; Moldovan, O.T., Kovac, L., Halse, S., Eds.; Springer Nature: Cham, Switzerland, 2018; pp. 41–67.

32. Ferreira, R.L.; de Oliveira, M.P.A.; Silva, M.S. Subterranean biodiversity in ferruginous landscapes. In *Cave Ecology*; Moldovan, O.T., Kovac, L., Halse, S., Eds.; Springer Nature: Cham, Switzerland, 2018; pp. 435–447.

33. Morris, R.C.; Ramanaidou, E.R. Genesis of the channel iron deposits (CID) of the Pilbara region, Western Australia. *Australian Aust. J. Earth Sci.* **2007**, *54*, 733–756. [CrossRef]

34. Danišík, M.; Evans, N.J.; Ramanaidou, E.R.; McDonald, B.J.; Mayers, C.; McInnes, B.I.A. (U–Th)/He chronology of the Robe River channel iron deposits, Hamersley Province, Western Australia. *Chem. Geol.* **2013**, *354*, 150–162. [CrossRef]

35. Mammola, S.; Giachino, P.M.; Piano, E.J.A.; Barberis, M.; Badino, G.; Isaia, M. Ecology and sampling techniques of an understudied subterranean habitat: The Milieu Souterrain Superficiel (MSS). *Sci. Nat.* **2016**, *103*, 1–24. [CrossRef] [PubMed]

36. Department of Water. *Ecological Water Requirements of the Lower Robe River*; Environmental Water Report Series 22; Department of Water: Perth, Australia, 2012; p. 38.

37. Edward, K.L.; Harvey, M.S. Short-range endemism in hypogean environments: The pseudoscorpion genera *Tyrannochthonius* and *Lagynochthonius* (Pseudoscorpiones: Chthoniidae) in the semiarid zone of Western Australia. *Invertebr. Syst.* **2008**, *22*, 259–293. [CrossRef]

38. Car, C.A.; Megan, S.; Huynh, C.; Harvey, M.S. The millipedes of Barrow Island, Western Australia (Diplopoda). *Rec. West. Aust. Mus. Suppl.* **2013**, *83*, 209–219. [CrossRef]
39. Harvey, M.S.; Leng, M.C. Further observations on *Ideoblothrus* (Pseudoscorpiones: Syarinidae) from subterranean environments in Australia. *Rec. West. Aust. Mus.* **2008**, *24*, 381–386. [CrossRef]
40. Sket, B. Can we agree on an ecological classification of subterranean animals? *J. Nat. Hist.* **2008**, *42*, 1549–1563. [CrossRef]
41. McKenzie, N.L.; van Leeuwen, S.; Pinder, A.M. Introduction to the Pilbara biodiversity survey, 2002–2007. *Rec. West. Aust. Mus. Supppl.* **2009**, *78*, 3–89. [CrossRef]
42. Page, T.J.; Hughes, J.M.; Real, K.M.; Stevens, M.I.; King, R.A.; Humphreys, W.F. Allegory of a cave crustacean: Systematic and biogeographic reality of *Halosbaena* (Peracarida: Thermosbaenacea) sought with molecular data at multiple scales. *Mar. Biodivers.* **2016**, *48*, 1185–1202. [CrossRef]
43. Des Châtelliers, M.C.; Juget, J.; Lafont, M.; Martin, P. Subterranean aquatic Oligochaeta. *Freshwater Biol.* **2009**, *54*, 678–690. [CrossRef]
44. Sorensen, J.P.R.; Maurice, L.; Edwards, F.K.; Lapworth, D.J.; Read, D.S.; Allen, D.; Butcher, A.S.; Newbold, L.K.; Townsend, B.R.; Williams, P.J. Using boreholes as windows into groundwater ecosystems. *PLoS ONE* **2013**, *8*, e70264. [CrossRef] [PubMed]
45. Halse, S.A.; Scanlon, M.D.; Cocking, J.S. Do Springs Provide a Window to the Groundwater Fauna of the Australian Arid Zone? In *Balancing the Groundwater Budget, Proceedings of an International Groundwater Conference, Darwin, Australia, 12–17 May 2002*; Yinfoo, D., Ed.; International Association of Hydrologists: Wallingford, UK, 2002; p. 12.
46. Sket, B.; Paragamian, K.; Trontelj, P. A census of the obligate subterranean fauna of the Balkan Peninsula. In *Balkan Biodiversity; Pattern and Process in the European Hotspot*; Griffiths, H.I., Kryštufek, B., Reed, J.M., Eds.; Springer: Dordrecht, The Netherlands, 2004; pp. 309–332.
47. Deharveng, L.; Gibert, J.; Culver, D.C. Biodiversity in Europe. In *Encyclopedia of Caves*, 3rd ed.; White, W.B., Culver, D.C., Pipan, T., Eds.; Academic Press: New York, NY, USA, 2019; pp. 136–145.
48. Zagmajster, M.; Polak, S.; Fišer, C. Postojna-Planina cave system in Slovenia, a hotspot of subterranean biodiversity and a cradle of speleobiology. *Diversity* **2021**, *13*, 271. [CrossRef]
49. Pipan, T.; Deharveng, L.; Culver, D.C. Hotspots of subterranean biodiversity. *Diversity* **2020**, *12*, 209. [CrossRef]
50. Karanovic, T. Subterranean copepods (Crustacea, Copepoda) from the Pilbara region in Western Australia. *Rec. West. Aust. Mus. Suppl.* **2006**, *70*, 1–239. [CrossRef]
51. Karanovic, T.; Djurakic, M.; Eberhard, S.M. Cryptic species or inadequate taxonomy? Implementation of 2D geometric morphometrics based on integumental organs as landmarks for delimitation and description of copepod taxa. *Syst. Biol.* **2016**, *65*, 304–327. [CrossRef]
52. Trotter, A.J.; McRae, J.M.; Main, D.C.; Finston, T.L. Speciation in fractured rock landforms: Towards understanding the diversity of subterranean cockroaches (Dictyoptera: Nocticolidae: *Nocticola*) in Western Australia. *Zootaxa* **2017**, *4232*, 361–375. [CrossRef]
53. Moore, G.I.; Humphreys, W.F.; Foster, R. New populations of the rare subterranean blind cave eel *Ophisternon candidum* (Synbranchidae) reveal recent historical connections throughout north-western Australia. *Mar. Freshwater Res.* **2018**, *69*, 1517–1524. [CrossRef]
54. Perina, G.; Camacho, A.I.; Huey, J.; Horwitz, P.; Koenders, A. Understanding subterranean variability: The first genus of Bathynellidae (Bathynellacea, Crustacea) from Western Australia described through a morphological and multigene approach. *Invertebr. Syst.* **2018**, *32*, 423–447. [CrossRef]
55. Coates, D.J.; Byrne, M.; Moritz, C. Genetic diversity and conservation units: Dealing with the species-population continuum in the age of genomics. *Front. Ecol. Evol.* **2018**, *6*. [CrossRef]
56. Wesener, T.; Voigtländer, K.; Decker, P.; Oeyen, J.P.; Spelda, J. Barcoding of Central European *Cryptops* centipedes reveals large interspecific distances with ghost lineages and new species records from Germany and Austria (Chilopoda, Scolopendromorpha). *ZooKeys* **2016**, *564*, 21–46. [CrossRef] [PubMed]
57. Halse, S. Conservation and impact assessment of subterranean fauna in Australia. In *Cave Ecology*; Moldovan, O.T., Kovacz, L., Halse, S., Eds.; Springer Nature: Gland, Switzerland, 2018; pp. 479–493.
58. Environmental Protection Authority. Report and Recommendations of the Environmental Protection Authority: Mesa A Hub Revised Proposal. Report 1640. Environmental Protection Authority, Perth, Australia. 2019. Available online: https://www.epa. wa.gov.au/sites/default/files/EPA_Report/Mesa%20A%20Hub%20Revised%20Proposal%20-%20EPA%20Report.pdf (accessed on 9 September 2021).

Editorial

An Overview of Subterranean Biodiversity Hotspots

David C. Culver [1,*], Louis Deharveng [2], Tanja Pipan [3] and Anne Bedos [2]

[1] Department of Environmental Science, American University, 4400 Massachusetts Ave. NW, Washington, DC 20016, USA

[2] Institut de Systématique, Évolution, Biodiversité (ISYEB), UMR7205, CNRS, Muséum National d'Histoire Naturelle, Sorbonne Université, EPHE, 45 Rue Buffon, 75005 Paris, France; dehar.louis@wanadoo.fr (L.D.); bedosanne@yahoo.fr (A.B.)

[3] Research Center of the Slovenian Academy of Sciences and Arts, Karst Research Institute, Novi Trg 2, SI-1000 Ljubljana, Slovenia; tanja.pipan@zrc-sazu.si

* Correspondence: dculver@american.edu; Tel.: +1-703-508-9509

Citation: Culver, D.C.; Deharveng, L.; Pipan, T.; Bedos, A. An Overview of Subterranean Biodiversity Hotspots. *Diversity* **2021**, *13*, 487. https://doi.org/10.3390/d13100487

Received: 22 August 2021
Accepted: 23 August 2021
Published: 6 October 2021

Publisher's Note: MDPI stays neutral with regard to jurisdictional claims in published maps and institutional affiliations.

1. Introduction

Riding a wave of interest in biodiversity patterns in surface-dwelling communities, in 2000, Culver and Sket [1] published a paper listing 20 caves and karst wells with 20 or more known species. At the time of their study, it was widely recognized that because of the extremely narrow distribution of most cave specialists, the overall species richness of a region or country was the result of the accumulation of species from a number of different caves, and that average single-site species richness (α-diversity) was a minor component of overall species richness (γ-diversity) [2]. This has since been confirmed in a number of large-scale studies of subterranean species diversity, especially in Europe [3–6]. Although the number of studies measuring both α-diversity (local diversity) and β-diversity (between-site diversity) has grown in the intervening 20+ years, they have largely been confined to Europe and, to a lesser extent, North America [3–7], mainly because it requires extensive datasets for about 100 sites for these to be estimated [6,8]. However, it is also well known that considerable subterranean species α-diversity lies outside these regions, including in Australia [9,10], Brazil [11,12], China [13], South East Asia [14], and Mexico [15], in part as a result of a search for additions to the list of hotspot caves reported by Sket and Culver.

The collection of papers in this Special Issue of *Diversity*—"Hotspots of Subterranean Biodiversity"—expands and enriches previous hotspot lists [2,14,16,17]. These hotspots may not capture all regions of high subterranean species richness; all of these hotspot caves are worthy of study and protection in their own right, and any protection strategy should include these exceptional sites as well as regional areas of species richness. In total, papers in this Special Issue cover 13 of 22 sites known to harbor at least 25 obligate subterranean species. Culver and Sket's original list of 20 has become too lengthy to be of much use; thus, we raised the bar to 25. One other cave is included in this Special Issue because of its special geographical position distant from any other hotspot site—Ganxiao Dong Cave in China [18]. However, we did not include it in our summary of hotspot caves.

2. Overview of Hotspot Subterranean Sites

In Table 1, we list the 22 known sites with at least 25 obligate subterranean species, and they are also shown on the map in Figure 1. Two of these sites (Ojo Guareña in Spain and Sistema Huautla in Mexico) have not been previously noted in any list as hotspot caves. There may be additional non-cave subterranean sites, particularly in hyporheic habitats, but these data have for the most part not been assembled into species lists for individual sites. In addition, there is the difficulty in deciding what constitutes a single site—is it just one Bou-Rouch pump site or a stream reach (e.g., a riffle)?

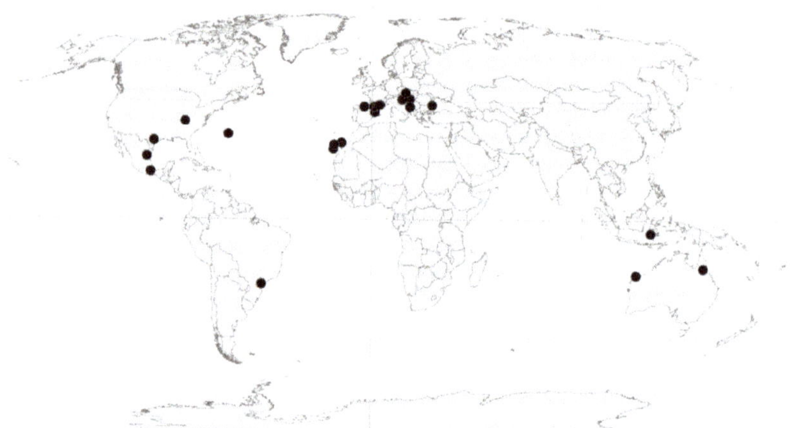

Figure 1. Map of hotspot sites listed in Table 1. Map prepared by Magda Aljančič, used with permission.

Table 1. All sites with 25 or more stygobionts plus troglobionts. Caves assessed in this special issue are marked with an asterisk.

	Stygo-bionts	Rank	Troglo-bionts	Rank	Total	Rank	No. Unde-scribed	Terrain and Hydrogeology	Latitude	Year Assessed	Reference
Temperate											
Postojna Planina System (SLO) *	72	1	45	1	117	1	8	karst	45.7	2021	[19]
Vjetrenica (B&H)	42	7	39	2	81	2	3	karst	42.9	2009	[20,21]
Križna Jama (SLO) *	32	10	28	6	60	4	5	karst	45.7	2021	[22]
Ojo Guareña (ESP) *	46	5	8		54	6	23	karst	43.0	2021	[23]
Logarček (SLO)	28		15		43	9	?	karst	43.5	2000	[1]
Mammoth Cave (USA) *	17		32	4	49	7	0	karst	45.9	2021	[24]
Coume Ouarnède (FRA) *	21		17		38		3	karst	43.0	2021	[25]
Pestera de la Movile (RO) *	13		25		38		4	karst, chemoautotrophic	37.2	2021	[26]
Cent fonts (FRA)	43	6	0		43	9	7	karst, phreatic	43.6	2006	[27]
Triadou aquifer (FRA)	34	9	0		34		?	karst, phreatic	43.7	2000	[1]
Baget system (FRA) [1]	24		9		33		?	karst, phreatic	43.0	2000	[1]
Subtemperate											
Walsingham Cave (BER) *	65	2	0		65	3	0	karst, anchialine	32.3	2021	[28]
San Marcos Artesian Well (USA) *	55	3	0		55	5	16	karst, phreatic, chemoautotrophic	29.9	2021	[29]
Jameos del Agua (ESP)	40	8	0		40		4	lava tube, anchialine	28.4	2018	[30]
Cueva de Felipe Reventon (ESP) *	0		38	3	38		0	lava tube	28.4	2021	[31]
Cueva del Viento-Sobrado (ESP) *	0		28	6	28		0	lava tube	28.4	2021	[31]
Sub-tropical											
Sistema Huautla (MEX) *	0		27	9	27		10	deep karst	18.1	2021	[32]
Undara Basalt Flow (AUS) *[2]	1		30	5	31		27	lava tube	−18.2	2021	[33]
Robe Valley wells (AUS) *[3]	48	4	0		48	8	18	karst, phreatic	−21.6	2021	[34]
Sistema Purificacion (MEX)	3		28	6	31		?	karst	23.8	2019	[14]
Areias system (BRA)	6		22		28		14	karst	−24.6	2016	[11]
Tropical											
Towakkalak (SUL) *	10		26	10	36		17	karst	−5.0	2021	[35]

[1] Unpublished records of L. Deharveng show 30 stygobionts and 11 troglobionts from the Baget Basin, including Grotte de Sainte-Catherine. Two species are undescribed. [2] Numbers are for the Undara Basalt Flow as a whole. Bayliss Cave, the largest cave, has 23 troglobionts and 1 stygobiont. [3] Clark et al. [34] describe a nearby terrestrial area but it includes both deep soil and MSS (milieu souterrain superficiel) components, and is not directly comparable to the other sites.

Several additional sites are worth mentioning because it is likely that if further collecting and identification is performed, they will have 25 stygobionts plus troglobionts. Trajano et al. [12] report a highly diverse troglobiotic fauna in a Brazilian karst region, with 29 troglobionts reported from 11 caves in a 25 km^2 region. This record is noteworthy both for its geographic position and the fact that the caves are in limestone, even though no one cave system has 25 troglobionts. No hotspot caves are reported from Africa, but

Deharveng and Bedos, in an ongoing study, have reported 30 troglobionts and 3 stygo-bionts known from an Algerian karst area—Djurdjura—in an area no more than 50 km in length. Especially noteworthy is that there are many species restricted to karstic holes with permanent snow above 2000 m, especially beetles and Collembola, which exhibit eye and pigment reduction, most living in caves as well. In Ganxiao Dong in China, a total of 19 troglobionts and 1 stygobiont have been reported [18], and it is likely that many more species will be found, especially because multi-taxa collection is in its infancy. The fauna is noteworthy because of the high level of morphological specialization, especially in Coleoptera and Myriapoda.

Outside of the Dinaric mountains, none of the sites were highly ranked (i.e., in the top ten) with respect to both their aquatic fauna and terrestrial fauna. Postojna Planina Cave system is by far the richest known subterranean site with respect to species richness; it ranks first with respect to both aquatic and terrestrial species richness [19]. Some sites had no terrestrial fauna (San Marcos well in Texas, Jameos del Agua in the Canary Islands, Walsingham Cave in Bermuda, Triadou Aquifer, and Robe Valley wells in Australia) and others had no aquatic fauna (Huautla in Mexico, Cueva de Felipe Reventón and Cueva del Viento-Sobrado in the Canary Islands). All continents are represented by at least one site, except for Africa.

When the aquatic and terrestrial fauna are considered separately, a different pattern emerges for the two habitats. For aquatic species, the top ten sites include six anchialine or deep phreatic sites, three sites in the Dinaric mountains, one site in the Cantabrian Mountains of Spain. The Dinaric and Cantabrian sites are within the ridge of high-diversity regions in Europe [4,6]. All of these sites are likely to be enhanced in available organic carbon and nutrients relative to other sites, but historical reasons in the case of the Dinaric sites may be important as well [4]. In particular, their proximity to the Mediterranean Sea, a source of colonists, especially during the Messinian salinity crisis, may be important.

For terrestrial species, the sites are more scattered outside the Dinaric mountains, which is the location of three of the five richest terrestrial caves (Table 1). The lava tubes on Tenerife in the Canary Islands are the site of two of the ten species-rich terrestrial caves, as are two caves in Mexico; the other top ten are scattered in Australia, Brazil, Sulawesi, and the United States. The mesas near the Robe Valley in Australia are very species rich, but the reported subterranean fauna includes the soil fauna [34], making comparisons impossible. It is difficult to determine what unites these terrestrial sites, because they are a combination of lava tubes and karst caves, with very different levels of organic carbon and nutrients. It is, however, noteworthy that tropical or subtropical sites are relatively more represented than for aquatic fauna.

The latitudinal distribution of hotspot sites shows an unexpectedly bimodal distribu-tion (Figure 2). The majority of sites are in the temperate zone, between 40 and 50 degrees north or south of the equator. This corresponds, at least in the northern hemisphere, to the region traditionally held to be the area of highest subterranean species richness, a result often attributed to the effects of repeated Pleistocene glaciations [36]. The second mode is in the sub-tropical and sub-temperate regions (between 20 and 30 degrees north and south of the equator). These sites are mostly lava tubes and wells connected to chemoautotrophic zones (Table 1). None of the sites are located in the arid tropics, also suggesting that food availability may be an important driver of species richness.

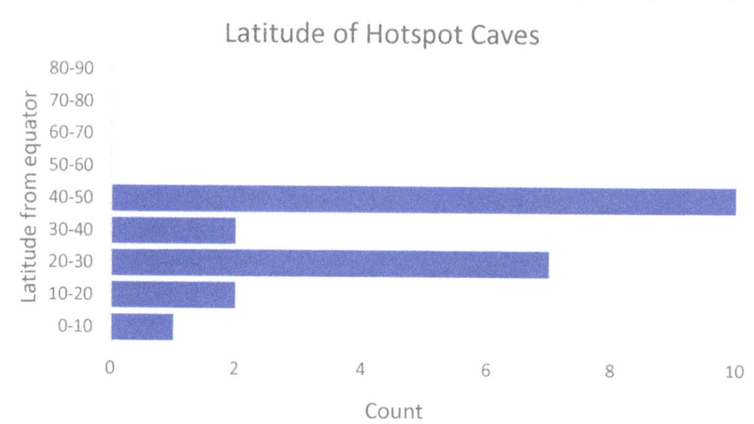

Figure 2. Distribution distances from the equator of the subterranean hotspot sites listed in Table 1.

3. Caveats and Challenges

3.1. Differences in Working Definitions of Troglobionts and Stygobionts

Although there are many nuances [14,37–39], troglobionts and stygobionts are considered to be species exclusively found in subterranean habitats, which usually, but not always, show convergent morphological modification to subterranean life, often called troglomorphies [40]. Troglomorphies include, but are not limited to, reduced or absent eyes and pigment. For areas and groups with a well-studied surface fauna, it is possible to detect species occurring in subterranean habitats with well-developed eyes and pigment, but absent from the surface. In many, perhaps most cases, these are recent isolates in subterranean habitats. The difficulty is that in areas where the surface fauna is not well studied, such as most of the tropics, it is impossible to ascertain that such non-troglomorphic species are troglobiotic (or stygobiotic) species. Therefore, numbers of troglobiotic and stygobiotic species may be underestimated or overestimated in these sites. Such species may be common and the uncertainty of the ecological assignment (e.g., troglobiont or troglophile) is a bias. Deharveng and Bedos [41] report that soil-inhabiting species (especially Collembola and Diplura) are commonly listed as troglobionts.

Some studies of poorly known fauna may be biased by the inclusion of too many species as troglobionts and stygobionts if the operational criteria for inclusion are eyelessness and lack of pigment. Many litter- and soil-dwelling species are eyeless and depigmented, and some major taxonomic groups, e.g., Symphyla or the highly diversified Onychiuridae (Collembola) species, are entirely eyeless and depigmented regardless of habitat [29,42]. Additionally, some troglophilic species have populations that have reduced eyes and pigment. Sometimes these are recognized as subspecies [43], but if not, they are usually not included in the lists.

A strong case could be made that the relevant communities to compare are the permanent, sustaining populations (troglobionts [stygobionts] and eutroglophiles [eustygophiles]) but such lists are rarely available. Eutroglophiles and eustygophiles (see [38] for definitions) may be hard to delineate from other non-troglobionts and non-stygobionts.

3.2. Differences in Taxonomic Groups, Habitats Sampled, and Sampling Techniques

Differences in taxonomic composition can be because of regional differences in the composition of the subterranean fauna or because of incomplete sampling and/or taxonomic description. The tropics and arid areas tend to be richer in Arachnida, with some minor arachnid orders entirely missing at higher latitudes [14,41]. Other cases are clearly ones of inadequate collection. The waters of epikarst are usually dominated by Harpacticoidea crustaceans, but they have not been sampled or inadequately sampled outside of Europe (but see [44]), and the fauna can be quite rich, with more than ten species in some Slovenian caves [45,46]. Among the species listed in Table 1, only Postojna Planina Cave System [19] and Ojo Guareña [23] have been thoroughly sampled for epikarst fauna. Other groups may be under-represented due to their small size. Among Collembola, the Neelidae have several species less than 0.5 mm long [41,47] and are probably much more common than existing records would indicate. Ostracoda have rarely been sampled, with only the Postojna Planina Cave System [19], Ojo Guareña [23] and San Marcos well [28] among the sites in Table 1 having been surveyed. Finally, there are often groups that are under-collected because they require special preservation techniques or there are no taxonomists to study them (the Racovitzan and Linnean shortfalls [48]). Planarians are a good example. Although many records of planarians are known, many, if not most, have not been identified to species, and few subterranean species have been described recently.

3.3. Described versus Undescribed Species

A particularly vexing problem in the analysis of species numbers in subterranean sites is the nearly universal problem of undescribed species. Except for Cueva de Felipe Reventón, Cueva del Viento-Sobrado, Walsingham Caves, and Mammoth Cave, all the hotspot caves listed in Table 1 include undescribed species, as do Bayliss Cave, Sistema Huautla, Towakkalak system and Ganxiao Dong included in this Special Issue. The percentage of undescribed species tends to be highest in tropical countries, with a rich fauna and a shorter tradition of taxonomic description, such as Brazil, but it is also high in countries such as Australia (Table 1).

The difficulty is that not all records reported as new species turn out to be new species. The West Virginia cave fauna provides an example of the problem. In 1976, Holsinger, Baroody, and Culver [49], listed 27 undescribed stygobionts and troglobionts. In a re-examination of the West Virginia cave fauna in 2007, Fong et al. [50] found that of the 27 undescribed species listed in 1976:

- 13 were described as new species;
- 6 were assigned to existing known stygobiotic and troglobiotic species;
- 8 remained unstudied.

This suggests a "discount rate" for undescribed species of 68% (13 of 19, because 6 were not new species after all), but this rate is likely to vary from region to region, and if it were applied to the data in Table 1, it would result in a bias against tropical regions and any other region with large numbers of undescribed species. Perhaps the best solution is to list both groups of species in any compilation of species richness but to indicate the number that remain undescribed, as we have done in Table 1.

Finally, there are questions about the quality of the taxonomic descriptions. This may be especially problematic for descriptions more than one century old, where the descriptions were very brief, and the range of diversity of different groups was largely unknown. This is, for example, a problem with mites described from Mammoth Cave in the 19th century, some of which have not been recollected [24].

4. Protection of Hotspot Caves

It is difficult to generalize, either about the threats or the level of protection for the hotspot caves listed in Table 1, at least beyond the existential threat that global warming poses for subterranean ecosystems [51,52]. In addition, a few generalities are possible. First, all sites are affected by surface processes. For caves, these inputs include infiltration through soil and epikarst, direct inputs from sinking streams, and direct human actions, including tourism.

Secondly, the area of the Earth's surface that impacts a subterranean fauna is greater for the aquatic fauna than for the terrestrial fauna. At one extreme is the artesian well at San Marcos, Texas, which accesses groundwater in the Edwards Aquifer at a depth of 59.5 m, and the Edwards Aquifer itself, which covers an area of over 10,000 km^2 [28]: water quality and quantity in the San Marcos well are potentially affected by events occurring through the Edwards Aquifer. At another extreme are the lava tubes in the Canary Islands. Most, if not all, of the organic input comes from tree roots as well as organic matter deposited from small cracks and crevices in the lava tube. In this case, the vulnerable zone is the projection of the cave onto the surface. In many cases, the limit of the vulnerable zone is unknown.

Thirdly, cave tourism is an important aspect of the state of affairs for a number of hotspot caves. Two caves in Table 1 have in excess of 100,000 visitors per year (Postojna Planina Cave System and Mammoth Cave). Others have smaller numbers of visitors, including wild cave tours (Walsingham Caves, Križna jama, Jameos del Agua, and Cueva del Viento Sobrado). Although cave tourism is often thought of in negative terms because of the attendant problems of light pollution, increases in CO_2, structural modifications of the cave, etc., it can have a positive impact in that the economic value of a tourist cave makes major disruption unlikely. Additionally, the tourist part of the cave is typically a small fraction of the entire cave.

Furthermore, many of the hotspot caves are protected by some form of designation as park land, or protected area. Even if there is no official designation, there are often laws protecting caves in general [53,54]. However, the efficacy of these regulations and designations is never absolute, and at some level all sites face threats. For example, even Mammoth Cave, a U.S. National Park, faces a number of problems, ones due to actions taken outside the park, as well as past actions by the NPS itself [55]. On the obverse side, some caves are largely protected by their difficulty of access, by their length, as well as by the secrecy of some of their entrances, such as Sistema Huautla, one of the deepest caves (>1 km) in the world [56] or Coume Ouarnède [25].

Finally, there is not one template for threats or one prescription for the protection of subterranean hotspots. What works for a site in the United States probably will not work for a site in Brazil. Each site must be considered separately and carefully, and upon threat evaluation. Threat levels are the basic criteria to be used to derive protection measures. All the sites listed in Table 1 are worthy of protection as well as a source of regional and national heritage.

Author Contributions: Data compilation: D.C.C., L.D., T.P., and A.B.; Manuscript writing: D.C.C.; Editing: L.D., T.P., and A.B. All authors have read and agreed to the published version of the manuscript.

Funding: This research received no external funding.

Institutional Review Board Statement: Not applicable.

Data Availability Statement: Not applicable.

Conflicts of Interest: The authors declare no conflict of interest.

References

1. Culver, D.C.; Sket, B. Hotspots of subterranean biodiversity in caves and wells. *J. Cave Karst Stud.* **2000**, *62*, 11–17.
2. Gibert, J.; Deharveng, L. Subterranean ecosystems: A truncated functional diversity. *Bioscience* **2002**, *52*, 473–481. [CrossRef]
3. Bregović, P.; Zagmajster, M. Understanding hotspots within a global hotspot—Identifying the drivers of regional species richness patterns in terrestrial subterranean habitats. *Insect Conserv. Biodivers.* **2016**, *9*, 268–281. [CrossRef]
4. Zagmajster, M.; Eme, D.; Fišer, C.; Galassi, D.; Marmonier, P.; Stoch, F.; Cornu, J.; Malard, F. Geographic variation in range size and beta diversity of groundwater crustaceans: Insights from habitats with low thermal seasonality. *Glob. Ecol. Biogeogr.* **2014**, *23*, 1135–1145. [CrossRef]
5. Malard, F.; Boutin, C.; Camacho, A.I.; Ferreira, D.; Michel, G.; Sket, B.; Stoch, F. Diversity patterns of stygobiotic crustaceans across multiple spatial scales in western Europe. *Freshw. Biol.* **2009**, *54*, 756–776. [CrossRef]
6. Culver, D.C.; Deharveng, L.; Bedos, A.; Lewis, J.J.; Madden, M.; Reddell, J.R.; Sket, B.; Trontelj, P.; White, D. The mid-latitude biodiversity ridge in terrestrial cave fauna. *Ecography* **2006**, *29*, 120–128. [CrossRef]
7. Christman, M.C.; Doctor, D.H.; Niemiller, M.L.; Weary, D.J.; Young, J.A.; Zigler, K.S.; Culver, D.C. Predicting the occurrence of cave-inhabiting fauna based on features of the Earth surface environment. *PLoS ONE* **2016**, *11*, e0160408. [CrossRef]
8. Dole-Olivier, M.J.; Castellarini, F.; Coineau, N.; Galassi, D.M.P.; Martin, P.; Mori, N. Towards an optimal sampling strategy to assess groundwater biodiversity comparison across six European regions. *Freshw. Biol.* **2009**, *54*, 777–796. [CrossRef]

9. Humphreys, W.H. Biodiversity patterns in Australia. In *Encyclopedia of Caves*, 3rd ed.; White, W.B., Culver, D.C., Pipan, T., Eds.; Academic Press: New York, NY, USA, 2019; pp. 109–126.

10. Guzick, M.T.; Austin, A.D.; Cooper, S.J.B.; Harvey, M.S.; Humphreys, W.F.; Bradford, T.; Eberhard, S.M.; King, R.A.; Leijs, R.; Muirhead, K.A.; et al. Is the Australian subterranean fauna uniquely diverse? *Invert. Syst.* **2010**, *24*, 407–418. [CrossRef]

11. Souza-Silva, M.; Ferreira, R.L. The first two hotspots of subterranean biodiversity in South America. *Subterr. Biol.* **2016**, *19*, 1–21. [CrossRef]

12. Trajano, E.; Gallão, J.E.; Bichuette, M.E. Spots of high diversity of troglobites in Brazil: The challenge of measuring subterranean diversity. *Biodivers. Conserv.* **2016**, *25*, 1805–1828. [CrossRef]

13. Latella, L. Biodiversity: China. In *Encyclopedia of Caves*, 3rd ed.; White, W.B., Culver, D.C., Pipan, T., Eds.; Academic Press: London, UK, 2019; pp. 127–135.

14. Deharveng, L.; Bedos, A. Biodiversity in the tropics. In *Encyclopedia of Caves*, 3rd ed.; White, W.B., Culver, D.C., Pipan, T., Eds.; Academic Press: London, UK, 2019; pp. 146–162.

15. Palacios-Vargas, J.G.; Juberthie, C.; Reddell, J.R. México. *Mundos Subterráneos* **2014–2015**, *25–26*, 1–101.

16. Culver, D.C.; Pipan, T. *Biology of Caves and Other Subterranean Habitats*, 1st ed.; Oxford University Press: Oxford, UK, 2009.

17. Culver, D.C.; Pipan, T. Subterranean ecosystems. In *Encyclopedia of Biodiversity*, 2nd ed.; Levin, S.A., Ed.; Academic Press: Waltham, MA, USA, 2013; Volume 7, pp. 49–62.

18. Huang, S.; Wei, G.; Wang, H.; Liu, W.; Bedos, A.; Deharveng, L.; Tian, M. Ganxiao Dong: A hotspot of cave biodiversity in northern Guangxi, China. *Diversity* **2021**, *13*, 355. [CrossRef]

19. Zagmajster, M.; Polak, S.; Fišer, C. Postojna Planina Cave System in Slovenia, a hotspot of subterranean biodiversity and a cradle of speleobiology. *Diversity* **2021**, *13*, 271. [CrossRef]

20. Lučić, I.; Sket, B. *Vjetrenica. Pogled U Dušu Zemlje*; Savez Speleologa Bosne I Hercegovine and Hrvatsko Biospeleološko Društvo: Zagreb, Croatia, 2003.

21. Ozimec, R.; Lučić, I. The Vjetrenica cave (Bosnia & Herzegovina)–one of the world's most prominent biodiversity hotspots for cave-dwelling fauna. *Subt. Biol.* **2009**, *7*, 17–24.

22. Polak, S.; Pipan, T. The subterranean fauna of Križna jama, Slovenia. *Diversity* **2021**, *13*, 210. [CrossRef]

23. Camacho, A.I.; Puch, C. Ojo Guareña, a hotspot of subterranean biodiversity in Spain. *Diversity* **2021**, *13*, 199. [CrossRef]

24. Niemiller, M.L.; Helf, K.; Toomey, R.S. Mammoth Cave: A hotspot of subterranean biodiversity in the United States. *Diversity* **2021**, *13*, 373. [CrossRef]

25. Faille, A.; Deharveng, L. The Coume Ouarnède system, a hotspot of subterranean biodiversity in Pyrenees (France). *Diversity* **2021**, *13*, 419. [CrossRef]

26. Brad, T.; Iepure, S.; Sarbu, S.M. The chemoautotrophically based Movile Cave groundwater ecosystem, a hotspot of subterranean biodiversity. *Diversity* **2021**, *13*, 128. [CrossRef]

27. Olivier, M.J.; Martin, D.; Bou, C.; Prié, V. Interprétation du suivi hydrobiologique de la faune stygobie, réalisé sur le système karstique des Cents Fonts lors du pompage d'essai. In *Système Karstique des Cent Fonts. Simulation de Scénarios d'Exploitation et de Gestion de la Resource*; Ladouche, B., Maréchal, J.C., Dörfliger, N., Lachassagne, P., Eds.; Bureau de Recherches Géologiques et Minières: Montpelier, France, 2006; pp. 127–142.

28. Iliffe, T.M.; Calderón Gutiérrez. Bermuda's Walsingham Caves: A global hotspot for anchialine stygobionts. *Diversity* **2021**, *13*, 352. [CrossRef]

29. Hutchins, B.T.; Gibson, J.R.; Diaz, P.H.; Schwartz, B.F. Stygobiont diversity in the San Marcos Artesian Well and Edwards Aquifer groundwater ecosystem, Texas, USA. *Diversity* **2021**, *13*, 234. [CrossRef]

30. Martínez, A.; Gonzalez, B.C. Volcanic anchialine habitats of Lanzarote. In *Cave Ecology*; Moldovan, O.T., Kováč, L., Halse, S., Eds.; Springer Nature: Cham, Switzerland, 2018; pp. 399–414.

31. Oromí, P.; Socorro, S. Biodiversity in the Cueva del Vientzo lava tube system (Tenerife, Canary Islands). *Diversity* **2021**, *13*, 226. [CrossRef]

32. Francke, O.F.; Monjaraz-Ruedas, R.; Cruz-López, J. Biodiversity of the Huautla Cave System, Oaxaca, Mexico. *Diversity* **2021**, *13*, 429. [CrossRef]

33. Eberhard, S.M.; Howarth, F.G. Undara basalt flow, a hotspot of subterranean biodiversity, Queensland, Australia. *Diversity* **2021**, *13*, 326. [CrossRef]

34. Clark, H.L.; Buzatto, B.A.; Halse, S.A. A hotspot of arid zone subterranean biodiversity: The Robe Valley in Western Australia. *Diversity* **2021**, *13*, 482. [CrossRef]

35. Deharveng, L.; Rahmadi, C.; Suhardjono, Y.R.; Bedos, A. The Towakkalak System, a hotspot of subterranean biodiversity in Sulawesi, Indonesia. *Diversity* **2021**, *13*, 392. [CrossRef]

36. Jeannel, R. *Les Fossiles Vivants des Cavernes*; Gallimard: Paris, France, 1944.

37. Trajano, E.; de Carvalho, M.R. Towards a biologically meaningful classification of subterranean organisms: A critical analysis of the Schiner-Racovitza system from a historical perspective, difficulties of its application and implications for conservation. *Subterr. Biol.* **2017**, *22*, 1–26. [CrossRef]

38. Sket, B. Can we agree on an ecological classification of subterranean animals. *J. Nat. Hist.* **2008**, *42*, 1549–1563. [CrossRef]

39. Culver, D.C.; Pipan, T. Ecological and evolutionary classifications of subterranean organisms. In *Encyclopedia of Caves*, 3rd ed.; White, W.B., Culver, D.C., Pipan, T., Eds.; Academic Press: London, UK, 2019; pp. 376–379.

40. Christiansen, K.A. Proposition pour la classification des animaux cavernicoles. *Spelunca* **2019**, *2*, 75–78.
41. Deharveng, L.; Bedos, A. Diversity of terrestrial invertebrates in subterranean environments. In *Cave Ecology*; Moldovan, O.T., Kováč, L., Halse, S., Eds.; Springer Nature: Cham, Switzerland, 2018; pp. 107–172.
42. Peck, S.B. Eyeless arthropods of the Galapagos Islands, Ecuador: Composition and origin of the cryptozoic fauna of a young, tropical, oceanic archipelago. *Biotropica* **1990**, *22*, 366–381. [CrossRef]
43. Schultz, G.A. Descriptions of new subspecies of *Ligidium elrodii* (Packard) comb. nov. with notes on other isopod crustaceans from caves in North America (Oniscoidea). *Am. Midl. Nat.* **1970**, *84*, 36–45. [CrossRef]
44. Pipan, T.; Christman, M.C.; Culver, D.C. Dynamics of epikarst communities: Microgeographic pattern and environmental determinants of epikarst copepods in Organ Cave, West Virginia. *Am. Midl. Nat.* **2006**, *156*, 75–87. [CrossRef]
45. Pipan, T. *Epikarst–A Promising Habitat*; Založba ZRC: Ljubljana, Slovenia, 2005.
46. Pipan, T.; Culver, D.C.; Papi, F.; Kozel, P. Partitioning diversity in subterranean invertebrates: The epikarst fauna of Slovenia. *PLoS ONE* **2018**, *13*, e0185991. [CrossRef]
47. Schneider, C.; Deharveng, L. First record of the genus *Spinaethorax* Papáč and Palacios-Vargas, 2016 (Collembola, Neelipleona, Neelidae) in Asia, with a new species from a Vietnamese cave. *Eur. J. Taxon.* **2017**, *363*, 1–20. [CrossRef]
48. Ficetola, C.F.; Canadoli, C.; Stoch, F. The Racovitzan impediment and the hidden biodiversity of unexplored environments. *Conserv. Biol.* **2019**, *33*, 214–216. [CrossRef]
49. Holsinger, J.R.; Baroody, R.A.; Culver, D.C. *The Invertebrate Cave Fauna of West Virginia*; West Virginia Speleological Survey: Barrackville, WV, USA, 1976; No. 7.
50. Fong, D.W.; Culver, D.C.; Hobbs, H.H., III; Pipan, T. *The Invertebrate Cave Fauna of West Virginia*, 2nd ed.; West Virginia Speleological Survey: Barrackville, WV, USA, 2007; No. 16.
51. Mammola, S.; Cardoso, P.; Culver, D.C.; Deharveng, L.; Ferreira, R.L.; Fišer, C.; Galassi, D.M.P.; Griebler, C.; Halse, S.; Humphreys, W.F.; et al. Scientists' warning on the conservation of subterranean ecosystems. *BioScience* **2019**, *69*, 641–650. [CrossRef]
52. Mammola, S.; Piano, E.; Cardoso, P.; Vernon, P.; Domìnguez-Viller, D.; Culver, D.C.; Pipan, T.; Isaia, M. Climate change going deep: The effects of global climatic alterations on cave ecosystems. *Anthr. Rev.* **2019**, *6*, 98–116. [CrossRef]
53. Niemiller, M.L.; Taylor, S.J.; Bichuette, M.E. Conservation of cave fauna, with emphasis on Europe and the Americas. In *Cave Ecology*; Moldovan, O.T., Kováč, L., Halse, S., Eds.; Springer Nature: Cham, Switzerland, 2018; pp. 451–478.
54. Halse, S.J. Conservation and impact assessment of subterranean fauna in Australia. In *Cave Ecology*; Moldovan, O.T., Kováč, L., Halse, S., Eds.; Springer Nature: Cham, Switzerland, 2018; pp. 479–496.
55. Olson, R.A. Environmental issues relevant to the Mammoth Cave area. In *Mammoth Cave. A Human and Natural History*; Hobbs, H.H., III, Olson, R.A., Winkler, E.G., Culver, D.C., Eds.; Springer Nature: Cham, Switzerland, 2017; pp. 265–275.
56. Steele, C.W.; Shifflett, T.E. Huautla cave system (Sistema Huautla), Mexico. In *Encyclopedia of Caves*, 3rd ed.; White, W.B., Culver, D.C., Pipan, T., Eds.; Academic Press: New York, NY, USA, 2019; pp. 527–536.

MDPI

St. Alban-Anlage 66

4052 Basel

Switzerland

Tel. +41 61 683 77 34

Fax +41 61 302 89 18

www.mdpi.com

Diversity Editorial Office

E-mail: diversity@mdpi.com

www.mdpi.com/journal/diversity